VOLUME FIFTY

ADVANCES IN INSECT PHYSIOLOGY

VOLUME FIFTY

Advances in Insect Physiology

Pine Bark Beetles

Edited by

CLAUS TITTIGER
University of Nevada, Reno, NV, United States

GARY J. BLOMQUIST
University of Nevada, Reno, NV, United States

ELSEVIER

AMSTERDAM • BOSTON • HEIDELBERG • LONDON
NEW YORK • OXFORD • PARIS • SAN DIEGO
SAN FRANCISCO • SINGAPORE • SYDNEY • TOKYO
Academic Press is an imprint of Elsevier

Academic Press is an imprint of Elsevier
125 London Wall, London, EC2Y 5AS, UK
The Boulevard, Langford Lane, Kidlington, Oxford OX5 1GB, UK
525 B Street, Suite 1800, San Diego, CA 92101-4495, USA
50 Hampshire Street, 5th Floor, Cambridge, MA 02139, USA

First edition 2016

ISBN: 978-0-12-802723-3
ISSN: 0065-2806

For information on all Academic Press publications
visit our website at https://store.elsevier.com/

Publisher: Zoe Kruze
Acquisition Editor: Kirsten Shankland
Editorial Project Manager: Helene Kabes
Production Project Manager: Vignesh Tamil
Designer: Vicky Pearson Esser

Typeset by SPi Global, India

CONTENTS

CONTRIBUTORS

M.N. Andersson
Lund University, Lund, Sweden

P.D. Batista
University of Alberta, Edmonton, AB; University of Northern British Columbia, Prince George, BC, Canada

G.J. Blomquist
University of Nevada, Reno, NV, United States

C.W. Gitau-Clarke
Department of Economic Development, Jobs, Transport and Resources, Rutherglen, Vic, Australia

D.P.W. Huber
Natural Resources and Environmental Studies Institute, University of Northern British Columbia, Prince George, BC, Canada

J.K. Janes
School of Environmental and Rural Science, The University of New England, Armidale, NSW, Australia

C.I. Keeling
Simon Fraser University, Burnaby, BC, Canada

K.F. Raffa
University of Wisconsin-Madison, Madison, WI, United States

J.A. Robert
Natural Resources and Environmental Studies Institute, University of Northern British Columbia, Prince George, BC, Canada

F. Schlyter
Swedish University of Agricultural Sciences, Alnarp, Sweden

B.T. Sullivan
USDA Forest Service, Southern Research Station, Pineville, LA, United States

M.R.E. Symonds
Centre for Integrative Ecology, School of Life and Environmental Sciences, Deakin University, Burwood, Vic, Australia

C. Tittiger
University of Nevada, Reno, NV, United States

PREFACE

Pine bark beetles include some of the most destructive forest pests of the Northern Hemisphere. Periodic outbreaks of destructive beetles have occurred in the past, but they pale in comparison to the current unparalleled destruction of lodgepole forests by the mountain pine beetle in the western United States and British Columbia. In addition to spectacular outbreaks of aggressive bark beetles that kill millions of trees and result in landscape-level changes, bark beetles also constantly work in the background initiating the process of recycling dead and dying trees. These remarkable insects present special challenges to those trying to understand them. They usually are only indirectly visible through their effects on their hosts. They are difficult to rear outside of their normal environments, and their lives are intimately linked to several factors—all of which are critical to their continued existence. Bark beetles are obligate parasites of their hosts, which protect and nurture them in an environment arguably so hostile that most organisms would not survive it. They rely on symbiotic microorganisms and fungi and on chemical-directed behaviour to survive and reproduce. And yet, steady effort by researchers investigating a variety of perspectives continues to fill the gaps in our knowledge and these interactions are beginning to be understood across the broadest range of scales, from molecular to landscape, and from current to evolutionary timeframes. It is an exciting time.

This book describes the physiology of bark beetles in the broadest sense of the word. There is information regarding factors influencing how beetles select and interact with host trees, how these factors are influencing population structures, and how the beetles continue to evolve. Given the central importance of pheromones in their life cycle, there are chapters describing their production, reception, continued evolution, and considerations impacting forest management (for one species). Finally, proteomic and genomic analyses are summarized with respect to responses to a variety of situations ranging from feeding to overwintering to pheromone production.

The editors have been fortunate to play a small part in advancing our understanding of bark beetle biology, especially in the area of aggregation pheromone production. They have had the good fortune to work with and know some of the leaders in the field whose work is covered and authored by the writers of this volume.

CLAUS TITTIGER
GARY J. BLOMQUIST

CHAPTER ONE

Host Selection by Bark Beetles: Playing the Odds in a High-Stakes Game

K.F. Raffa*, M.N. Andersson†, F. Schlyter‡
*University of Wisconsin-Madison, Madison, WI, United States
†Lund University, Lund, Sweden
‡Swedish University of Agricultural Sciences, Alnarp, Sweden

Contents

Advances in Insect Physiology, Volume 50
ISSN 0065-2806
http://dx.doi.org/10.1016/bs.aiip.2016.02.001

Abstract

Bark beetles face challenges and trade-offs during host selection, imposed by lethal tree defences, lower nutrition and higher competition in less well-defended trees, scarcity and ephemeral distribution of susceptible hosts, limitation of suitable hosts to one beetle generation, and relatively short lifespan and vulnerability of adults during host searching. Beetles contend with these challenges by using multiple, integrated sensory modalities, and sequential decision making. They incorporate both negative and positive information at multiple scales to locate susceptible hosts across large and heterogeneous landscapes. Some of the olfactory sensory neurons that convey non-host signals are relatively broadly tuned, whereas those that underlie intraspecific communication and host quality assessment are more specific, an arrangement that maximizes the use of antennal space while retaining high fidelity of detection channels strongly linked to survival and reproduction. The pertinent co-localization of neurons within sensilla provides the means for odour mixture processing in the periphery and enhances odour source discrimination and evaluation of chemical ratios in host tissue. Bark beetles show high behavioural plasticity in their orientation sequences, which allows them to adjust to environmental variability. This plasticity is modulated by environmental, genetic, and gene × environment drivers. Behavioural plasticity allows individuals to best realize the benefits that can be derived from pheromone-mediated cooperative attacks when populations are high. Cross-scale linkages among neurons, sensilla, orientation sequences, populations, and habitat structure underlie the landscape-scale environmental and socioeconomic impacts bark beetles exert. They also underlie rapid outbreaks in response to human-induced environmental alterations, such as climate change, habitat manipulation, and global transport.

ABBREVIATIONS

CSP chemosensory protein
GC-EAD gas chromatography–electroantennographic detection
GLV green leaf volatiles
GR gustatory receptor
IR ionotropic receptor
NHV non-host volatiles
OBP odorant-binding protein
OR odorant receptor

OSN olfactory sensory neuron
SDH semiochemical diversity hypothesis
SNMP sensory neuron membrane protein
SSR single sensillum recordings

1. CHALLENGES AND TRADE-OFFS POSED BY THE RESOURCE: THE HOUSE USUALLY WINS

Most of our knowledge about bark beetles comes from information gathered during large-scale outbreaks. This makes sense from the perspectives of both resource management and scientific inquiry. Limited resources must be allocated to where and when they are most urgently needed, and conducting adequately replicated studies on low-density populations in large forested ecosystems is expensive, difficult, and often unrewarding. Yet the overall consequence of this imbalance is that we have a largely distorted view of bark beetle behaviour and ecology. In fact, only a small proportion of bark beetle species undertake the risk of entering live trees (Lindgren and Raffa, 2013). Of these, only a few undertake the heightened risk of entering healthy trees and consequently cause intermittent outbreaks. And among these few species, most generations are not spent during their outbreak phase, but rather during extended periods when reproductive gains within plants are offset by losses during host finding and colonization (Kausrud et al., 2011; Raffa et al., 2008; Safranyik et al., 1975).

Upon closer examination, being a bark beetle is a difficult way to make a living. They must locate host trees across complex and ever-shifting landscapes, and host location must be refined not just to species but also to appropriate age and size categories, and further yet to a relatively narrow range of physiological conditions (Wood, 1972). Accepting the wrong tree is punished with death, due to trees' highly sophisticated defence systems (Bohlmann and Gershenzon, 2009; Franceschi et al., 2005; Raffa and Berryman, 1983). Not accepting the right tree is punished with lost opportunity, ie, failure to reproduce (Berryman, 1979; Pope et al., 1980). When beetles do succeed in colonizing the right host, their means of overcoming tree defence, pheromone-mediated mass attacks, is exploited by predators, some of which are even more strongly attracted than the bark beetle to its own pheromone (Hulcr et al., 2005; Raffa et al., 2007; Wood, 1982). Herbivores also exploit bark beetle aggregation pheromones. These so-called secondary species usually cannot colonize live trees but are superior competitors in dead trees, including those killed by the primary bark beetle (Allison et al., 2001; Lindgren and Raffa, 2013; Poland and Borden, 1998b). To add insult to injury, some late-arriving competitors then produce their own mating pheromones, which attract more predators that subsequently feed on both the secondary and primary beetles (Boone et al., 2008). Examination of life table data (Amman, 1984; Berryman, 1973; Coulson, 1979), especially during the non-outbreak periods that typify most generations, suggests a bark beetle's odds of success are worse than ours at most casinos.

1.1 Ecological and Socioeconomic Importance, Diversity, and Life History of Bark Beetles

Despite the many challenges confronting bark beetles, these insects exert important effects on native and managed ecosystems, cause serious socioeconomic impacts, and pose resource managers with difficult, often value-laden decisions (Grégoire et al., 2015). Moreover, the impacts of some species are intensifying with anthropogenic inputs, most notably climate change, transport associated with global commerce, and habitat manipulation (Bentz et al., 2010; Jactel and Brockerhoff, 2007; Ryall and Fahrig, 2005; Simard et al., 2012).

Bark beetles play valuable roles in many ecological processes that are important to the functioning of forest ecosystems (Griffin et al., 2011; Müller et al., 2008). For example, they contribute to nutrient cycling by initiating the process of decomposition. Bark beetles are often the first colonizers of dead, aged, or declining trees, which they mechanically fragment and inoculate with microorganisms. Initial colonization launches a successional sequence that includes a rich diversity of wood-boring beetles, flies, wasps, and other arthropods. Avian and other vertebrate species feed on them and construct nests in the decaying trees. Mortality to sporadic trees across the landscape fosters canopy thinning, gap formation, and structural heterogeneity (Harvey et al., 2013). During outbreaks, bark beetles contribute to landscape-scale successional processes, which they may either accelerate or reverse, depending on other factors.

Bark beetles likewise cause substantial socioeconomic challenges (Embrey et al., 2012). Their resource, subcortical tissues, is vital to tree health, which places them into direct competition with humans. We derive multiple benefits from trees and invest large sums in their production, protection, and utilization (Cosanza et al., 1997). The economic hardships resulting from tree mortality often fall disproportionately on rural communities, and large-scale outbreaks threaten to convert sustainable resource-based industries into boom-and-bust bubbles. At an elemental level, bark beetles and people are both in the business of converting trees into homes. So conflict is inevitable.

The socioeconomic challenges, and also emerging environmental problems, posed by bark beetles are increasing due to anthropogenic activities that alter climate, forest structure, and species distributions. Recent climatic changes have resulted in more frequent outbreaks, larger outbreaks, and higher synchronicity of outbreaks among multiple species than in recorded history (Raffa et al., 2008). Warming temperatures benefit bark beetles by reducing their overwintering mortality, decreasing their development time, and adding transpiration stress on trees, which reduces their defences (Creeden et al., 2014; Logan and Powell, 2009; Sherriff et al., 2011). Severe drought compromises tree resistance by interfering with the mobilization and transport of defence chemicals through resin canals (Gaylord et al., 2013). Recent climate-aided outbreaks have occurred over such a large scale in North America that they transformed large areas from carbon sinks to carbon sources, potentially fostering additional warming (Ghimire et al., 2015; Kurz et al., 2008).

Warming temperatures have also allowed several species to extend their ranges to higher elevations and latitudes (Cullingham et al., 2011; Marini et al., 2012). For example, *Dendroctonus ponderosae* Hopkins historically impacted relatively lower-elevation lodgepole pines (*Pinus contorta* Doug. ex Loud.), its primary host. During intermittent warming trends, beetles dispersed into higher elevations dominated by whitebark pines (*Pinus albicaulis* Engelm.), but these episodes were relatively short lived and terminated with the return of normal cold conditions. In recent years, however, *D. ponderosae* has caused sustained mortality to whitebark pine, posing significant environmental threats (Logan et al., 2010). This keystone species is important for high-elevation vegetation processes, wildlife, hydrology, and soil retention. Further, ecosystems dominated by this slow-growing, late-maturing tree lack the resilience of those dominated by the main historical host, fast-growing, early cone-setting lodgepole pine. *D. ponderosae* has also spread northward in British Columbia, Canada, reaching such high numbers that it breached the historical eastern geophysical barrier of the Rocky Mountains and spread into previously uninfested Albertan forests (Lusebrink et al., 2013; Safranyik et al., 2010). Lodgepole pine hybridizes with jack pine (*Pinus banksiana* Lamb.) in this region, creating an unprecedented connection with the pine forests of eastern North America. Other examples of northward spread include the southern pine beetle, *Dendroctonus frontalis* Zimmerman, which spread beyond its historical range of the southeastern United States to New Jersey, New York, and Connecticut (Niemiec et al., 2014; Ungerer et al., 1999).

Global trade has also fostered the movement of bark beetles and their microbial symbionts. Once in a new region, beetles encounter unadapted trees and may exhibit much more destructive dynamics than in their native range. Recent examples include the introduction of *Dendroctonus valens* LeConte and its fungal symbiont *Leptographium procerum* into China (Sun et al., 2013), and ongoing widespread movement of coffee berry borer (Burbano et al., 2011). There are also linkages between climate change and human transport. For example, the large-scale pine mortality in the western United States exceeds local processing capabilities, creating economic incentives to transport killed timber to midwestern mills and markets, and with that increased likelihood of beetle or microbial introduction. Habitat manipulations that increase tree species, age, and genotype homogeneity alleviate many of the obstacles bark beetles face in natural ecosystems (Jactel and Brockerhoff, 2007). Likewise, forest fragmentation can impede predators, which are less likely than are beetle prey to disperse to new habitat patches to find resources (Ryall and Fahrig, 2005).

As a subfamily, the Scolytinae shows a high degree of diversity in the parts, taxonomy, ages, and physiological conditions of the plants they utilize (Hulcr et al., 2015; Kirkendall et al., 2015). Thus, no review can address host selection across this range. Therefore, we will focus on one group. Bark beetles that colonize the inner bark of main stems of conifers in temperate forests exert the greatest landscape-scale impacts, exert the greatest impacts on humans, are showing the most dramatic changes in response to anthropogenic activities, and comprise our greatest source of knowledge (Raffa et al., 2015). Therefore, we will emphasize this group, particularly *Dendroctonus*, *Ips*, and *Scolytus*. We will provide additional examples where useful.

The general life cycle of the above bark beetles is deceptively simple, given the complexities of their behaviours, population dynamics, and ecological relationships. Adults emerge from the tree in which they developed, and disperse. They land on trees, and if appropriate stimuli are present, bore into the cambium (Fig. 1). The host-selecting gender varies with genus. As beetles bore through the bark, they emit pheromones that attract both sexes in most species (Blomquist et al., 2010). These pheromones are often closely linked with host tree chemicals, which can serve as precursors, synergists, or stimulants of de novo synthesis (Blomquist et al., 2010; Seybold et al.,

Fig. 1 Life cycle of bark beetles: (A) Adult North American spruce beetle, *Dendroctonus rufipennis*, on outer bark during host selection. (B) Brood adult *Ips typographus* preparing to emerge from *Picea abies* to seek new hosts. Note larval tunnels that begin at egg niches along ovipositional gallery in phloem, radiate outward, and terminate in pupal chambers. Note also staining by microorganisms, some of which facilitate host colonization and utilization, and others that inhibit brood development. *Panel (A): Photo by Y. Cardoza, UW-Madison. Panel (B): Photo by SLU, Alnarp.* (See the colour plate.)

1995; Wood, 1982). Mating occurs under the bark, with *Dendroctonus* and *Scolytus* typically monogamous and *Ips* polygamous. Females excavate tunnels along which they oviposit. The larvae tunnel away from the ovipositional galleries, except in species that conduct solitary adult attacks where larvae feed in colonial chambers. Larvae feed on phloem and on fungi that the adults introduce into the tree (Bleiker and Six, 2007). Pupation and development to teneral adults occur in the phloem. Mature adults exit the tree and repeat the process. The number of generations per year varies among species, and with temperature within species.

1.2 The Lethal Barrier of Tree Defence

Conifers are equipped with highly sophisticated, multicomponent, and integrated defences against subcortical attack (Franceschi et al., 2005; Lombardero et al., 2000; Raffa and Berryman, 1983; Raffa et al., 2005). These include both constitutive and inducible components. The relative importance of each component varies among tree genera and species (Berryman, 1972; Matson and Hain, 1983). These defences can kill, repel, or inhibit attacking insects and microorganisms (Fig. 2) (Krokene, 2015; Matson and Hain, 1983; Schiebe et al., 2012).

Bark beetles have several morphological and behavioural adaptations that enable them to penetrate trees' first line of defence, ie, tough outer bark, such as conically shaped heads, compressed antennae, spined tarsi, and orientation to thin bark crevices. When beetles penetrate live tissue, they are confronted with a copious flow of resin. Resin is stored in various structures such as glands and is transported through canals and ducts (Berryman, 1972). This sticky material can entomb the beetles, thus terminating the attack, or it can delay their progress, allowing time for other defences to activate (Raffa, 2001; Raffa and Berryman, 1983). Resin consists of a matrix of carbohydrates, toxic monoterpenes, diterpenes, and phenolics, and also sesquiterpenes of unknown biological activity.

Fig. 2 Adult bark beetles killed by trees during failed colonization attempts. (A) Red turpentine beetle, *Dendroctonus valens* in *Pinus resinosa*. (B) Mountain pine beetle, *Dendroctonus ponderosae* in *Pinus contorta*. (C) European spruce beetles, *Ips typographus* in *Picea abies. Panel (A): Photo by K. Raffa. Panel (B): Photo by A. Lerch, UW-Madison. Panel (C): Photo by F. Schlyter.* (See the colour plate.)

Conifer-induced reactions include two major components, hypersensitivity and biosynthesis (Berryman, 1972). Trees undergo a rapid autonecrotic reaction that confines the beetle and its associated fungi within lesions (Krokene, 2015; Reid et al., 1967). Concentrations of monoterpenes, diterpenes, phenylpropanoids, and phenolics increase rapidly within this lesion, soon reaching concentrations above the tolerance limits of the beetle, its brood, and its symbionts (Raffa et al., 2005). These materials are repellant, antifeedant, inhibitory to pheromonal attraction, adulticidal, ovicidal, inhibitory to mycelial growth and spore germination, and bactericidal (Erbilgin et al., 2007; Faccoli and Schlyter, 2007; Klepzig et al., 1996; Manning and Reid, 2013). Typically no one compound inhibits all aspects of beetle and microbial development, but conversely, no aspect of beetle and microbial development is unaffected by all of the bioactive compounds present. In addition to quantitative increases, the proportions of various compounds may change over time, placing the beetle in the difficult situation of having selected one environment but finding itself in a very different one (Raffa, 1991).

Conifer–bark beetle interactions differ from most plant–herbivore systems in being largely all-or-nothing relationships. When folivorous insects feed on a plant within their host species range that has high concentrations of defence compounds, they most commonly experience delayed growth, reduced growth, and lowered fecundity (Delvas et al., 2011; Hwang and Lindroth, 1997; Martemyanov et al., 2015). Also, externally feeding insects can migrate to neighbouring plants (Hunter and Lechowicz, 1992; Stoyenoff et al., 1994), something bark beetle larvae cannot do. When bark beetle adults encounter effective defences they may exit, die, or overcome the defences by attracting enough conspecifics. When pheromone-mediated mass attacks are successfully elicited, often little of the tree's prior potential to reduce beetle performance is manifested among any of the attackers' reproductive success (Raffa and Berryman, 1983). Whether these cooperative attacks succeed or fail depends not only on a tree's defensive capability but also on a higher scale factor, the beetle's stand-level population density (Boone et al., 2011).

At least four factors of bark beetle life history and ecology necessitate that their host selection decisions must be consistently correct, rapid, and responsive to a shifting environment. First, most species that colonize living trees deposit all or most of their eggs in a single tree. This precludes substantial bet-hedging by partitioning eggs among different hosts. Second, most bark beetles are generally short lived as adults (Coulson, 1979). While outside trees, they are exposed to generalist predators, weather, and rapid depletion of their energy reserves (Atkins, 1969; Kinn et al., 1994; Williams and Robertson, 2008). Third, bark beetles engage in intense scramble competition. As each susceptible tree in the landscape is located and colonized, it is eliminated from the available pool. At the tree scale, aggregation is within days terminated by the release of antiaggregation pheromones (Wood, 1982). Fourth, even though the likelihood of attack success increases with the number of colonizers, the available phloem per individual declines with the number of colonizers (Anderbrant et al., 1985). Further, a tree is typically available for only one or, in some cases, two beetle generations. So successful colonization permanently removes a tree from the resource, and the better-defended trees remain.

1.3 Trade-offs in Host Selection Affecting Beetle Fitness

One factor that bark beetles can exploit to tip the balance in their favour is the direct relationship between environmental stress and tree susceptibility. Almost all forms of severe stress can reduce tree defences against bark beetles. These include biotic processes such as defoliation (Wallin and Raffa, 2001), root disease (Klepzig et al., 1991), and competition (Raffa and Berryman, 1983), and abiotic processes such as severe drought (Breshears et al., 2009; Gaylord et al., 2013; Netherer et al., 2015), lightning (Coulson et al., 1983), and fire injury (Davis et al., 2012; Hood and Bentz, 2007; Powell and Raffa, 2011; Powell et al., 2012). Stresses can be acute or chronic such as due to old age (Safranyik et al., 1975). So restricting host acceptance to highly stressed trees has some clear advantages.

Colonizing severely stressed trees also incurs some disadvantages, however (Lindgren and Raffa, 2013; Raffa and Berryman, 1987). First, such trees are relatively uncommon, and highly unpredictable in space and time (Atkins, 1966), and as each stressed tree is colonized, it is removed from the supply. So strict reliance on stressed trees increases the risk of opportunity costs, ie, failure to reproduce. Second, highly stressed trees are available not only to the primary bark beetle species but also to many other bark beetles, wood borers, and saprogenic fungi. These species are typically superior competitors to tree-killing bark beetles (Amman and Safranyik, 1985; Rankin and Borden, 1991; Safranyik et al., 2000). Third, trees that are under stress often have thinner phloem or may be lower in nutrients, thus providing a lower-quality substrate.

In general, then, bark beetles face a difficult trade-off. Healthy trees are risky to enter and require high numbers of conspecifics to overcome, but they are abundant, provide high substrate quality, and are less available to interspecific competitors. Severely stressed trees pose less risk, but they are less abundant and more scattered, provide relatively lower substrate quality, and are laden with interspecific competitors. Complicating their decision is that in nature host selection is made under no-choice conditions. That is, a beetle can either enter or leave the tree on which it landed, not select among concurrent alternatives as in some laboratory bioassays.

1.4 Some General Solutions to Winning a Game of Chance in an Uncertain Environment

The following sections provide some of the specific mechanisms by which bark beetles perform host tree selection. These behaviours include some general features that can be better understood within the above context of multiple and conflicting selective pressures. These include:

Using multiple, integrated sensory modalities allows beetles to obtain complementary information. These sources of information include volatile and labile chemical cues, visual cues, tactile cues, and internal cues such as age and lipid content.

Making decisions in a stepwise and non-deterministic fashion. Orientation sequences provide opportunities to alter decisions at multiple junctures. Bark beetles choose to play blackjack, not roulette.

Incorporating both negative (eg, non-host volatiles (NHV) and antiattractive pheromone compounds) and positive information at multiple scales fosters host location across landscapes. Broader tuning of sensory neurons to convey non-host vs narrow tuning to access within-host variation can maximize efficiency. Pertinent co-localization of neurons within sensilla can enhance exploitation of complex chemical mixtures.

Incorporating plasticity to track a changing environment. This allows beetles to adjust to environmental variability, some of which arises from feedbacks to their own reproduction, by making context-dependent decisions. This plasticity can foster host-switching and rapid responses by bark beetles to human-induced alterations in their environment.

Our objective is to link the physiological mechanisms by which bark beetles recognize and orient to host plants with the ecological and evolutionary context in which they perform these decisions. However, we must admit to some immediate caveats. First, we have more information, especially about neurophysiological mechanisms, for some species than others. For example, much more work has been done on chemoreception of *Ips typographus* L. than other species, especially concerning detection of non-pheromonal compounds. In some aspects, this imbalance in mechanistic studies between the European and North American literatures hinders direct comparisons and syntheses, because these regions also have differences in forest structure, climate, and management regimes. Also, the most important and hence most studied North American species are in the genus *Dendroctonus*, compared to *Ips* in Europe. Several *Dendroctonus* species are more capable of surviving and reproducing in a defended, standing tree, while most *Ips* in both Europe and North America appear more dependent on highly stressed or even dead trees, such as wind fells (Kärvemo and Schroeder, 2010). Second, there is no one species that has been studied extensively across all scales, further impeding synthesis. Third, genomic, transcriptomic, and proteomic approaches are just now beginning to be completed for some bark beetle species. No one would characterize these insects as 'model species' for controlled mechanistic studies, given that they cannot be easily reared in artificial media and instead are most suitably reared in logs from mature trees, cannot develop into normal individuals without a specific flora of external symbionts, cannot be easily observed during feeding, and have very slow generation times. Finally, the rationale to study bark beetles primarily because of their ecological and economic importance has fostered emphasis on a few species that are not particularly representative of most members of the subfamily.

2. MECHANISMS: MAKING THE MOST OF THE HAND YOU ARE DEALT

2.1 Multiple Senses in the Host Selection Sequence

Finding rare suitable host trees in ever-changing forest landscapes dominated by unsuitable trees in combination with antagonizing factors, such as intra- and interspecific competitors, predators, tree defences, and other external threats, represents a tremendous challenge and a high-stake game for host-seeking bark beetles. Host selection mistakes, such as accepting the wrong tree or deciding to not colonize a suitable tree, are associated with high costs that have to be paid in the currency of reduced survival and reproduction, or the complete lack thereof. Hence, host selection imposes strong selection pressures on the sensory systems that bark beetles can utilize for detecting and interpreting external cues and signals from hosts and other ecological sources in their environment. Only with a fine-tuned sensory make-up can a bark beetle accomplish the difficult task of finding a host tree of the right species, proper age and size, appropriate stress level, sufficient nutrient content, and one without too many competitors and enemies or too few potential mates. Thus, bark beetles, like other animals, integrate information from multiple senses to make advantageous decisions as to which hosts to select and in which habitats to search for them (Campbell and Borden, 2006b). The chemical senses have received the most attention; however, visual cues are also of great importance. For example, tree-killing bark beetles seem to be more attracted to vertical objects resembling standing tree trunks as compared to horizontal objects (Lanier, 1983). In addition, several species of bark beetles are more attracted to black traps than to white traps, probably because the visual spectrum of black traps more closely resembles that of the dark stems of host trees (Campbell and Borden, 2006b).

It is likely that different long-range sensory modalities (vision and olfaction) are used during different steps of host selection, in a sequential process (Schlyter and Birgersson, 1999). These steps include the localization of (i) the correct habitat, (ii) host species, and (iii) a suitable host individual (Schlyter and Birgersson, 1999; Zhang and Schlyter, 2004). Vision might contribute in the habitat localization step, for instance, to discriminate between forests and fields or between conifer forests and deciduous forests. Vision has also been shown to influence the decision of bark beetles to land at close range (<2 m) (Campbell and Borden, 2006a). Thus, within a suitable habitat visual cues are likely integrated with olfactory information to aid the bark beetles when searching for a suitable tree species to colonize.

In this section, however, we will focus on the olfactory sense, which is of utmost importance for long-range attraction to hosts and for which several recent studies have advanced our knowledge of the physiological mechanisms that underlie bark beetle host selection behaviour. Over decades of research, it has become apparent that probably all bark beetle species use a combination of positive (attractive) and negative (antiattractive or repellent) olfactory signals to select their hosts. Attractive signals include aggregation pheromones and volatiles released by the host trees (Byers, 2004), while antiattractive signals include con- and heterospecific pheromones (Schlyter and Birgersson, 1999), NHV from angiosperm plants (Zhang and Schlyter, 2004), as well as certain volatile compounds that likely are included in the host tree's arsenal of chemical defences (Andersson et al., 2010; Schiebe et al., 2012).

2.2 Attraction to Hosts

The relative importance of host compounds vs aggregation pheromones in host attraction varies among bark beetle species. For instance, species exhibiting so-called primary attraction are attracted to host compounds (eg, monoterpenes and ethanol) even without pheromones, whereas species with secondary attraction use aggregation pheromones as the main long-distance attractant for host colonization (Byers, 2004). All species that are able to successfully colonize and kill healthy trees exhibit secondary attraction and coordinate mass attacks using aggregation pheromones (reviewed in Byers and Zhang, 2011). However, it is still debated how the pioneering individuals of these species find the host trees at a distance. One hypothesis is that they land on trees at random (Byers, 1996; Saint-Germain et al., 2007; Wood, 1982) and then decide which trees to enter based on contact chemical and physical cues (see also Section 5.1). Host acceptance is then followed by the production and release of aggregation pheromones, which attract the majority of the attacking individuals. This is compatible with the view that host compounds may attract beetles to the right habitat, but not to specific trees (Saint-Germain et al., 2007). Major host monoterpenes can also synergize pheromone attraction, both in species with primary and secondary attraction (Erbilgin et al., 2007; Hulcr et al., 2006; Renwick and Vité, 1969; Rudinsky et al., 1972). It has also been hypothesized that the major monoterpenes released by more or less all conifer tree species might constitute a common attractive habitat signal for coniferous bark beetles in general, whereas specific minor host compounds or compound ratios (cf. Bruce et al., 2005) might be important in the selection of the correct species (Kalinová et al., 2014; Tømmerås and Mustaparta, 1987). In addition, the released amounts of the minor compounds might depend on the trees' age and stress and defence level (Andersson et al., 2010; Blažytė-Čereškienė et al., 2015; Kalinová et al., 2014; Schiebe, 2012; Schiebe et al., 2012). Thus, relatively less abundant compounds might be crucial in the selection of suitable host individuals, especially for the pioneering individuals that are not guided by the aggregation pheromone. In addition, relatively less abundant host compounds that can inform the beetle about the stress or defence status of a potential host might be especially important during endemic population levels when only weakened trees can be successfully colonized, as these trees have to be distinguished from the vigorous ones that are likely to repel or kill its attackers.

Irrespective of the modality and interspecific variability of host selection sequences, the antennae of all studied bark beetle species can detect a substantial number of major and minor host volatiles, suggesting that host compounds play significant roles in host selection (Andersson et al., 2010; Blažytė-Čereškienė et al., 2015; Kalinová et al., 2014; Schiebe, 2012). Indeed, bark beetles possess several classes of olfactory sensory neurons (OSNs) specific for host monoterpenes (Andersson, 2012; Andersson et al., 2009; Schiebe, 2012; Tømmerås, 1985), suggesting that specific detection of these compounds is important for their fitness, and sufficiently so to be maintained over the course of evolution (cf. Andersson et al., 2015).

2.3 Modulation by Antiattractive Signals and the Semiochemical Diversity Hypothesis

A meta-analysis showed that there is a general reduction in herbivory by oligophagous insects as the biodiversity of forests increases (Jactel and Brockerhoff, 2007). The semiochemical diversity hypothesis (SDH), as postulated by Zhang and Schlyter (2003), suggests that this decrease might be due to the presence of antiattractive volatiles from non-host plants, reducing the searching efficiency of specialist and oligophagous insects. Thus, bark beetles may actively avoid diverse habitats because they worsen the odds of finding a potential host as compared to searching in conifer-dominated habitats. Numerous studies on bark beetles have shown that NHV reduce the attraction to individual point sources (reviewed in Zhang and Schlyter, 2004). In addition, several studies have also indicated area-wide or 'habitat-scale' inhibitory effects of NHV in bark beetles and other forest insects (Jactel et al., 2011; Schlyter, 2012).

By using passive (unbaited) traps around point sources baited with aggregation pheromone and NHV, the 'active inhibitory radius' of NHV mixtures on *I. typographus* was estimated at 2–4 m, suggesting that the effect of NHV is not restricted to individual point sources (Zhang and Schlyter, 2003). In addition, odour source spacing experiments, in which single pheromone dispensers were separated from single NHV dispensers on multiple-funnel 'Lindgren' traps, indicated that one point source of an NHV mixture can reduce the attraction of *I. typographus* to a pheromone released >1 m away from the NHV source (Andersson et al., 2011). It was also shown that eight NHV dispensers, again releasing a mixture of compounds, positioned in a circle around a central pheromone trap reduced trap catches up to a radius of 2–3 m (Andersson et al., 2011). While these studies all suggest that individual dispensers releasing a mixture of NHV compounds inhibit pheromone attraction over a few metres distance, they do not conclusively demonstrate that inhibition scales up to habitat-scale effects. However,

Schiebe et al. (2011) treated groups of susceptible Norway spruce trees (*Picea abies* (L.) Karst) at forest edges with NHV mixtures (ie, no synthetic attractant was used), which diverted *I. typographus* attacks from the experimental zones to untreated trees 15–30 m away. This observation indicates habitat-scale effects of NHV on bark beetle host selection behaviour. Avoiding searching for a suitable host in habitats where the chances of finding one are slim seems adaptive, as it likely reduces host selection costs, such as energy expenditure, predation risk, and time that otherwise could have contributed to reproduction by the short-lived adults. The variation in the observed active distance of NHV in the different studies is likely due to the use of aggregation pheromone as attractant, which is a very strong signal, vs the absence of such a signal when applying NHV to non-attacked trees. This difference also highlights the profound difference between the decisions facing pioneering individuals, which must select optimal hosts in the absence of conspecific cues, vs the responding beetles that can ignore negative olfactory signals from the host or surrounding plants in the presence of cues indicating success by pioneers.

It is possible that the different constituents of the NHV mixtures are active at different steps in the host selection sequence. It has been hypothesized that general green leaf volatiles (GLV, mostly C_6-alcohols), primarily released by angiosperm trees, might serve as a negative signal at the habitat level (Schlyter and Birgersson, 1999), whereas more specific NHV from the bark (eg, C_8-alcohols and the spiroacetal *trans*-conophthorin) might represent unsuitable host species signals (Zhang and Schlyter, 2004). Antiattractive pheromonal compounds, such as verbenone, produced primarily by yeasts in bark beetle galleries (Leufvén et al., 1984), might signal host unsuitability at the host individual level (Zhang and Schlyter, 2004). Bark beetles are thought to use these compounds to avoid trees with high intraspecific competition and low nutritional value (Schlyter et al., 1989). In addition, it was recently shown that the amounts of certain host compounds, including 1,8-cineole and *p*-cymene, increase in *P. abies* heavily attacked by *I. typographus* (Andersson et al., 2010). 1,8-Cineole (and *p*-cymene to some extent) was shown to strongly inhibit pheromone attraction of *I. typographus* and thus might also serve as a negative signal in host suitability assessment (Andersson et al., 2010). Schiebe et al. (2012) demonstrated that 1,8-cineole is present in significantly higher amounts in *P. abies* that were resistant to bark beetle attack (ie, trees in which pioneering individuals were unsuccessful), as compared to those that were successfully attacked and killed. Thus, a bark beetle is likely to increase its chances of survival and reproduction by avoiding trees with strong chemical defences as signalled and/or mediated by 1,8-cineole (Andersson et al., 2010; Binyameen et al., 2014). Similarly, the host phenylpropanoid 4-allylanisole (also known as estragole), which inhibits growth of bark beetle-associated fungi, inhibits pheromone attraction of several North American *Ips* and *Dendroctonus* spp. (Hayes and Strom, 1994; Hayes et al., 1994). In addition, a mixture of major host monoterpenes interrupted pheromone attraction of the Asian larch bark beetle *Ips subelongatus* Motschulsky (Zhang et al., 2007). It was hypothesized that the high release rates of monoterpenes that were used might have been indicative of a healthy tree—too vigorous to succumb to bark beetle colonization. Thus, host-derived compounds can serve as both attractants and antiattractants, with their exact behavioural effects being context dependent and varying among species, sexes, physiological state, and release rate. Finally, inhibition of attraction to aggregation pheromones by heterospecific pheromone compounds is common across Scolytinae (Birch et al., 1980; Byers, 1993; Poland and Borden, 1998a). This inhibition is likely to represent a mechanism for preventing interspecific competition and/or mating mistakes.

2.4 Odour Coding Mechanisms: Recent Advances

As outlined earlier the chemical ecology of bark beetles is complex, involving an entire suite of attractive and antiattractive volatiles of different ecological origins and meanings. Presumably, bark beetles need to be able to detect and interpret a large variety of volatiles and volatile mixtures from different ecological sources in order to respond appropriately to external cues and contend with the difficult trade-offs imposed by host selection. For a compound to be behaviourally active, the insect must possess sensory neurons that detect it. Especially in recent years, electrophysiological studies from individual OSNs (ie, single sensillum recordings, SSR) in combination with field trapping experiments on *I. typographus* have provided mechanistic insight into the peripheral coding of the signals that underlie bark beetle host selection behaviour. Studies have identified specific OSN classes for a large number of attractive and antiattractive odorants, and some OSN classes more broadly tuned to several antiattractants with similar chemical structure. Strong links have been established between the physiological response profiles of the neurons, or the frequency of certain OSN classes on the antenna, and the behavioural effects of the active compounds in the field. In addition, recent studies on bark beetles have revealed mechanisms of peripheral odour coding that are likely to be generally applicable to the Insecta class. We return to these discoveries in Section 2.5, after summarizing both the early discoveries as well as the recent advances that have been made with respect to peripheral detection of positive and negative olfactory signals, and their links to behavioural responses.

2.4.1 Pioneering Studies: Pheromone Responses

A large number of SSR studies were conducted on bark beetles during the 1970s and 1980s, largely focusing on the detection of pheromonal compounds (attractive and antiattractive, from con- and heterospecifics), and also a few major host volatiles (reviewed in Andersson, 2012). Because Andersson (2012) provides an extensive compilation of the early data, we will only provide a brief summary of these results here. OSNs responding specifically to pheromonal compounds were identified in *I. typographus* (Mustaparta et al., 1984; Tømmerås, 1985; Tømmerås and Mustaparta, 1987; Tømmerås et al., 1984), *Ips pini* (Mustaparta et al., 1977, 1979, 1980, 1984), *Ips paraconfusus* Lanier (Mustaparta et al., 1980), *Ips grandicollis* (Eichhoff) (Ascoli-Christensen et al., 1993), *Dendroctonus pseudotsugae* Hopkins (Dickens et al., 1984, 1985), *D. frontalis* (Dickens and Payne, 1977; Payne et al., 1982), *Dendroctonus micans* Kugelann (Tømmerås et al., 1984), and *Trypodendron lineatum* (Olivier) (Tømmerås and Mustaparta, 1989). Additionally, a few neuron classes responding to host compounds were found in *I. typographus* (Tømmerås, 1985), *I. pini* (Mustaparta et al., 1979), *D. pseudotsugae* (Dickens et al., 1984), *D. frontalis* (Dickens and Payne, 1977), *T. lineatum* (Tømmerås and Mustaparta, 1989), and a neuron class for the non-host compound benzyl alcohol in *Tomicus destruens* Woll. (Guerrero et al., 1997).

A major take-home message from the early SSR studies is that bark beetle OSNs that respond to attractive or antiattractive pheromonal compounds generally are highly specific, responding strongly to only one or a few structurally related compounds. A similarly high specificity is seen among the majority of the OSNs that detect sex pheromone components in moths, implying that high receptor specificity has been evolutionarily favoured in both sex- and aggregation-pheromone communication systems across the Insecta (Andersson et al., 2015). High OSN specificity of pheromone neurons is expected to provide high fidelity in the mate recognition channels, preventing costly heterospecific matings and contributing to reproductive isolation. In addition, bark beetle pheromone neurons often show high selectivity for a certain enantiomer or isomer of a compound, including OSNs specific for either the (+)- or (−)-enantiomer of *cis*-verbenol, *trans*-verbenol, verbenone, ipsdienol, and ipsenol in *Ips* spp. (Mustaparta et al., 1980; Tømmerås, 1985), and (+)- or (−)-frontalin in *Dendroctonus* spp. (Dickens et al., 1985; Payne et al., 1982). This is consistent with the specific, sometimes opposing, behavioural responses to different enantiomers of these pheromone compounds (Birch et al., 1980; Wood, 1982). Furthermore, the frequency (or abundance) of enantiomer-specific OSNs on the antennae seems to correlate with the attractiveness of the compound, such as *I. paraconfusus* being attracted to (+)-ipsdienol having more neurons tuned to the (+)-enantiomer than to the (−)-enantiomer, whereas *I. pini* that is attracted to (−)-ipsdienol shows the opposite pattern (Birch et al., 1980; Mustaparta et al., 1980). That is, enantiomer-dependent behavioural responses to pheromones are reflected in the peripheral olfactory system, both in terms of OSN specificity and OSN frequency on the antennae.

2.4.2 Responses to Host- and Non-Host Volatiles

Since the early SSR studies, our knowledge of the chemical ecology of bark beetles has improved substantially, perhaps especially due to the identification of the antiattractive volatiles that are released by non-host plants (Zhang and Schlyter, 2004). Hence, Andersson et al. (2009) screened the olfactory sensilla of *I. typographus* with an expanded test odour panel comprised of similar numbers of pheromone-, host-, and non-host compounds, all with ecological relevance. In this study, the high specificity of pheromone OSNs was confirmed from the previous studies. Additionally, several new classes of OSNs responding to major or minor host compounds as well as non-host compounds were characterized (Table 1). For some of these OSN classes, additional active ligands have been identified more recently by Schiebe (2012) (Table 1). The newly discovered plant odour-responding neurons showed a variety of response specificities, with several being as specific as the pheromone neurons and others being more broadly tuned (Andersson et al., 2009; Schiebe, 2012). Several links between the response profiles of OSNs and the behavioural response to the active compounds could be established and are discussed below. Since *I. typographus* is the most well-studied bark beetle species in terms of olfactory detection by individual OSNs, we will use this species as a case study in the coming sections describing the recent advances in chemoreception physiology that underlie bark beetle host selection.

Three non-host GLV-alcohols, 1-hexanol, (*E*)-2-hexenol, and (*Z*)-3-hexenol, all reduce the attraction by *I. typographus* to its aggregation

Table 1 Strongly Responding (>100 Hz to Screening Dose, Generally 10 μg) Olfactory Sensory Neuron (OSN) Classes of *I. typographus*

Ligand Source	Number	OSN Class[a]	Secondary Responses[b]	Physiological or Behavioural Activity in *D. ponderosae*[c]	References
Beetle	1	(−)-*cis*-Verbenol	(−)-Verbenone	+	1, 3, 4
	2	(+)-*trans*-Verbenol		+	4
	3	(−)-Verbenone		+	1, 4
	4	2-Methyl-3-buten-2-ol			1, 4
	5	(+)-Ipsdienol	(−)-Ipsdienol (±)-Ipsenol Amitinol	+	1, 4, 5, 6
	6	(−)-Ipsdienol	(+)-Ipsdienol		1, 4, 5, 6
	7	(−)-Ipsenol	(+)-Ipsenol		1, 4
	8	Amitinol	(±)-Ipsdienol		1, 4
	9	Phenylethanol		+	4
Host	10	(+)-α-Pinene/ (−)-α-pinene	(−)-β-Pinene (−)-*cis*-Verbenol	+	1
	11	Myrcene	Terpinolene	+	1, 4
	12	Myrcene/*p*-cymene	Terpinolene	+	1
	13	*p*-Cymene	γ-Terpinene 3-Carene (±)-Limonene (±)-Carvone Estragole	+	1, 3
	14	1,8-Cineole		−	1
	15	3-Carene		+	1
	16	Pinocarvone/ (±)-camphor/ pinocamphone	(−)-β-Pinene	+	3, 4
	17	*trans*-4-Thujanol	Terpinene-4-ol 3-Octanol		3
	18	Styrene			3
Non-host	19	GLV-OHs: 1-Hexanol *Z*3-Hexenol *E*2-Hexenol	Hexanal *E*2-Hexenal (±)-1-Octen-3-ol (±)-Chalcogran	+	1
	20	(±)-3-Octanol	(±)-1-Octen-3-ol (+)-3-Octanol (−)-1-Octen-3-ol *Z*3-Hexenol 1-Hexanol	−	1, 2

Table 1 Strongly Responding (>100 Hz to Screening Dose, Generally 10 μg) Olfactory Sensory Neuron (OSN) Classes of *I. typographus*—cont'd

Ligand Source	Number	OSN Class	Secondary Responses	Physiological or Behavioural Activity in *D. ponderosae*	References
	21	(−)-1-Octen-3-ol	(±)-1-Octen-3-ol (±)-3-Octanol (+)-3-Octanol	+	1
	22	(*S*,*S*)-*trans*-Conophthorin	(±)-Chalcogran (±)-*exo*-Brevicomin[d] (*R*,*R*)-*trans*-Conophthorin Dehydro-conophthorin	+	1, 7

[a]OSN classes are named according to the compound(s) eliciting the primary response.
[b]Only compounds eliciting intermediate to strong (>70 Hz) secondary responses at the high screening dose are listed.
[c]From Andersson et al. (2013; see their Additional file 5). +, activity supported by positive data; −, negative data for activity; empty cells, no data.
[d]Tømmerås et al. (1984) found OSNs for *exo*-brevicomin, but since *trans*-conophthorin and chalcogran were not tested, it is likely that these neurons belong to the *trans*-conophthorin class.
References: 1, Andersson et al. (2009); 2, Andersson et al. (2012b); 3, Schiebe (2012); 4, Tømmerås (1985); 5, Mustaparta et al. (1984); 6, Tømmerås et al. (1984); 7, Unelius et al. (2014).
Ecological source of primary ligands, ligands eliciting secondary responses, and indication of semiochemical activity in *D. ponderosae* are also listed.

pheromone. These three compounds are behaviourally redundant; ie a mixture of all three compounds can be replaced by, eg, 1-hexanol alone with the same antiattractive effect given that the total release rate is kept constant (Unelius et al., 2014; Zhang and Schlyter, 2003). This behavioural redundancy is likely to be explained by the fact that the three GLVs strongly activate only one class of OSN (Table 1), and this OSN class has essentially the same sensitivity to all three compounds. Maintaining such 'selective broad tuning' (Andersson et al., 2015) of this OSN class seems adaptive because the three compounds basically confer the same message to a coniferous insect specialist, ie, non-host (Andersson, 2012). Thus, having an OSN class that responds to all three compounds would likely increase the probability that an important non-host signal is detected, without having to 'waste antennal space' by expressing several different GLV receptors in a larger number of sensilla. Several different types of GLV neurons are commonly found among angiosperm feeding insects, with some being specific for a single compound (Andersson et al., 2012a; Hansson et al., 1999; Larsson et al., 2001). An analogy is seen among certain OSNs for pheromone antagonists in moths, which have quite broad tuning, detecting several antagonizing heterospecific pheromone compounds (Andersson et al., 2015; Koutroumpa et al., 2014). Moreover, antiattractive compounds that synergize the inhibitory effect of the GLVs (Unelius et al., 2014; Zhang and Schlyter, 2003), such as verbenone and *trans*-conophthorin, are detected by different neurons that do not respond to GLVs (Andersson et al., 2009; Tømmerås, 1985) (Table 1). Thus, the synergistic effects of these antiattractants are likely to be a result of central integration of odour input from several OSN classes. Likewise, the synergistic effect of aggregation-pheromone components is also likely due to central integration, because individual pheromone components are detected by different and specific OSN classes across bark beetle species (Andersson, 2012 and references therein).

Consistent with the multiple roles of semiochemicals in bark beetle orientation, chalcogran, an aggregation-pheromone component of the sympatric competitor *Pityogenes chalcographus* (L), also reduces the attraction of *I. typographus* to its aggregation pheromone (Byers, 1993). Interestingly, the OSN class of *I. typographus* that primarily responds to the (5*S*,7*S*)-enantiomer of *trans*-conophthorin is also highly sensitive to chalcogran (Table 1), suggesting that the same neuronal channel mediates the inhibitory effect of both compounds, which are structurally similar spiroacetals (Andersson et al., 2009). Thus, this OSN class represents yet another example of selective broad tuning, allowing single neurons to detect several antiattractive compounds, ie, increasing perceived chemosensory space using a fixed number of OSNs 'to make the most of the hand you're dealt'. In addition, the *trans*-conophthorin neuron is more sensitive to both chalcogran and *exo*-brevicomin (both tested as racemic mixtures) than to the (5*R*,7*R*)-enantiomer of *trans*-conophthorin (Andersson et al., 2009). The >100-fold difference in sensitivity to the two enantiomers demonstrates that high enantiomer selectivity is not restricted to the pheromone communication system. The lack of electroantennogram responses to the *R*,*R*-enantiomer in *I. typographus* (Zhang et al., 2002) is consistent with the low OSN sensitivity to this compound and suggests that only the *S*,*S*-enantiomer is behaviourally active. Comparative molecular and SSR studies on species that utilize chalcogran or *exo*-brevicomin as attractants in their pheromone blends (ie, *Pityogenes* and *Dendroctonus* spp.) are needed to reveal whether detection of *trans*-conophthorin and its negative behavioural effect in *I. typographus* represent the ancestral state or a derived character.

Finally, high concentrations of α-pinene were shown to synergize the attraction to low concentrations of aggregation pheromone in *I. typographus* (Erbilgin et al., 2007). Both Tømmerås (1985) and Andersson et al. (2009) identified an OSN class that primarily responded to this compound (Table 1). Interestingly, the somewhat low sensitivity of this OSN class (ie, a sensitivity threshold at ca. one order of magnitude higher dose than for most other OSNs to their key ligands) (Andersson et al., 2009) might explain why release rates above 500 mg/day of α-pinene was required to increase the attraction to the pheromone. It is unknown whether the lower OSN sensitivity to this compound represents an ecological adaptation to constantly high α-pinene concentrations in coniferous habitats, or if it simply indicates that this OSN class is more sensitive to yet unidentified compounds.

2.4.3 Responses to Volatiles from Weakened vs Vigorous Hosts

During non-epidemic population densities, semiaggressive species such as *I. typographus* only attack weakened trees and prefer older over younger, better-defended hosts (Grodzki et al., 2006). Thus, one would expect such species to have an olfactory capacity to discriminate weakened from vigorous trees, the latter being likely to kill the attackers if too few conspecifics join in on the attack. Indeed, gas chromatography–electroantennographic detection (GC-EAD) recordings have shown that the antennae of *I. typographus* respond to several compounds associated with tree stress or age (Blažytė-Čereškienė et al., 2015; Kalinová et al., 2014; Schiebe, 2012). Oxygenated terpenes are present in very small amounts in the bark of healthy trees, but increase after bark beetle attack (Azeem et al., 2013; Pettersson and Boland, 2003) and also after artificial defence induction using the hormone methyl jasmonate (Martin et al., 2003). Thus, oxygenated monoterpenes may provide the beetle with crucial information about the health and defence status of a potential host tree. In *I. typographus*, strong GC-EAD responses were elicited by several stress-related compounds, eg, 1,8-cineole, *trans*-4-thujanol, terpinene-4-ol, camphor, and pinocarvone (Blažytė-Čereškienė et al., 2015; Schiebe, 2012). The amount of *trans*-4-thujanol was shown to decline with tree age and induced repellent responses in laboratory bioassays, thus possibly representing a signal of host suitability related to age (Blažytė-Čereškienė et al., 2015). The two non-terpenoid compounds estragole (synonym 4-allylanisole) and styrene also elicited strong GC-EAD activity (Schiebe, 2012). The antennally active compounds were subsequently tested by SSR, and specific OSN classes for some of them were identified (Table 1). One class responded primarily to *trans*-4-thujanol with secondary responses elicited terpinene-4-ol and a few other alcohols. A second OSN class responded specifically to styrene, whereas a third class responded primarily to pinocarvone, but was almost as strongly activated by camphor (Schiebe, 2012). In this study, estragole elicited intermediate responses in the previously identified OSN class responding primarily to *p*-cymene; however, a recent screening also identified an OSN class primarily tuned to this compound (D. Kandasamy, M.N. Andersson, et al., unpublished data). An OSN class responding specifically to 1,8-cineole had previously been identified (Andersson et al., 2009) (see also Section 2.5). Apart from 1.8-cineole, that strongly inhibits pheromone attraction (Andersson et al., 2010; Binyameen et al., 2014), the behavioural effects of the other compounds on *I. typographus* remain to be tested in the field.

2.4.4 OSN Frequency and Antennal Distribution

Andersson et al. (2009) showed that an unexpectedly high frequency of OSNs on the antenna of *I. typographus* is tuned to antiattractive NHV. This result highlights the ecological importance of these negative signals; ie, colonizing an angiosperm plant results in zero reproductive output, and searching for a host in a habitat dominated by non-hosts is costly and perhaps ultimately unsuccessful. As many as 23% of the responding neurons responded primarily to NHV, and these OSNs could be grouped into four OSN classes based on their response spectra (Andersson, 2012; Andersson et al., 2009) (Table 1). In comparison, 42% of the neurons (seven classes) responded to con- and heterospecific pheromone compounds, whereas 35% (six classes) were tuned to host compounds. The latter observation further strengthens the idea that host compounds are important for species that exhibit both primary and secondary attractions (see also, eg, Blažytė-Čereškienė et al., 2015; Kalinová et al., 2014), though their individual or combinatory effects on behaviour largely remain elusive and are likely to be context dependent and variable among species.

The olfactory sensilla of *I. typographus* and other bark beetle species (eg, Chen et al., 2010) are localized in three main areas (or bands), previously named A, B, and C (from proximal to distal), on the ventral surface of the flattened antennal club (Hallberg, 1982) (Fig. 3A). The different OSN classes of *I. typographus* are not randomly distributed across the three areas. Instead, a certain OSN class tends to be found either in both A and B areas, but not in the distal C area, whereas other classes are found only in the C area. The OSN classes responding to the aggregation-pheromone component *cis*-verbenol, the major host compound α-pinene, and the minor host compound 1,8-cineole, respectively, were mainly found in area C (Fig. 3B), whereas the majority of the other classes were found in both the A and B areas, but not in area C (Andersson et al., 2009). These two distribution patterns of OSNs correspond to the distribution of two morphological

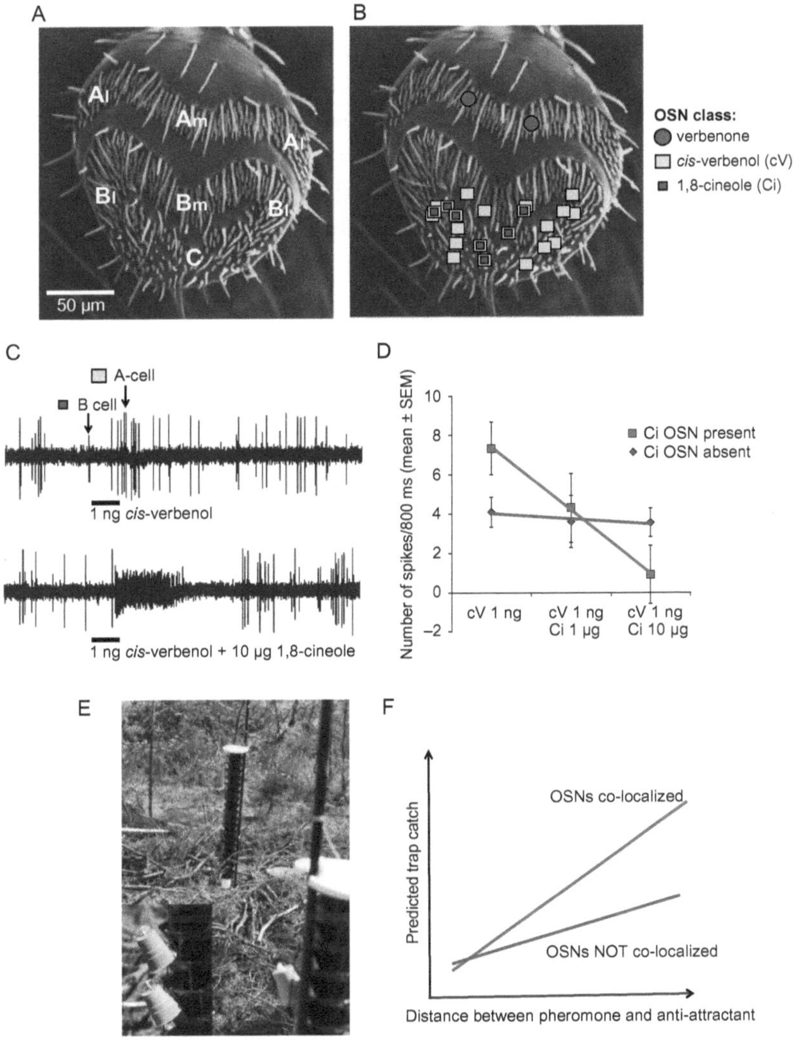

Fig. 3 (A) Ventral surface of the antennal club of *Ips typographus* showing the arrangement of olfactory sensilla in three main areas (A, B, and C). Based on the presence of certain functional classes of olfactory sensory neurons (OSNs) exclusively on the lateral sides of the A and B areas, these areas are subdivided into lateral (A_l and B_l) and medial (A_m and B_m) regions. (B) Approximate locations of sensilla containing OSNs for verbenone (filled *red circles*), *cis*-verbenol (cV, *large yellow squares*), and 1,8-cineole (Ci, *small blue squares*), the latter OSN class always co-localized with cV-OSNs. Ca. half of the cV-OSNs are co-localized with an uncharacterized OSN. (C) Raw spike trains showing the large amplitude cV A-neuron and the small amplitude Ci B-neuron in the same sensillum. *Upper trace* shows the excitation of the cV-neuron in response to stimulation with

(Continued)

types of olfactory sensilla; ie, the shorter and thinner single-walled sensillum type 1 (SW 1) occurs predominantly in areas A and B, whereas the longer and slightly thicker SW 2 is mainly present in area C (Hallberg, 1982). However, a recent study that aimed to identify additional OSN classes among the antennal sensilla of *I. typographus* suggests that the spatial organization of OSN classes on the antenna is more fine scaled than previously suggested; ie, a few newly discovered OSN classes were exclusively found on the lateral zones of areas A and B (D. Kandasamy, M.N. Andersson, et al., unpublished data; Fig 3A). Whether the spatial organization of OSNs across the antennae represents an ecological or physiological adaptation, or simply is due to neurodevelopmental constraint, remains unknown. Likewise, the functional significance, if any, of utilizing two morphological types of sensilla (SW 1 and SW 2) that seem to harbour OSNs for both pheromonal and plant volatiles also remains to be investigated.

2.5 Integration of Odour Blends by Co-Localized Neurons: Bark Beetle Research Revealed Novel Coding Mechanisms

As outlined earlier, minor host compounds are likely to be important, in a variety of ways, for bark beetles when selecting a suitable host tree to attack. One example is the monoterpene-ether 1,8-cineole, a putative spruce defence compound (Schiebe et al., 2012), that strongly reduces the attraction of *I. typographus* to its aggregation pheromone (Andersson et al., 2010; Binyameen et al., 2014). Interestingly, the OSN class that specifically

Fig. 3—cont'd 1 ng cV. *Lower trace* shows the inhibition of the cV response in the A-neuron and the simultaneous excitation of the Ci B-neuron due to stimulation with a binary mixture of cV and Ci. *Black horizontal bars* indicate the 0.5 s odour pulse. (D) Mean response of the cV A-neuron to 1 ng cV presented alone or in binary combinations with Ci ($N = 10$–12). The cV response is inhibited by Ci only in sensilla in which the cV OSN is paired with the Ci OSN (*blue line*). (E) Lindgren multiple-funnel traps with vertically spaced pheromone and antiattractant dispensers (grey cups) used in the odour source spacing experiment (Binyameen et al., 2014). (F) Predicted trap catch increase in response to odour source spacing of pheromone and antiattractants detected by co-localized (*blue line*) vs non-co-localized OSNs (*red line*), respectively. *Panels (A) and (B): Modified from Andersson, M.N., Larsson, M.C., Schlyter, F., 2009. Specificity and redundancy in the olfactory system of the bark beetle* Ips typographus*: single-cell responses to ecologically relevant odors. J. Insect Physiol. 55, 556–567. Panel (B): Micrograph provided by E. Hallberg. Panels (C) and (D): Redrawn from Andersson, M.N., Larsson, M.C., Blaženec, M., Jakuš, R., Zhang, Q.-H., Schlyter, F., 2010. Peripheral modulation of pheromone response by inhibitory host compound in a beetle. J. Exp. Biol. 213, 3332–3339. Panel (E): Photo by F. Schlyter.* (See the colour plate.)

responds to 1,8-cineole (OSN with small spike amplitude; referred to as B cell) is always found co-localized in the same sensilla as the OSN class that specifically responds to *cis*-verbenol (large amplitude, A cell), although ca. half of the *cis*-verbenol neurons are co-localized with B cells with yet unknown response specificity (Andersson et al., 2009) (Fig. 3B). Thus, this organization of pheromone neurons being co-localized with neurons for host compounds in bark beetles is different from the organization in moths where neurons for pheromone compounds and host compounds generally are grouped in different sensilla (Berg et al., 2005; Hansson et al., 1986). This might be a reflection of the fact that host and mate localization in bark beetles is an integrated process that involves the detection of both pheromone compounds and host volatiles (Andersson et al., 2009, 2010).

In insects in general, the assessment of specific ratios of volatile mixtures is often crucial for behavioural decisions, and it is thought that co-localization of neurons in sensilla improves the ratio assessment of mixtures, facilitated by coincidence detection of odorants by the co-localized OSNs (Bruce et al., 2005). Furthermore, it has been hypothesized that co-localization allows for signal modulation in the periphery via passive interactions between responding OSNs sharing the same extracellular compartment (Vermeulen and Rospars, 2004). It has also been suggested that co-localization improves the discrimination of closely separated odour sources, again because co-localization allows for coincidence detection of odour filaments. This hypothesis was derived from wind tunnel experiments on male noctuid moths, showing that the disruptive effect of a pheromone antagonist was significantly reduced when the antagonist was released from a source 1 mm from the pheromone source, corresponding to 0.001 s temporal separation of the odour filaments from the two sources (Baker et al., 1998; Fadamiro et al., 1999). It was hypothesized that such fine-scale resolution could only be achieved if the detectors for the two compounds were located at the same point in space, ie, are co-localized in the same sensilla (Baker et al., 1998). In fact, the well-characterized olfactory system of *I. typographus* has been exploited to find experimental support for both peripheral OSN interactions and improved odour source discrimination via co-localization of OSNs. These studies suggest that mixture processing by the peripheral olfactory system might represent an important mechanism that aids beetles in the challenging task of finding a host in which they are likely to survive and successfully propagate.

Andersson et al. (2010) found that the spontaneous activity of the *cis*-verbenol neuron of *I. typographus* was completely inhibited upon stimulation with 1,8-cineole. This inhibition occurred only in the sensilla that contained the B cell for 1,8-cineole in addition to the *cis*-verbenol A cell, ie, not in the *cis*-verbenol sensilla that contained a non-responsive B cell. Thus, the inhibition of the *cis*-verbenol A cell that occurred during the B cell response to 1,8-cineole suggests that the response of the B cell inhibits the A cell, rather than ligand–receptor interactions in the *cis*-verbenol cell (Fig. 3C). Furthermore, not only the spontaneous activity but also the responses to 1–10 ng *cis*-verbenol were reduced by simultaneous stimulations with high doses (1–10 μg, in binary *cis*-verbenol/1,8-cineole mixtures) of 1,8-cineole and in a dose-dependent manner (Fig. 3D). These observations support the hypothesis that odour blend responses can be modulated in the periphery via interactions between responding, co-localized, OSNs. With the genetic toolbox available for *Drosophila melanogaster*, peripheral OSN interactions have now been more conclusively demonstrated with potentially strong effects on odour-mediated behaviours (Su et al., 2012). Thus, it is likely that the inhibitory effect of 1,8-cineole on pheromone attraction of *I. typographus* is, at least partly, mediated via the decreased firing rate of the *cis*-verbenol neuron when 1,8-cineole is released from the same point source as the aggregation pheromone (Andersson et al., 2010). Thus, in this scenario, bark beetles might become anosmic to their own aggregation pheromone when 1,8-cineole is present above a certain concentration threshold. Such a mechanism would relieve the brain from processing certain olfactory mixtures, thus representing yet another means of 'making the most of the hand a bark beetle is dealt'.

Binyameen et al. (2014) provided the first experimental support (Fig. 3E) for the hypothesis that OSN co-localization in insects improves the discrimination of closely separated odour sources. High odour source discrimination capability is likely to be important for bark beetles when homing in on a suitable tree that is surrounded by unsuitable neighbours, or when deciding exactly where to land within a tree to avoid heavily colonized or well-defended parts. From the hypothesis, one can predict that the inhibitory effect of a pheromone antagonist that is detected by an OSN co-localized with a pheromone OSN would be reduced at shorter separation distance from the pheromone source as compared to the effect of an antagonist detected by an OSN not co-localized with pheromone OSNs. In other words, trap catches should increase more rapidly with odour source separation distance if the antagonist and pheromone compounds are detected by co-localized OSNs (Fig. 3F). Hence, Binyameen et al. (2014) performed a field trapping experiment where they separated (0–48 cm) a source releasing the aggregation pheromone (*cis*-verbenol and 2-methyl-3-buten-2-ol) of *I. typographus* from a source releasing 1,8-cineole. The resulting trap catches were compared with those obtained from the same separations between sources of pheromone and verbenone—an antiattractant detected by OSNs not co-localized with pheromone OSNs (Fig. 3B). Consistent with the prediction, trap catch increased significantly more with odour source separation distance for 1,8-cineole than for verbenone (Binyameen et al., 2014). In fact, a strong effect of separation distance for 1,8-cineole on trap catch was found already between 4 and 8 cm, whereas similar effects for verbenone were found only at the largest (48 cm) separation distance (Binyameen et al., 2014). In addition to the co-localization of the two neurons per se, it is possible that the inhibition of the *cis*-verbenol neuron by the 1,8-cineole neuron also contributes to the larger increase in trap catch with separation distance for 1,8-cineole as compared to verbenone. As soon as the pheromone source is spatially separated from the 1,8-cineole source, odour filaments from the two sources will activate the two neurons asynchronously, which partly might release the *cis*-verbenol neuron from the inhibition imposed by the 1,8-cineole neuron. This hypothesis should be tested in future electrophysiological experiments.

Assuming that stereotypic pairing of OSNs in sensilla is adaptive (Baker et al., 1998; Bruce et al., 2005) and not due to neurodevelopmental constraints, the results of Andersson et al. (2010) and Binyameen et al. (2014) raise the question as to why the neuron for 1,8-cineole has been selected to be co-localized with a neuron for an aggregation-pheromone component, whereas the neuron for verbenone has not? First, one explanation could be that attacking a tree with high levels of 1,8-cineole is associated with a higher fitness cost as compared to attacking a tree from which verbenone is released. If a high amount of 1,8-cineole means a high risk of dying or reproduction failure (as suggested by Schiebe et al., 2012), attempting to colonize such a tree would be costlier than colonizing a tree releasing verbenone, which is thought to represent a signal of over-crowding and low nutritional value (Schlyter et al., 1989). In the latter case, there is still a chance to reproduce and some offspring might survive despite the competition from conspecifics (Pureswaran et al., 2006). Second, the specific co-localization of OSNs might have been selected due to the spatiotemporal release patterns of *cis*-verbenol and 1,8-cineole compared to the release of verbenone (or other antiattractive NHV) (Binyameen et al., 2014). For instance, since 1,8-cineole is released from both attacked and non-attacked trees, it is present as a background odour when *cis*-verbenol is released from attacked trees (Andersson et al., 2010; Schiebe et al., 2012). Thus, it might be important for attacking bark beetles to differentiate between the 1,8-cineole in the background and the combined pheromone/1,8-cineole signal emanating from an attacked tree. In comparison, verbenone is not present as a background odour and it is mainly released from the same point sources as *cis*-verbenol, ie, the entrance holes of bark beetle galleries (Birgersson and Bergström, 1989; Byers, 1983b; Leufvén and Birgersson, 1987). However, verbenone is not present (or is present only in small amounts) in the initial attack phase, and when the release of verbenone peaks in later attack phases, the release of *cis*-verbenol has declined (Birgersson et al., 1984). The antiattractive NHV from angiosperms are never released from the same source as *cis*-verbenol, since coniferous bark beetles do not colonize such trees, nor can they produce aggregation pheromone from them. Hence, both the temporal and spatial differences in the release of 1,8-cineole, verbenone, and NHV in relation to the release of the aggregation pheromone might provide a possible explanation for why the neuron for 1,8-cineole, but not the neurons for the other antiattractants, has been selected to be paired with the neuron for *cis*-verbenol (Binyameen et al., 2014).

2.6 Bark Beetle Chemosensory Genes and Proteins

It is when odour molecules bind to receptors in the cell membrane of OSN dendrites that the chemical information in the environment is translated into an electrical signal that can be interpreted by the insect nervous system. Thus, studies of the chemosensory gene families and the function of the proteins they encode are of paramount importance to fully understand the cellular and molecular mechanisms that underlie host selection behaviour in bark beetles as well as the evolution of host specialization.

The insect receptors for airborne molecules are coded by three different multigene families, ie, the odorant receptors (ORs), ionotropic receptors (IRs), and gustatory receptors (GRs) (Benton et al., 2009; Kwon et al., 2007; Robertson and Kent, 2009; Robertson et al., 2003; Sato and Touhara, 2009; Touhara and Vosshall, 2009; Vosshall and Stocker, 2007). The latter family notably includes receptors for carbon dioxide in addition to receptors for non-volatile tastants. The ORs detect volatile compounds (both pheromones and 'general' odorants), while IRs are involved in both olfaction (antennal IRs) and taste (divergent IRs) (Abuin et al., 2011; Croset et al., 2010; Touhara and Vosshall, 2009). In *Drosophila*, the antennal IRs have different response profiles as compared to the ORs. They primarily detect nitrogen-containing compounds, aromatics, and acids (Croset et al., 2010). Additional gene families encoding proteins with roles in chemosensation include the odorant-binding proteins (OBPs), chemosensory proteins (CSPs), and sensory neuron membrane proteins (SNMPs) (Benton et al., 2007; Pelosi et al., 2006; Sánchez-Gracia et al., 2009). OBPs are thought to solubilize hydrophobic odour molecules and transport them across the hydrophilic sensillar environment to the receptors (Pelosi et al., 2006; Sánchez-Gracia et al., 2009), and they have in some (but not all) cases been shown to affect odour responses, ie, they might act as a prereceptor filter (Leal, 2012). The roles of CSPs are multiple but less known. Some members bind odorants and might, thus, have similar roles as the OBPs (Pelosi et al., 2006). SNMPs (ie, SNMP1) are important for proper pheromone responses in *Drosophila* and moths, though studies provide somewhat conflicting results of their exact role (necessary for response, increasing sensitivity of the response, or affecting temporal response dynamics) (Benton et al., 2007; Li et al., 2014; Pregitzer et al., 2014). The rapid technical and economical advances currently being made with respect to DNA and RNA sequencing have resulted in a tremendous increase in the number of insect species for which chemosensory genes have been identified from genomes and transcriptomes. However, Coleoptera in general and bark beetles, in particular, remain an underrepresented group in this regard. To our knowledge, members of the different chemosensory gene families have so far only been satisfactory identified in three species of coniferous bark beetles, namely from antennal transcriptomes of *I. typographus*, *D. ponderosae*, and *D. valens* (Andersson et al., 2013; Gu et al., 2015). The numbers of identified chemosensory genes from the different families differ substantially among species (Table 2), which is likely due to technical reasons rather than true biological differences.

Among the identified chemosensory genes, the three conserved carbon dioxide receptors were identified from the antennal transcriptome of *I. typographus* (only two receptors identified in the antennal transcriptome of *D. ponderosae*), suggesting that the bark beetle antennae detect carbon dioxide (Andersson et al., 2013). To our knowledge, possible roles for carbon dioxide in bark beetle host selection have not been investigated. In addition, all of the olfactory antennal IRs that are conserved across insect orders were recovered from *D. ponderosae*. The low numbers of GRs and IRs of the divergent class (likely involved in taste (Croset et al., 2010)) identified in all three bark beetle species suggest that receptors for taste are mainly expressed in tissues other than the antennae. To date, a full genomic repertoire of bark

Table 2 Numbers of Chemosensory Genes Identified from Transcriptomes of Three Species of Bark Beetles

	Species		
Gene Family	***Ips typographus*[1]**	***Dendroctonus ponderosae*[1]**	***Dendroctonus valens*[2]**
OR	43	49	22
GR	6	2	4
IR	7	15	3
OBP	15	31	22
CSP	6	11	6
SNMP	3	3	4

Abbreviations: *CSP*, chemosensory protein; *GR*, gustatory receptor; *IR*, ionotropic receptor; *OBP*, odorant-binding protein; *OR*, odorant receptor; *SNMP*, sensory neuron membrane protein.
References: 1, Andersson et al. (2013); 2, Gu et al. (2015).

beetle chemosensory genes has yet to be identified, and a good place to begin investigations would be in the currently available genome sequence of *D. ponderosae* (Keeling et al., 2013). It would be interesting to compare the genomic OR repertoire size of bark beetles with that of the flour beetle *Tribolium castaneum* (Herbst), which is unexpectedly large with 259 seemingly intact OR genes (Engsontia et al., 2008).

Phylogenetic comparisons of the bark beetle ORs with the ones from *T. castaneum* (Tenebrionidae) and the longhorn beetle *Megacyllene caryae* (Gahan) (Cerambycidae) have shown that the majority of the bark beetle ORs group in bark beetle-specific expanded lineages (Andersson et al., 2013; Gu et al., 2015). These patterns of receptor lineage expansions in bark beetles probably reflect their specialization to conifer environments of the three species studied, in combination with their intricate and specialized pheromone communication systems. However, some of the bark beetle ORs show a closer relatedness to receptors in *M. caryae*, which colonizes hickory, and a few also to receptors in *T. castaneum*, suggesting that some olfactory functions might have been conserved across Coleoptera (Andersson et al., 2013; Gu et al., 2015).

Within the bark beetle-specific lineages, there are several examples of orthologous receptors shared by different bark beetle species (Andersson et al., 2013; Gu et al., 2015). The relatively large number of orthologous receptors might partly be due to the species' evolutionary relatedness in comparison to the two non-bark beetle species, but it might also be explained by the fact that the different bark beetles share 'semiochemical space', which is largely dominated by conifer monoterpenes and oxygenated versions of such. Indeed, a large number (>50) of semiochemicals are produced, detected, and/or behaviourally active in both *I. typographus* and *D. ponderosae* (see Additional file 5 in Andersson et al., 2013).

Although a few bark beetle-specific expansions also are present among the OBPs, a higher degree of conservation is seen among members in this family than in the OR family in Coleoptera. Indeed, dendrograms show that there are several subfamilies that contain OBP members from all of the above-mentioned beetle species (Andersson et al., 2013; Gu et al., 2015). Although overall sequence similarity might not reflect similar protein functions, the difference in conservation between the ORs and OBPs suggests that the OR family diverges to a larger extent as a result of ecological specialization as compared to the OBP family. However, in order to study functional divergence or conservation, ligand affinity and specificity of the proteins must be determined. To our knowledge, none of the bark beetle ORs or OBPs has been functionally characterized, but doing so should be prioritized in future studies. Functional studies will be facilitated by the large number of known semiochemicals and are needed to further our knowledge of bark beetle chemical ecology and host selection behaviour down to the level of the molecular players, which are the ultimate responders to natural selection.

3. PLASTICITY: HEDGING YOUR BETS

3.1 A Highly Variable, Largely Unpredictable, and Density-Dependent Host Resource

Bark beetles must function in a continuously and rapidly changing environment, particularly in regard to the availability, quality, and optimal choice of host trees. The background against which individuals must make host selection decisions includes a diverse set of higher-scale factors, fine-scale factors such as the beetles' internal physiology, and both negative and positive feedbacks arising from their own population densities. Each of these sources of variation provides both opportunities and challenges and exhibits its own spatiotemporal patterns, magnitude of change, and degree of predictability.

Higher-scale factors include weather, abiotic stressors on trees such as wildfire and lightning, biotic stressors such as defoliators and root pathogens, and ongoing changes in forest structure. Temperature extremes affect beetles directly, increasing or decreasing their overwintering survival and development rates (Bentz et al., 2010). These processes drive beetle population size (Powell and Bentz, 2009), which in turn influences the range of tree physiologies that can be overcome by cooperative attacks. Relationships between temperature-driven increases and attack behaviour are particularly pronounced in species that show 'adaptive seasonality', threshold-based developmental responses to degree-day accumulation that generate relatively synchronous adult emergence despite a wide range of starting points (Logan and Bentz, 1999). Temperature anomalies commonly occur over relatively large spatial areas, but also show strong microsite effects arising from elevation and exposure. They are largely stochastic and often operate on a scale of one to several years. Precipitation anomalies can suddenly increase the availability of susceptible trees on a large scale (Breshears et al., 2009). These events can be highly spatially variable however and are strongly influenced by microsite effects such as soil type and aspect. Cumulative effects of inter-annual drought can be particularly important. Wildfire is likewise relatively stochastic, but is highly correlated with weather. Burn injury can reduce tree defences, thereby increasing the pool of available trees (Powell and Raffa, 2011). However, only a relatively narrow range of burn intensities provide an optimal condition for tree-killing bark beetles, and this tends to be a 'pulsed resource' that is rapidly exhausted (Davis et al., 2012; Hood and Bentz, 2007; Powell and Raffa, 2011). Burn-injured trees tend to be a spatially highly aggregated resource.

Resource availability arising from tree felling by storms is the most stochastic factor in time and in Europe is generally considered as the main purveyor of breeding material and the most common factor for starting bark beetle eruptions, particularly in spruce (Marini et al., 2013; Mezei et al., 2014; Økland and Bjørnstad, 2003). Likewise, in North America, windstorms affecting spruce stands are commonly followed by outbreaks of *Dendroctonus rufipennis* (Kirby) (Reynolds and Holsten, 1994). In Europe, it is likely that beetle outbreaks in spruce will increase in frequency and size with particular temperature, wind speed, and precipitation variabilities, as reviewed by Kausrud et al. (2012). In northern Europe especially, there is a high likelihood of more rapid beetle population growth after disturbances, when voltinism increases from one to two or even three generations per year (Jönsson et al., 2009, 2011).

Lightning is likewise an acute stress that impairs tree defences and, unlike chronic stresses, can provide beetles with trees that are both highly susceptible and have thick phloem (Blanche et al., 1985). Although lightning may be sporadic in space and time locally, it is abundant on a landscape basis, especially in some regions (Coulson et al., 1983). Defoliators can reduce tree defences against bark beetles, often quickly enough to render trees susceptible while they still have thick phloem (Wallin and Raffa, 2001). Some folivore species show intermittent, temperature-responsive outbreaks, but others are quite cyclic in nature due to system-intrinsic processes such as time-delayed feedbacks with natural enemies (Allstadt et al., 2013; Haynes et al., 2009). Defoliator outbreaks tend to build up over a couple of years, reach high intensity, and extend over large areas during their peaks, and then collapse. Root insect and disease infestations, in contrast, are often much more spatially aggregated, continuous, and chronic as they extend from epicentres (Aukema et al., 2010). Root-infested trees often show lower growth rates prior to their being colonized by bark beetles.

A bark beetle's internal environment is likewise changing during its adult lifespan. Flight is energetically costly and depletes their lipid stores. Adults are quite literally racing a biological clock. Many species have a relatively short-lived lifespan outside trees during their dispersal, host-seeking phase.

Bark beetles exert strong feedbacks on the availability on their host tree resource. Each successful colonization removes that host from the available food base. So there can be no successive utilization by multiple generations, as is commonly seen in other feeding guilds. This depletion exerts an important source of negative feedback on populations. On the other hand, each population increment that arises from successful colonization also expands the size of the available cohort to respond to pheromones emitted by beetles entering a tree in the next generation. This provides an important source of positive feedback on populations. Tree susceptibility then is relative to local beetle density, and dynamic, not absolute. How this relates to host selection behaviour is just beginning to be understood.

3.2 The Value of Making Sequential Rather than Absolute Decisions (Why Card-Counters Win)

One means by which bark beetles contend with the challenges posed by host resource trade-offs, environmental uncertainties, and near-total investment in single hosts imposed by their life history is by making a series of sequential, rather than terminal, decisions. Each of these decisions can be reversed prior to oviposition. In this regard, their host selection sequence is somewhat analogous to animal courtship rituals, which may either proceed to the next step or be terminated at any stage prior to copulation. Continually gathering and incorporating new information into subsequent choice, and hedging final bets for as long as possible, tilt the odds in favour of a deft gambler.

During flight, responses to chemical and optical stimuli can be mediated by beetle fat reserves (Atkins, 1969; Graham, 1959; Williams and Robertson, 2008). Initial landing is mediated by the visual silhouette and volatile plumes emanating from trees (Shepherd, 1966; Strom et al., 1999). Landing, however, does not guarantee that a beetle will enter a tree. In fact, several studies have shown that *D. ponderosae*, an 'aggressive' tree-killing species, lands on trees that are subsequently entered or not entered with equal frequency (Hynum and Berryman, 1980; Moeck et al., 1981; Raffa and Berryman, 1980). The result is that almost all trees within the host species range (and even some outside the host species range) are landed upon each flight season. Once on a tree, beetles perceive gustatory and short-range olfactory cues, which provide information about host condition (Elkinton et al., 1981; Raffa and Berryman, 1982). Walking *D. ponderosae* are attracted to odours from the host limber pine (*Pinus flexilis* James) but not the non-host bristlecone pine (*Pinus longaeva* Bailey) (Gray et al., 2015). Thigmotactic stimulation of the pronotum, provided by orienting under bark flaps and crevices, is commonly required for these chemical stimuli to elicit response (Borg and Norris, 1969). The decision of whether or not to enter is based on both the composition and concentration of host compounds, of which monoterpenes play a particularly important role (Wallin and Raffa, 2000). Those beetles that do not enter resume flight, land on new trees, and repeat the process. The risk of not entering is that dispersal exposes beetles to generalist predators, exhausts their energy supplies, and costs valuable time as susceptible trees become taken by intraspecific competitors. It is not entirely a scramble competition, however, because dispersing beetles may encounter pheromones produced by entering beetles and then join in a mass attack, a consideration we address in Section 3.4.

Once a beetle begins to tunnel, it may still abandon a colonization attempt if host responses are rapid and intense, that is, if the likelihood of success is lower than initially perceived. The mechanisms by which bark beetles partition the behaviours of initial entry and continued tunnelling involve behavioural repertoires that incorporate multiple host chemicals. For example, non-polar extracts of *P. contorta* elicited initial feeding by *D. ponderosae*, but polar extracts were needed to elicit sustained feeding (Raffa and Berryman, 1982). Interestingly, the extent to which beetles fed on extracts from a particular tree did not predict which trees would be killed in the field, again indicating the importance of multiple stimuli and modalities and context dependence. Similarly, specific monoterpenes had different effects on two distinct components of the host selection sequence, ie, whether or not to enter and if so how far to tunnel, in both *I. pini* and *D. rufipennis* (Wallin and Raffa, 2000, 2001, 2004).

3.3 Sources and Modalities of Plasticity

The extent, sources, and mechanisms by which various bark beetle species incorporate behavioural plasticity into host selection comprise an essential yet poorly understood aspect of their biology. Few species have been studied in detail, and differing methodologies preclude rigorous comparisons among systems. However, the studies conducted to date confirm that there is high intraspecific variation in host selection behaviour, and this variation arises from environmental, genetic, and gene × environment factors.

Beetle age, and also the number of times a beetle had previously rejected a potential substrate, influences host acceptance (Wallin et al., 2002). That is, older *I. pini* males (the host-selecting gender) were more likely to enter media amended with high concentrations of monoterpene than were younger beetles. Moreover, as individual beetles rejected high monoterpene substrate, their likelihood of eventually accepting a substrate amended with that same concentration increased, even when corrected for age. In human terms, this is analogous to sampling the environment and correcting the initial expectation about what is available by incorporating real-time data. Although personalized terms such as 'becoming frustrated' or 'settling' are not mechanistically appropriate, it does seem reasonable that an insect having such a short adult life span would lower the threshold concentration of allomones above which it will not enter, when the alternative may be a replacement rate of zero.

There is also a hereditary component. Again with *I. pini*, an insect that typically does not enter healthy trees in nature, adult males varied in whether or not they entered substrate amended with concentrations of monoterpenes emulating healthy trees. This variability was then used as the basis for artificially selecting different breeding lines in the laboratory. Over three generations, these breeding lines diverged significantly, yielding one that would enter substrate amended with high concentrations, an unselected control that showed no difference from the initial population, and one that would only enter substrate amended with the very lowest concentrations of monoterpenes (Wallin and Raffa, 2002; Wallin et al., 2002). Such breeding experiments, while demonstrating heritability, do not indicate the mode of inheritance, for which backcrossing experiments are needed, nor do they identify molecular markers, for which genomic and proteomic analyses are required. *D. rufipennis* also shows a genetic component to its host selection behaviour. Specifically, offspring emerging from felled and live trees from a wide range of sites across western North America yielded females that showed different host acceptance behaviour in laboratory assays. Further, there was significant mother–daughter correlation in the concentration of monoterpenes that females (the host-selecting gender) will accept. Wallin and Raffa (2004) collected brood adults from the hosts in which they developed, established adults on logs in a common environment, collected progeny females, conducted bioassays with a concentration of α-pinene mimicking live trees, established independent female–male breeding lines, conducted the same bioassay with progeny females, and computed correlations between mothers and daughters. Such designs, while improving ecological realism, cannot totally eliminate potential maternal effects however.

One of the most important sources of behavioural plasticity is the relationship between host selection decisions made by individuals and stand-level population density in shaping optimal choice, because it relates directly to the ability of bark beetles to transform entire landscapes. As population density increases, the physiological range of trees that irruptive bark beetle species can colonize broadens. This generates positive feedback, across scales. Observational reports by early forest entomologists consistently describe how beetles become 'more aggressive' during outbreaks, not only attacking more trees in relation to their increased numbers but also attacking physiologically different trees (Evenden et al., 1943; Rudinsky, 1962; Schwerdtfeger, 1955; Thalenhorst, 1958). Some examples include Keen (1938) who stated: 'Endemic populations select weaker, less vigorous trees for attack, but no such selection is apparent during epidemic conditions', and Beal (1939) who stated: 'When the Black Hills beetle is not numerous it breeds in weakened trees or those injured by lightning or in some other way. … During outbreaks this insect attacks vigorous healthy trees' and even 'shows a slight preference for the more vigorous, rapidly growing trees'. More recently, Lewis and Lindgren (2002) reported that *D. rufipennis* showed a strong association with root disease during its endemic but not irruptive phase. Likewise, Powell et al. (2012) found that *D. ponderosae* attacked uninjured *P. contorta* following wildfire when their local populations were in the irruptive phase, but only attacked injured trees when beetles were in their endemic phase. Similarly, Wallin and Raffa (2004) found in a manipulative experiment that *D. rufipennis* attacked live standing *Picea* spp. when local populations were in their irruptive but not endemic phase. Likewise, Boone et al. (2011) found that *D. ponderosae* selected increasingly well-defended *P. contorta* as their populations increased, and actually switched their relative attack rates from relatively less well defended to relatively better defended trees, relative to the population mean (Fig. 4). A driving force behind this putatively anomalous shift is that better-defended trees are also larger and hence provide a superior food base for beetle development (Graf et al., 2012); ie Beal was right. Using the regression between *D. ponderosae* pupal production and *P. contorta* diameter derived by Graf et al. (2012), the increased sizes of trees overcome during outbreaks relative to endemic periods (Boone et al. 2011) yield approximately 20% more brood on a per square metre basis and 45% more on an absolute basis. A second driver is that as beetle populations rose, poorly defended trees were progressively culled from the host population (Boone et al., 2011). Hence, if beetles were hard wired to never enter healthy trees, they would be at a disadvantage because poorly defended trees no longer remain in the stand. Genotypes associated with high preference for stressed trees would likely be maintained at some level during outbreaks, however, because these beetles could still join in mass attacks.

The extent to which bark beetles expand the physiological range of trees they will accept varies among species. For example, the onset of all three outbreaks examined by Raffa et al. (2008), *D. ponderosae*, *D. rufipennis*, and *Ips confusus* (LeConte), was released by weather anomalies, specifically high temperatures and severe drought. But the former two species actually caused their highest tree mortality after the weather anomalies had subsided, indicating that their populations had risen past a threshold after which they became self-driving. In contrast, when the drought that elicited the *I. confusus* outbreak subsided, populations crashed: these insects were tracking the pool of severely stricken trees rather than expanding onto healthy trees. Likewise, Aukema et al. (2010) observed that most attacks by *I. pini* on *Pinus resinosa* Rudolf were preceded by fungal and herbivore infestation of roots and basal stems. When these stands experienced drought, the pine engravers exploited this additional source of stressed trees and their populations rose. When the drought subsided, however, *I. pini* populations crashed, and were again almost entirely limited to the root-diseased resource. Additionally, an experimental study with artificially water deprived host trees of *I. typographus* shows that water stress not only lowers tree defence/increases beetle success but also indicates that extreme stress might lower host acceptance (Netherer et al., 2015).

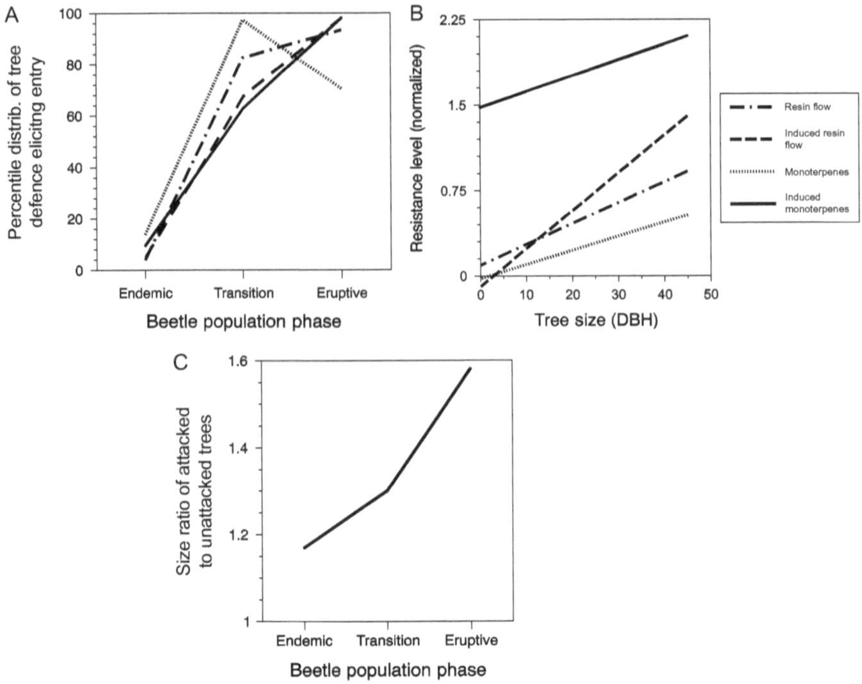

Fig. 4 From bottom feeder to top predator: plasticity of host selection by *D. ponderosae*. (A) The defensive capacity of *P. contorta* trees entered by *D. ponderosae* relative to other *P. contorta* in stand increases with beetle population size. Data show four metrics of tree defence. When populations are at endemic levels, the median defensive capacity of entered trees is far below the stand average (eg, 5 percentile for constitutive resin flow, 9.5 for induced monoterpenes), even though this insect is typically thought of as an 'aggressive' species. But as beetles pass through the transition threshold they attack progressively more vigorous, better-defended trees, and during their eruptive phase even preferentially attack well-defended trees (eg, 97.9 percentile for constitutive resin flow, 98.1 for induced monoterpenes). (B) Well-defended trees (using the same four metrics) are also the largest trees, which are well known to produce more beetle brood if they can be killed. Outbreaks are correlated with large-diameter trees, a relationship validated in panel (C). But a common inference from that association, that large-diameter trees are more susceptible is incorrect, inverts cause and effect. Instead, large populations can overcome large trees, which yield more beetles, which allow populations to colonize even better-defended trees. *Redrawn from Boone, C.K., Aukema, B.H., Bohlmann, J., Carroll, A.L., Raffa, K.F., 2011. Efficacy of tree defense physiology varies with bark beetle population density: a basis for positive feedback in eruptive species. Can. J. For. Res. 41, 1174–1188.*

Thus, it is critical to incorporate interspecific differences in behaviour when evaluating plasticity, management, and landscape-scale inferences associated with bark beetles. It is also critical when inferring relationships between higher scale drivers, such as weather, to beetle population dynamics to (1) quantify beetle numbers (as opposed to numbers of dead trees) and do so at the relevant scale at which beetles actually interact with trees, (2) incorporate the critical distinction between the onset and growth of outbreaks, and (3) have information on the physiology and biochemistry of trees prior to their attack, to distinguish whether the insects truly 'killed' trees or merely occupied trees already in the process of dying. Unfortunately, this information is often lacking, whereas indirect, postthreshold, or coarse-scale data are more easily acquired, and incorrect inferences of process from pattern can arise.

The mechanisms by which bark beetles can expand the range of tree physiologies that elicit their entry behaviours are not well understood. However, they appear to incorporate a gene × environment interaction. Wallin and Raffa (2004) found that contrary to their predictions, individual *D. rufipennis* females from irruptive populations were not more likely to enter monoterpene-amended media than females from endemic populations. There was a key difference between beetles from these sources, however, in how they responded to the presence of other beetles. Endemic beetles made entry decisions based solely on monoterpene concentrations, independent of whether other beetles were present in their assay units. In contrast, the aversion that irruptive individuals showed to high monoterpene concentrations declined in the presence of other beetles. The mechanism by which this communication occurs is unknown, as they did not find evidence of it being mediated by aggregation pheromones. It may be that short-distance signalling, chemical or otherwise, is important in mediating this link between on-bark behaviour and stand-level population densities.

The population density- and age-dependent plasticity in bark beetle host selection behaviour may be underpinned by plasticity in their chemosensory systems, which might occur at both central and peripheral levels (Claudianos et al., 2014; Kromann et al., 2015; Sandoz, 2012; Saveer et al., 2012; Wilson et al., 2004). While no studies have addressed this possibility in bark beetles, there are several examples of context-dependent olfactory modulation from other insect groups. For instance, plasticity can be achieved by differential expression of OR genes, which occurs in mosquitoes after blood feeding (Fox et al., 2001) and in honeybees depending on experienced odour environment (Claudianos et al., 2014). Olfactory plasticity can also be obtained by neuromodulatory hormones acting centrally on olfactory neurons in the antennal lobe (Ignell et al., 2009), sometimes depending on internal state (ie, hunger level) (Root et al., 2011), or by modulating the responses of OSNs (Flecke and Stengl, 2009). Future molecular and physiological studies should target these questions using bark beetles captured in their natural environment to compare OR gene expression levels, OSN sensitivities, and OSN class frequencies in individuals from endemic vs epidemic populations, and pioneering individuals vs pheromone-responding attackers. Beetles reared under controlled laboratory conditions should also be investigated for possible signs of olfactory plasticity related to age and physiological status.

3.4 Why Not Cheat?

Given the complexities, uncertainties, and risks involved in selecting suitable hosts for brood development, and the ability of bark beetles to collectively overcome tree defences through pheromone-mediated mass attacks, several researchers have raised the question of whether cheating, ie, avoiding unattacked trees but instead only entering in response to pheromone signals from entering beetles, might not be a viable strategy? Indeed most entries probably occur by beetles that arrived at trees in response to pheromones, rather than beetles that initiated the first entry. We currently lack the information to answer whether some genotypes within a population might simply respond to insect-produced rather than host-produced cues. Therefore, we can only approach this problem by identifying features of the system that could influence the viability of cheating.

The major advantage to cheating is risk avoidance. Initial colonizers of 'aggressive' species such as *D. ponderosae* and *D. rufipennis* enter a tree that will soon kill them by undergoing rapidly induced defences, unless they are joined by enough other beetles. If successful in eliciting aggregation, the initial beetle's chances of reproducing are extremely high. This seems a good time to join an attack, because it provides an opportunity to derive benefits without incurring costs, to place a wager after the outcome is known.

There are several disadvantages, however, to relying solely on insect signals. First, aggregation on a tree generally lasts only a few days, because pheromone signalling in bark beetles is intimately tied to immediate tree physiology. *Dendroctonus* beetles exploit tree defence compounds as precursors and synergists of their aggregation pheromones, so attraction ceases shortly after tree defences succumb. At this point, beetles also oxidize their aggregation pheromones into antiaggregation pheromones that reduce host attractiveness. Second, many bark beetles live for only short periods as adults, so there can be severe opportunity costs to ignoring susceptible trees, if after passing up on this chance a beetle does not encounter aggregation pheromones. This risk seems particularly high when populations are low, which encompasses most beetle generations. Third, in some systems late arrivers produce less brood than early arrivers due to intraspecific crowding within the fixed carrying capacity of the phloem (Anderbrant et al., 1985; Berryman and Stark, 1962; Wagner et al., 1987, 1988). Fourth, predators exploit bark beetle pheromones as kairomones, so late-arriving beetles may experience relatively higher predation. This has been validated in laboratory assays (Abuin et al., 2011; Aukema and Raffa, 2004), but has not been tested in the field.

At least three other considerations weigh on the relative advantages of cheating in this system. First, the risks (and energy expenditures) involved in host selection may be mitigated by the sequential process that beetles use. That is, early stages of host acceptance can be terminated at any time, landing requires little energy, and beetles may even gain some energy during bark tasting (Borg and Norris, 1969). Second, the advantages of cheating may be density dependent, similar to the advantages of relative host selection behaviours. That is, plasticity in response to insect vs host signals may be more adaptive than any uniform strategy (Birgersson et al., 1988; Schlyter and Birgersson, 1999). Third, in some systems beetles that land during the mid-stages of aggregation may experience better reproductive fitness than either those arriving late or entering first (Pureswaran et al., 2006). However, beetles that arrived at different stages did not vary in body size or the amounts of pheromones they produce, suggesting that stabilizing selection contributes to the synchrony of attacks needed to overcome defence (Pureswaran et al., 2006).

4. CHOOSING AMONG HOST SPECIES IN A CHANGING LANDSCAPE: THERE'S MORE THAN ONE GAME IN TOWN

Like other herbivorous insects, most bark beetles can feed on several species of plants, and this capability can increase their host selection options. In general, bark beetle species that typically colonize only dead or highly stressed trees, such as *Orthotomicus caelatus* Eichhoff and *Dryocoetes autographus* (Ratzeburg), often colonize a broader taxonomic range of trees than those that undergo large-scale outbreaks on healthy trees, such as *D. ponderosae*, *D. rufipennis*, *D. frontalis*, and *I. typographus* (Raffa et al., 2015). The former often colonize three or more genera, whereas the latter typically colonize one tree genus. Some of the moderately aggressive species, such as *Dendroctonus pseudotsugae* Hopkins, *Dryocoetes affaber* Mannerheim, and on occasion *I. pini*, commonly utilize two genera. Thus, host range in bark beetles is best understood as a manifold encompassing both tree taxonomy (and its underlying interspecific chemical distinctions) and tree physiology (and its underlying intraspecific variability in defensive capability). That is, in addition to being specialists or generalists in terms of tree species, each bark beetle can also be viewed as a relative specialist or generalist in terms of tree condition.

Although bark beetle species are often limited to a specific conifer genus, they can usually colonize all or most of the sympatric tree species within that genus (Huber et al., 2000; Wood, 1982). There are exceptions, such as *Dendroctonus jeffreyi* Hopkins, which is largely specific to *Pinus jeffreyi* A. Murray, a tree with unusual chemistry that is not attacked by most other bark beetle species. As with other herbivores, bark beetles may show relative behavioural preferences for particular tree species within a genus. For example, *D. frontalis* most commonly exploits *Pinus taeda* L. and *Pinus echinata* Mill., frequently colonizes several other species, and less frequently colonizes *Pinus strobus* L. and *Pinus palustris* Mill (Hain et al., 2011). Likewise, *D. rufipennis* exerts the highest mortality to Lutz spruce in Alaska (a hybrid of Sitka and white spruce, sometimes designated *Picea lutzii* Little) but likewise colonizes a broad range of *Picea*, such as *Picea glauca* (Moench) Voss and *Picea sitchensis* (Bong) Carr., and appears to have relatively low affinity for *Picea mariana* (Mill. B.S.P.) (Werner et al., 2006), and in other regions colonizes and *Picea engelmannii* and *Picea rubens* (Sarg.).

4.1 Historical Ranges: Regional Variation and Experience

Although a bark beetle species may exploit several host species, varying levels of tree mortality cannot necessarily be extrapolated to differences in actual behavioural preference, because different tree species possess different levels of postentry defence, and also because related conifer species generally show low spatial overlap. Each conifer species has its unique ecophysiological niche, and so is more likely to dominate a particular region that is characterized by a particular temperature, precipitation, and soil regime. Further, different tree species are commonly associated with a particular stocking density, age-class distribution, size-class distribution, understory composition, and abundance and proximity of non-host species. Conifer biomes are often characterized by relatively large expanses that are dominated by a single species, and transition into neighbouring large expanses dominated by another. Within-site intermingling of different species from the same genus is not typically high. Thus, it often cannot be separated whether different tree mortality patterns reflect true behavioural preference on the part of the beetle, the extent to which ecophysiological requirements of the beetle overlay those of different tree species, whether the forest structure and site conditions associated with a tree species are more likely to predispose its communities to outbreaks than other species, differences in predator or competitor loads, differences in biotic and abiotic disturbance agents, some other variable, or various combinations thereof (West et al., 2014).

The difficulty of ascribing field patterns of bark beetle attack to actual behavioural preference likewise cannot be as readily addressed using controlled bioassays under laboratory conditions as in other herbivore guilds. Assays with logs and phloem strips greatly increase host acceptance by bark beetles, and so can obscure actual behavioural differences for live trees of different species (Huber et al., 2009; Raffa et al., 2013; Safranyik and Linton, 1983). This differs from the much more useful discriminating power of detached-leaf assays, which often simulate field behaviour of folivores with substantial accuracy. Another approach is to use tree mortality data, but this likewise has limited value. First, tree mortality only indicates those trees that were successfully mass attacked, rather than those that were entered. Those values are often dissimilar and can actually be in opposite directions in interspecific rankings among tree species (Bentz et al., 2015; Raffa et al., 2013). Also, bark beetles change their host selection behaviour as they transition from their endemic to outbreak phase, but tree mortality data only become available after that transition occurs. Other approaches that have proven useful, although difficult, include quantifying natural beetle entry rates in regions where different tree species comingle and caging beetles onto live trees (Raffa, 1988). But these assays likewise have limitations, specifically an implicit assumption that the trees are under equal levels of vigour/stress despite interspecific differences in ecophysiological requirements, the susceptibility of caging assays to uncontrolled weather factors and the reduced sensitivity that comes with any no-choice bioassay (Raffa, 1988).

Many herbivorous insects show changes in host species preference based on experience. The general life history of bark beetles largely limits this possibility. That is, a beetle often oviposits her entire egg clutch within one tree (although there are exceptions), so the opportunity for host species learning at this level is limited. Another possibility is for learning to occur based on the host species in which an individual developed. A version of this scenario was developed with bark beetles specifically in mind, and is termed 'Hopkins Host Selection Principle'. As originally devised by Hopkins, it does not necessitate that learning occurs during the larval stage as is frequently ascribed to this hypothesis, but rather learning could also occur during or immediately after adult emergence (Barron, 2001). Even with the broader interpretation, however, there is not strong evidence for host-based learning in bark beetles (McKee et al., 2015; West et al., 2014).

Host species range can expand during bark beetle outbreaks. At high beetle populations, host acceptance broadens as the breadth of tree physiological conditions that can be overcome by mass attacks increases (Lindgren and Raffa, 2013). During the later stages of an outbreak, this relaxation of selection behaviour can become extreme, such as when beetles attack very small trees that yield little to no brood adults even though the tree is killed. Beetles may also spread into tree genera that are typically considered non-hosts during these conditions (Huber et al., 2009). This host range expansion may occur on a micro-scale, attacking trees adjacent to mass-attacked pines (Amman et al., 1990), or on a stand level, such as *D. ponderosae* attacking and reproducing in *Picea* (Huber et al., 2009; McKee et al., 2013, 2015). Under most conditions these range expansions are temporary, and historical host restrictions resume after the outbreak subsides. However, given that some reproduction occurs, it is possible that such episodes formed the basis for earlier evolutionary species diversifications among bark beetles and as such provide insights into host-driven sympatric speciation (Huber et al., 2009).

4.2 Ranges Expanding with Climate Change

All of the above interactions are occurring on a changing template. As climates warm due to anthropogenic inputs, bark beetles are expanding their geographic ranges to higher latitudes and elevations, resulting in new host–beetle associations and altered prevalence of existing associations. For example, *D. frontalis* has established in several historical and naïve *Pinus* species in New Jersey, Long Island New York, and Connecticut (Niemiec et al., 2014). In western North America, *D. ponderosae* has expanded to regions of British Columbia north of its historical outbreaks and then spread eastward across the Rocky Mountains where it now colonizes *P. contorta* × *P. banksiana* Rudolph and Laidly hybrids and *P. banksiansa* in Alberta (Clark et al., 2010; Cudmore et al., 2010; Cullingham et al., 2011; Safranyik et al., 2010). This poses a threat to midwestern and eastern forests of North America, because they are now connected by host type with *D. ponderosae* for the first time in recorded history. Previously, midwestern forests were separated from *D. ponderosae* by the Great Plains to the east and cold temperatures to the north. In addition to *P. banksiana*, forests in midwestern North America contain native *P. resinosa* Rudolf and *P. strobus* Wendel and Smith, which have no known evolutionary history with *D. ponderosae*. These species appear susceptible based on their chemistry (Raffa et al., 2015) and beetle performance in laboratory assays (Cale et al., 2015; Taft et al., 2015). In the Rocky Mountains of western North America, conifer biomes show pronounced elevational gradients, with *P. ponderosa* at the lowest elevations, then the predominant host *P. contorta* at mid-elevations, then non-host species *Picea* and *Abies* at higher elevations, followed by five-needled pines such as *P. albicaulis* at the highest elevations, reaching tree line. Historically, cold winter temperatures, coupled with the buffer of non-host species, reduced access of *D. ponderosae* to whitebark pine (*P. albicaulis*) stands. During occasional warming spells beetles were able to survive at these elevations and would kill some trees, but these short duration episodes would subside when the return of normal temperatures caused beetle populations to crash. Now, climatic conditions support the continual survival and development of *D. ponderosae* in *P. albicaulis* habitat (Logan et al., 2010).

Despite the potentially severe environmental and economic threats posed by climate-driven geographic range expansions, we currently have little understanding of how beetles will respond to the hosts they encounter in new habitats. Difficulties in assessing this potential include the above-stated limitations of detached-part bioassays with bark beetles and the obvious hazards of moving beetles (and their associated microorganisms) into new habitats for purposes of bioassay. However, *D. ponderosae* is readily attacking *P. banksiana* in western Canada (Cullingham et al., 2011) and has attacked planted trees of all three native midwestern *Pinus* species in western arboreta (Furniss and Schenk, 1969). In mixed-pine stands in the western United States, *P. albicaulis* attacked by *D. ponderosae* are more likely to be successfully mass attacked and killed than are *P. contorta* attacked in the same stands (Bentz et al., 2015). This is consistent with their underlying physiological defences. *P. contorta*, which has experienced more sustained and intense selective pressure by *D. ponderosae* than *P. albicaulis*, undergoes more substantial induced chemical defences in response to attack (Raffa et al., 2013). However, *D. ponderosae* retains its preference for *P. contorta*, likewise reflecting its evolved responses at the behavioural and chemoreception levels. In mixed stands, *D. ponderosae* attacks higher relative proportions of *P. contorta* than *P. albicaulis* (Bentz et al., 2015; Raffa et al., 2013). This is consistent with the higher concentrations of the putative host recognition kairomone β-phellandrene (Huber et al., 2000; Jost et al., 2008; Miller and

Borden, 1990) in *P. contorta* than *P. albicaulis* (Bentz et al., 2015; Raffa et al., 2013). This preference is relative, not absolute, however. Preference for *P. contorta* declines as the composition of *P. albicaulis* in a stand increases, and beetles show equally high entrance rates in bark strips in controlled no-choice assays (Bentz et al., 2015; Raffa et al., 2013). As anthropogenic changes improve climatic suitability, it does not appear that host acceptance behaviours will substantially limit bark beetle range expansions in regions harbouring congeneric tree species.

4.3 New Geographic Ranges and Host Associations Arising from Human Transport

As global trade increases, the incidence of accidentally transporting insects into new geographic regions is likewise increasing, and cryptic insects such as bark beetles are particularly likely to avoid detection (Aukema et al., 2010). These introductions expose naïve trees to bark beetles against which they had not previously been exposed, and vice versa. As with climate-driven expansions, we have little knowledge upon which to predict impacts and population trends. Host selection behaviour will contribute one important component, but it will interact with numerous other factors. As with introductions of other insect feeding guilds, bark beetles show the full range of responses, including failure to establish, relatively innocuous blending into existing communities, and unprecedented outbreaks. We briefly illustrate three examples here, with emphasis on host selection behaviour.

The pine shoot beetle (*Tomicus piniperda* L.) is an important pest of *Pinus sylvestris* L. plantations in Europe. Upon first detection in the United States in Michigan in the 1990s, this insect raised multiple concerns about its potential impacts to commercial Christmas tree stands, adjoining native stands of *P. resinosa*, *P. banksiana*, and *P. strobus*, and more distant North American *Pinus* species that might be accessed via subsequent human transport. Initial host range experiments, which were necessarily conducted in logs for quarantine reasons, found that *T. piniperda* could enter and survive in every *Pinus* species tested, including representatives from midwestern, southern, and western forests (Lawrence and Haack, 1995). Subsequent studies focusing on midwestern species found successful reproduction in logs of all three native pines plus *P. sylvestris*, with variable evidence of interspecific preferences and performances under these conditions (Siegert and McCullough, 2001, 2003; Ye et al., 2005). However, in the field, *T. piniperda* continues to behave as a secondary species, attacking only severely stressed or dead native species. It continues to be a noteworthy but manageable economic pest of Christmas tree *P. sylvestris* farms (Mccullough and Sadof, 1998; Morgan et al., 2004).

The eastern five-spined ips, *I. grandicollis* (Eichhoff), is a minor pest of eastern North American pines, associated primarily with trees experiencing physiological stress. This insect was accidentally introduced into Australia in the early 1900s. It rapidly began colonizing *Pinus radiata* D. Don, a western North American species with which it had no evolutionary history. Interestingly, the host utilization behaviour of *I. grandicollis* changed in Australia, where it often conducts separate feeding and breeding attacks (Yousuf et al., 2014a,b). This change in behaviour can increase the likelihood of spreading the symbiotic fungus *Ophiostoma ips* (Rumbold) Nannfeldt and further predisposing trees to attack. Although *I. grandicollis* made the requisite host species switch to establish in Australia, it remains limited to physiologically stressed hosts. Despite its behavioural distinctions, *I. grandicollis* can be managed in Australia using modified silvicultural, sanitation, and semiochemical procedures employed in North America coupled with classical biological control. However, Australian *P. radiata* stands suffer significant economic losses from another introduced species, the woodwasp, *Sirex noctilio* F. (Hymenoptera Siricidae). Major strides have recently been made against this aggressive tree killer, using trap logs laden with entomopathogenic nematodes. However, *I. grandicollis* can disrupt this biological control tactic, providing an example of important indirect negative effects that can be exerted by introduced bark beetles (Yousuf et al., 2014b).

The red turpentine beetle, *D. valens* LeConte, was introduced from North America to China during the early 1980s. This insect is a relatively minor pest in North America, where it colonizes the base or stumps of trees. In the former case, it can predispose trees to subsequent attack by tree-killing bark beetle species. In China, however, *D. valens* readily attacks *Pinus tabuliformis* Carrière, including healthy trees. It has become a major pest, killing tens of millions of trees over hundreds of thousands of hectares (Sun et al., 2013). It causes significant economic and environmental impacts, including posing a threat to the endangered *Pinus bungeana* Zuccarini ex Endlicher. The reasons behind the drastically different dynamics of *D. valens* between its native and introduced ranges are not entirely clear, but appear to include a combination of altered semiochemical communication that facilitates mass attacks analogous to those of species that kill healthy trees in their native range, higher susceptibility of Chinese than North American *Pinus* to fungi vectored by *D. valens*, rapid genetic changes in microbial symbionts, and interactions between tree-fungal and tree-beetle dynamics, specifically elevated emission of 3-carene in response to inoculation with *L. procerum*, resulting in heightened attraction of *D. valens* (Lu et al., 2010, 2011; Sun et al., 2013). These drivers have been exacerbated by changing climate (Sun et al., 2013). Overall, *D. valens* represents a dramatic example of how a relatively minor bark beetle pest can become a major threat once transported to a new region.

In each of the earlier examples, natural host species range did not pose a major barrier to establishment on novel hosts within the same genus. The nature and extent of impact, however, could not be explained solely based on host selection mechanisms. Rather the plasticity of host selection responses among bark beetles appears to provide the potential for severe outbreaks, but whether the realized impacts are minor or large scale, manageable or largely out of control, depends on additional factors such as host tree defence, symbiotic associations, and climate.

5. INTEGRATING MECHANISTIC KNOWLEDGE OF HOST SELECTION BEHAVIOUR WITH THE LANDSCAPE ECOLOGY AND MANAGEMENT OF BARK BEETLES

5.1 Integrating Ecology and Physiology, Pattern, and Process

Despite the limitations that bark beetles may pose as models for basic neurophysiological studies, they provide outstanding, and perhaps unequalled, models for linking fundamental sensory and neurointegrative mechanisms with natural selection in complex habitats, landscape-scale impacts of herbivores on natural, managed, and socioeconomic systems, and management decisions and opportunities. Understanding the multicomponent and sequential nature of host selection by bark beetles can help resolve some of the long-standing and ongoing disputes about their behaviour, ecology, and coevolution with trees and forest ecosystems. For example, vigorous debates have centred around whether bark beetles initially land on trees in a 'directed' vs 'random' fashion. This debate raises some interesting mechanistic questions, but is less fruitful when 'landing' and 'host selection' are used interchangeably as sometimes happens. That is, movement towards a source and landing are 'directed', in that they can be elicited by visual stimuli and host compounds. However, landing on a particular tree can be 'random' relative to whether or not that tree is ultimately 'selected', ie, whether it elicits entry and oviposition. Thus, evidence that a particular compound orients beetles to a point source is consistent with its role in one phase of the overall orientation sequence, but should not be extrapolated to encompass the broader repertoire of host selection. Likewise, first-principle arguments that natural selection would be unlikely to yield 'random host selection' are compelling. But when extrapolated to infer that landing must therefore be non-random, such arguments incorrectly conflate 'landing' with 'host selection' and are inconsistent with the evidence for some species.

A second area in which understanding the sequential nature and mechanisms of host selection is useful concerns the role of tree defence physiology in bark beetle population dynamics. Although the appearance of dead beetles in pitch tubes and lesions is a dramatic illustration of host defence (Fig. 2), the frequency at which this is actually observed is relatively low. Persisting in attacks that are unlikely to succeed because tree responses are outpacing the arrival of additional beetles is maladaptive. So beetles often leave after initial biting, as evidenced by the equal landing rates on subsequently entered and not entered trees, observations, and bioassays. However, departure leaves little if any signature. Hence bark beetle 'life tables', which are almost always within-tree partial life tables, often show a low component attributable to tree defence. Within-tree life table data are valuable for a number of purposes, but can be misapplied when extrapolated to infer full replacement rates. A more holistic estimate of the full role of tree defence should include the proportion of adults that emerge from brood trees but never oviposit, ie, opportunity costs. The available data indicate that such losses during host seeking are high (Berryman, 1979; Pope et al., 1980).

Finally, it is important to incorporate the diversity of behaviours among bark beetle species when evaluating different phases of host selection. A useful example of understanding the unique behaviours of various species is illustrated by the classic review of Rudinsky (1962). He proposed that the dispute over 'directed' vs 'random' landing could be partially resolved by recognizing differences between those species that never (or rarely) kill healthy trees (ie, most scolytines) from those that do. He hypothesized the former rely more heavily on volatiles associated with severe tree stress and microbial metabolism, whereas the latter rely more heavily on the land-and-taste strategy. His theory appears to have stood the test of time. More recently, landscape approaches to disturbances caused by bark beetles, and their underlying drivers, have substantially improved our understanding of ecosystem processes. In some cases however, the coarseness of taxonomic resolution has allowed spurious correlations and obscured important drivers.

Integrating pattern with process is a major challenge to understanding any complex biological system. This gap can be partially resolved by introducing the elements of spatial and temporal scale, and feedback, which can have valuable bridging attributes. We propose a conceptual model in Fig. 5 that overlays what is known about various modalities, signals, and feedbacks involved in host selection onto gradients of space and time. Comparative approaches, in which common methods are employed across a range of genetic, chemosensory, behavioural, and ecological studies conducted on several key species, could employ this template to facilitate understanding. Such comparisons should include diversities of geographic region, genera, and population dynamics.

5.2 Management Implications

On a global scale, forests are facing both increasing demands from a growing human population and increasing threats due to anthropogenic feedbacks (Lugo, 2015). We currently lack both the fundamental scientific knowledge

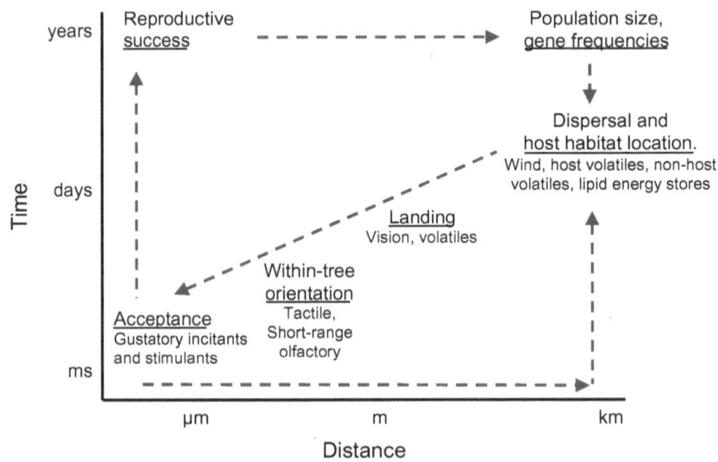

Fig. 5 Conceptual overview of host selection sequence by bark beetles. Beetles disperse upon emergence, and orient using anemotactic and chemical cues to locate host habitats. They land in response to vertical silhouettes and general host compounds such as monoterpenes and ethanol. Upon landing, they orient under bark crevices and in cracks based on thigmotactic stimuli that also elicit responsiveness to short-distance olfactory stimuli and biting. Tunnelling is elicited by chemosensory stimuli perceived in sensilla, with different chemical classes, compounds, and concentrations interacting to process host recognition and host acceptance reactions. The sequence may be interrupted (red—negative feedback) at any stage prior to oviposition. Reproductive success is determined by the interaction of host defences, beetle arrival in response to pheromones, host nutritional quality, and competition. Relative reproductive success influences population size and has genetic consequences, which jointly feed back into host acceptance mechanisms (green). (See the colour plate.)

and the technical management tools to economically and environmentally address this challenge. Because bark beetles and their responses to anthropogenic change are at the centre of many of these challenges, it is worthwhile to examine how an understanding of host selection behaviour can assist forest managers. Many detailed reviews have focused entirely on this subject, so our goal here is to highlight only some general principles, examples, and novel future approaches that emerge from our prior discussion of host selection behaviour.

An overarching principle is that knowledge of underlying mechanisms can directly improve management decisions. As an example, it has long and correctly been noted that tree mortality to some bark beetle species is correlated with large-diameter trees. However, this has sometimes been extrapolated to infer that larger trees are more susceptible and more highly preferred. This, in turn, has led to management recommendations to remove large trees. But in fact larger trees are better defended and less preferred by *D. ponderosae* than suppressed trees during endemic periods (Boone et al., 2011). The correlation arises because once an extraneous factor releases populations, they can become high enough to colonize large trees, which in turn produce more beetles, which continues the outbreak. The important distinction is that some other agent initially released the populations to the point where beetles can colonize large trees, not that the large trees cause outbreaks. This example illustrates the critical need for data at the relevant scale of resolution. It likewise illustrates that 'inferring process from pattern' can be useful when employed to designate and subsequently conduct emergent experiments, but misleading when extrapolated inferences are accepted per se. A second overarching principle is that tactics that manipulate the behaviour of pests have the potential to also adversely affect natural enemies, a consideration that must always be investigated and weighed.

The SDH predicts that lower numbers of specialist herbivores will be present in mixed habitats diverse in semiochemicals (Zhang and Schlyter, 2003), which has been confirmed at the stand level in a meta-analysis (Jactel and Brockerhoff, 2007). Recently, two larger landscape-level studies in Sweden (Kärvemo, 2015; Kärvemo et al., 2014) confirmed a negative influence by presence of *Betula* spp. stands on attack probability of *I. typographus* on *P. abies* at two levels of scale. While such functional influence of biodiversity can also be due to mechanisms of lower host availability and higher predator abundance (Risch et al., 1983; Root, 1973), both are partly mediated by chemical signalling. Stand-scale experiments and observations of bark beetles (Schlyter, 2012) and processionary moth (Dulaurent et al., 2012; Jactel et al., 2011) indicate that the SDH could act as a major factor for the abundance and distribution of forest insects and may act both at the stand-level proper and across stand-level borders (Schiebe et al., 2011). Interestingly, the Swedish Board of Forestry is slowly shifting their policy from historical conifers—only to advice on higher tree diversity in production plantations (Bergqvist, 2015). This is especially important, in the face of the uncertainty and increasing storm and insect damage due to global change over the last decades. The loss of diversity combined with global change was recently highlighted as a future problem for Fennoscandian and other boreal forests (Gauthier et al., 2015).

A more direct use of the SDH is to use dispensers with NHV and/or verbenone or other antiattractants to lower the density of potential landers and attackers at the tree or stand level. It should be noted that while NHV in both *Ips* and *Dendroctonus* come from the same source and has a similar general message of 'unsuitable habitat', the verbenone in *Dendroctonus* is a pheromonal compound, ie, largely produced by the insect (Schlyter and Birgersson, 1999), while only found in smaller amounts produced by microbial conversion or autoxidation (Byers, 1983a; Conn et al., 1984; Hunt et al., 1989), giving the more fine-scale message of 'unsuitable or overcrowded patch'. In contrast, in *Ips* the verbenone is a cue mainly produced by commensal yeasts late in the attack giving a pronounced 'old patch' signal (Schlyter and Birgersson, 1999). Correspondingly, the field activity of verbenone alone can be comparatively strong for *Dendroctonus* and it is commercially available for forest protection, while the verbenone alone is a relatively weaker signal and only shows commercial grade efficacy against *I. typographus* when used in combination with NHV blends (Zhang and Schlyter, 2004). A meta-study of *I. typographus* and *D. ponderosae* field tests with these two signals showed an overall strong effect (Schlyter, 2012), not different between species, but with large variation in effect among experiments within species due to tree status and beetle population levels (Fig. 6).

Little is known about bark beetle responses to metabolites produced by their associated fungi and bacteria, although this has been demonstrated in some cases. It is possible that low levels of fungal compounds might be used as attractants (eg, pheromone synergists), or higher levels as cues to avoid colonizing an old, already overcrowded host, similar to verbenone. Thus, future studies should attempt to identify such volatiles, investigate their possible detection and coding by olfactory neurons, and test active

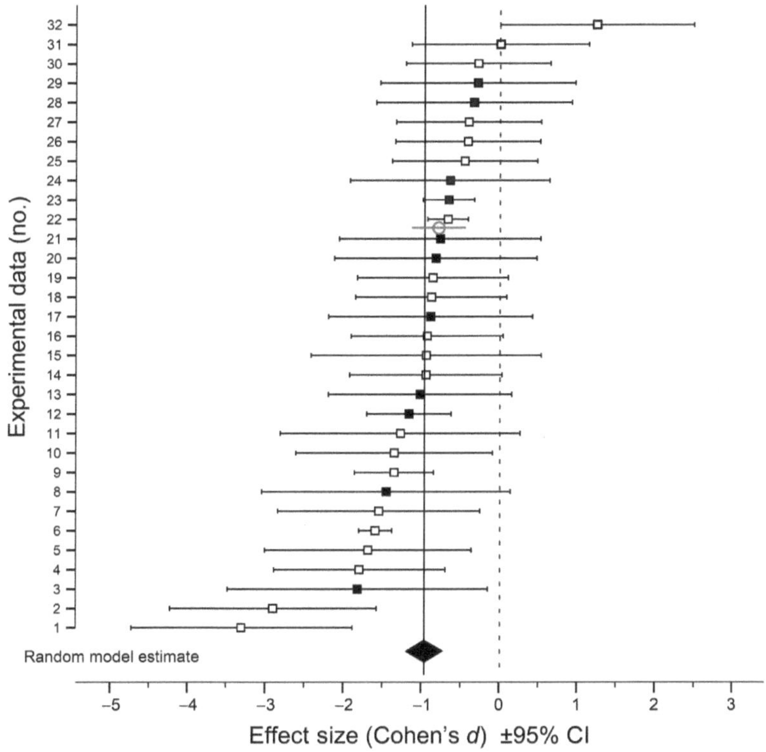

Fig. 6 Tree protection efficiency of antiattractant dispensers in *Ips typographus* and *Dendroctonus ponderosae*, tested by a meta-analysis comparing treated and untreated plots from a range of experimental results published until 2011. The efficiency is gauged by the effect size, where the difference between treatment and control is scaled by division of their pooled standard deviations, with a strong effect often considered to be at |0.8| or larger (Nakagawa and Cuthill, 2007). Thus, the overall effect is here rather large with a mean of ca. −1, as shown by the 'diamond' on last row giving the mid-point and width of the 95% CI of grand mean effect size. The graph is an ordered plot of individual study effect sizes and their ±95% confidence intervals; known as a 'Forest plot'. *Vertical line*: ——, estimated overall effect size; - - - -, effect size = 0 or no effect. Numbers on the dependent axis correspond to the 'Expt nr' in Appendix of Schlyter (2012). ■, *I. typographus*; □, *D. ponderosae*; ◯, data point summarizing a recent 7-year study of high-mountain whitebark pine protection against *D. ponderosae* (Perkins et al., 2015). *Redrawn from Schlyter, F., 2012. Semiochemical diversity in practice: anti-attractant semiochemicals reduces bark beetle attacks on standing trees—a first meta-analysis. Psyche J. Entomol. 2012, Article ID 268621.* (See the colour plate.)

compounds for behavioural effects. Incorporating active compounds from third partners in the 'bark beetle host game' could possibly provide synergies for semiochemical-based control. Such antiattractants may have their most cost-effective use in trees that are of particular high individual value, owing to their location, use, or potential for other desired properties such as putative resistance to invasive pathogens (Perkins et al., 2015).

Interruption of bark beetle orientation sequences can also be achieved by manipulating features of forest structure (Fettig et al., 2014). For example, thinning managed stands can enhance the resistance levels of remaining trees by reducing competition for water, nutrients, and sun. Thinning also causes changes in microclimate that interfere with arrestment. Further, thinning has also been proposed to create canopy gaps that can alter the movement and reduce the functioning of aggregation pheromones (Fares et al., 1980). The available evidence suggests that thinning can partially prevent, but not halt, bark beetle outbreaks. Also, procedures such as thinning and rapid rotation must be consistent with overall management objectives to be worthwhile. It has also been suggested that the orientation by bark beetles towards crevices, by which thigmotactic stimuli are integrated with chemical cues (eg, Borg and Norris, 1969; Bunt et al., 1980; Paynter et al., 1990), might also scale up to the whole-tree level and so influence tree mortality rates (Ferrenberg and Mitton, 2014), although this has not been tested.

Many invasions can be effectively managed by semiochemical monitoring and mass trapping when programmes are well organized and when efficient attractants are available. Experiences in New Zealand provide effective examples (El-Sayed et al., 2006). The newer technologies of SSR and functional studies to de-orphanize ORs (Section 2) promise to deliver new cards to be dealt from the semiochemical deck here. An altogether new set of tools consist of sentry dogs that can locate pheromone from distances of >100 m from trees attacked by *I. typographus*. Presently, this method is foremost employed for rapid location and removal of attacked trees as required by law (Schlyter and Johansson, 2010). This approach differs from trapping schemes that disclose landscape-level existence or may lower populations down to Allee-effect levels, but will make eradication feasible by locating the last remaining attacked trees before they release dispersing offspring. Two unproven methods under development include interferences with OR reception (Chen and Luetje, 2012; Tauxe et al., 2013) and sniffer dog detection of tree stress markers such as 1,8-cineole.

Rapid advances in sequencing and molecular techniques now allow for quick and inexpensive identification of OR sequences in non-model species, and subsequent functional testing to identify active ligands of ORs expressed in heterologous systems. To date, there are no functionally characterized ORs (or other olfactory proteins) in bark beetles. However, such studies are needed to understand the mechanics and evolution of odour detection in bark beetles at the cellular and subcellular levels. Not only will such studies provide mechanistic knowledge, but they may also lead to novel, potentially more efficient, pest control strategies. For instance, if the receptors for aggregation pheromones or antiattractive semiochemicals are identified, they can be screened using massive-throughput screening platforms (Tauxe et al., 2013) to identify natural or synthetic antagonists (Chen and Luetje, 2012, 2013) or agonists that have a higher affinity to the OR than the natural odorant. Indeed, c.500,000 compounds were recently screened in silico to identify antagonists and agonists for the CO_2 receptor in the malaria- and yellow fever mosquitoes, establishing this receptor as a target for mosquito-vectored disease control (Tauxe et al., 2013). If similar compounds can be identified for crucial ORs in bark beetles, they could be incorporated into Integrated Pest Management strategies in combination with pheromones and antiattractants. Since different bark beetle species share many semiochemicals, it is possible that the same compounds could be used for multiple pests, as indicated by the presence of conserved OR orthologs in relatively unrelated species, such as *I. typographus* and *D. ponderosae* (Andersson et al., 2013). In addition, some of the NHV, such as *trans*-conophthorin, are expensive to synthesize. Thus, if less expensive ligands for this OR can be identified, management using NHV will become more economically feasible.

There is evidence for some host compounds, such as 1,8-cineole, to be both antiattractive semiochemicals and tree defence moieties, and together with other oxygenated terpenes potential volatile markers and cues for levels of tree resistance or susceptibility (Kalinová et al., 2014; Schiebe, 2012). One could train sniffer dogs to such cues, which are chemically similar to the compounds in *I. typographus* pheromone (oxygenated terpenoids), allowing location of single trees or groups of trees of different tolerance from a distance. In particular, groups of susceptible trees could be identified for proactive removal during sanitation following storms or normal thinning operations in production forestry.

6. CONCLUSIONS

1. Bark beetles are confronted with complex decisions during host selection. Factors that shape these decisions include tree defensive responses that can kill the adults and their brood, trade-offs between host susceptibility and host quality, life history constraints such as relatively short adult life spans and often total oviposition within a single tree, commonly

a restriction to only one generation within a given host, and a largely stochastic environment that exerts strong higher scale inputs on survival and reproductive opportunities.

2. Bark beetles contend with these challenges by using multiple, integrated sensory modalities, making host selection decisions in a sequential fashion that includes several junctures for potential termination, and exhibiting plasticity to better react to a changing environment, including their own population densities.
3. Bark beetles use a mixture of fine-tuned and more broadly tuned neurons to identify host habitats, host species, and susceptible individuals within-host populations. Some neurons for negative signals such as NHV and certain heterospecific pheromones respond to several compounds with related structure, possibly a mechanism to improve the chances of detecting non-hosts or unsuitable host and habitat signals. Detection of positive as well as some negative signals involves fine-tuned neurons for predominant host compounds, rare host compounds, and pheromones.
4. Antennae provide spatially structured landscapes of neurons that optimize beetle responses to heterogeneous signals arising from complex forest landscapes. Specific co-localization of OSNs in sensilla allows for odour mixture processing by the peripheral nervous system and also improves spatial discrimination of odour sources to maximize the efficiency with which bark beetles conduct host selection.
5. Beetle–host associations are being altered by climate change, management practices, and global transport. The extent to which bark beetles establish and impact new hosts will be driven by many factors, but host range at the genus level does not appear to pose a major barrier.
6. Bark beetles provide an important model system for linking neurophysiological processes with landscape-scale ecological, environmental, and socioeconomic patterns and outcomes.
7. Understanding mechanisms of host selection can provide several approaches to managing populations of pest species and preventing impacts of invasive species. This understanding can likewise lead to the development of new strategies based on specific interactions with receptors.

ACKNOWLEDGEMENTS

K.F.R. was supported by the University of Wisconsin College of Agricultural and Life Sciences and UW Graduate School Beers-Bascom Fellowship. M.N.A. was funded by the Swedish Research Council FORMAS (#217-2014-689). F.S. was partly funded by

Formas, a Linnaeus initiative grant (#217-2006-1750) 'IC-E3' and by Department of Plant Protection Biology, Alnarp, SLU. The critical review by Staffan Lindgren (University of Northern British Columbia) is greatly appreciated. M.N.A. is grateful to his parents Hans and Anne-Marie Andersson for babysitting during manuscript preparation.

REFERENCES

Abuin, L., Bargeton, B., Ulbrich, M.H., Isacoff, E.Y., Kellenberger, S., Benton, R., 2011. Functional architecture of olfactory ionotropic glutamate receptors. Neuron 69, 44–60.

Allison, J.D., Borden, J.H., McIntosh, R.L., De Groot, P., Gries, R., 2001. Kairomonal response by four *Monochamus* species (Coleoptera: Cerambycidae) to bark beetle pheromones. J. Chem. Ecol. 27, 633–646.

Allstadt, A.J., Haynes, K.J., Liebhold, A.M., Johnson, D.M., 2013. Long-term shifts in the cyclicity of outbreaks of a forest-defoliating insect. Oecologia 172, 141–151.

Amman, G.D., 1984. Mountain pine beetle (Coleoptera: Scolytidae) mortality in three types of infestations. Environ. Entomol. 13, 184–191.

Amman, G.D., Safranyik, L., 1985. Insects of lodgepole pine: impacts and control. In: Baumgartner, D.M., Krebill, R.G., Arnott, J.T., Weetnam, G.F. (Eds.), Lodgepine—The Species and Its Management. May 8–10, 1984, Spokane, Washington, USA and repeated May 14–16, 1984, Vancouver, British Columbia, Canada. Cooperative Extension, Washington State University, Pullman, pp. 107–124.

Amman, G.D., McGregor, M.D., Dolph Jr., R.E., 1990. Mountain Pine Beetle, Forest Insect and Disease Leaflet 2. USDA Forest Service, Washington, DC.

Anderbrant, O., Schlyter, F., Birgersson, G., 1985. Intraspecific competition affecting parents and offspring in the bark beetle *Ips typographus*. Oikos 45, 89–98.

Andersson, M.N., 2012. Mechanisms of odor coding in coniferous bark beetles: from neuron to behavior and application. Psyche J. Entomol. 2012. Article ID 149572.

Andersson, M.N., Larsson, M.C., Schlyter, F., 2009. Specificity and redundancy in the olfactory system of the bark beetle *Ips typographus*: single-cell responses to ecologically relevant odors. J. Insect Physiol. 55, 556–567.

Andersson, M.N., Larsson, M.C., Blaženec, M., Jakuš, R., Zhang, Q.-H., Schlyter, F., 2010. Peripheral modulation of pheromone response by inhibitory host compound in a beetle. J. Exp. Biol. 213, 3332–3339.

Andersson, M.N., Binyameen, M., Sadek, M.M., Schlyter, F., 2011. Attraction modulated by spacing of pheromone components and anti-attractants in a bark beetle and a moth. J. Chem. Ecol. 37, 899–911.

Andersson, M.N., Larsson, M.C., Svensson, G.P., Birgersson, G., Rundlöf, M., Lundin, O., Lankinen, Å., Anderbrant, O., 2012a. Characterization of olfactory sensory neurons in the white clover seed weevil, *Apion fulvipes* (Coleoptera: Apionidae). J. Insect Physiol. 58, 1325–1333.

Andersson, M.N., Schlyter, F., Hill, S.R., Dekker, T., 2012b. What reaches the antenna? How to calibrate odor flux and ligand–receptor affinities. Chem. Senses 37, 403–420.

Andersson, M.N., Grosse-Wilde, E., Keeling, C.I., Bengtsson, J.M., Yuen, M.M., Li, M., Hillbur, Y., Bohlmann, J., Hansson, B.S., Schlyter, F., 2013. Antennal transcriptome analysis of the chemosensory gene families in the tree killing bark beetles, *Ips typographus* and *Dendroctonus ponderosae* (Coleoptera: Curculionidae: Scolytinae). BMC Genomics 14 (198), 1–16.

Andersson, M.N., Löfstedt, C., Newcomb, R.D., 2015. Insect olfaction and the evolution of receptor tuning. Front. Ecol. Evol. 3, 53.

Ascoli-Christensen, A., Salom, S.M., Payne, T.L., 1993. Olfactory receptor cell responses of *Ips grandicollis* (Eichhoff) (Coleoptera: Scolytidae) to intra- and interspecific behavioral chemicals. J. Chem. Ecol. 19, 699–712.

Atkins, M.D., 1966. Behavioural variation among scolytids in relation to their habitat. Can. Entomol. 98, 285–288.
Atkins, M.D., 1969. Lipid loss with flight in the Douglas-fir beetle. Can. Entomol. 101, 164–165.
Aukema, B.H., Raffa, K.F., 2004. Gender- and sequence-dependent predation within group colonizers of defended plants: a constraint on cheating among bark beetles? Oecologia 138, 253–258.
Aukema, B.H., Zhu, J., Moeller, J., Rasmussen, J., Raffa, K., 2010. Interactions between below- and above ground herbivores drive a forest decline and gap-forming syndrome. For. Ecol. Manag. 259, 374–382.
Azeem, M., Rajarao, G.K., Nordenhem, H., Nordlander, G., Borg-Karlson, A.K., 2013. *Penicillium expansum* volatiles reduce pine weevil attraction to host plants. J. Chem. Ecol. 39, 120–128.
Baker, T.C., Fadamiro, H.Y., Cossé, A.A., 1998. Moth uses fine tuning for odour resolution. Nature 393, 530.
Barron, A.B., 2001. The life and death of Hopkins' host-selection principle. J. Insect Behav. 14, 725–737.
Beal, J.-A., 1939. The Black Hills beetle: a serious enemy of Rocky Mountain pines. In: Farmer's Bulletin, USDA, Washington, DC, pp. 1–22.
Benton, R., Vannice, K.S., Vosshall, L.B., 2007. An essential role for a CD36-related receptor in pheromone detection in *Drosophila*. Nature 450, 289–293.
Benton, R., Vannice, K.S., Gomez-Diaz, C., Vosshall, L.B., 2009. Variant ionotropic glutamate receptors as chemosensory receptors in *Drosophila*. Cell 136, 149–162.
Bentz, B.J., Régnière, J., Fettig, C.J., Hansen, E.M., Hayes, J.L., Hicke, J.A., Kelsey, R.G., Negrón, J.F., Seybold, S.J., 2010. Climate change and bark beetles of the western United States and Canada: direct and indirect effects. Bioscience 60, 602–613.
Bentz, B.J., Boone, C., Raffa, K.F., 2015. Tree response and mountain pine beetle attack preference, reproduction and emergence timing in mixed whitebark and lodgepole pine stands. Agric. For. Entomol. 17, 421–432.
Berg, B.G., Almaas, T.J., Bjaalie, J.G., Mustaparta, H., 2005. Projections of male-specific receptor neurons in the antennal lobe of the oriental tobacco budworm moth, *Helicoverpa assulta*: a unique glomerular organization among related species. J. Comp. Neurol. 486, 209–220.
Bergqvist, J., 2015. Skapa skogar med variation (Create forests with variation). Skogsaktuellt 30, 26–27. Swedish Forestry Agency Newsletter (in Swedish).
Berryman, A.A., 1972. Resistance of conifers to invasion by bark beetle-fungus associations. Bioscience 22, 598–602.
Berryman, A.A., 1973. Population dynamics of the fir engraver, *Scolytus ventralis* (Coleoptera: Scolytidae). I. Analysis of population behavior and survival from 1964 to 1971. Can. Entomol. 105, 1465–1488.
Berryman, A.A., 1979. Dynamics of bark beetle populations: analysis of dispersal and redistribution. Mitt. Schw. Entomol. Ges. 52, 227–234.
Berryman, A.A., Stark, R., 1962. Studies on the effects of temperature on the development of *Ips confusus* using radiographic techniques. Ecology 43, 722–726.
Binyameen, M., Jankuvova, J., Blaženec, M., Jakuš, R., Song, L., Schlyter, F., Andersson, M.N., 2014. Co-localization of insect olfactory sensory cells improves the discrimination of closely separated odour sources. Funct. Ecol. 28, 1216–1223.
Birch, M.C., Light, D.M., Wood, D.L., Browne, L.E., Silverstein, R.M., Bergot, B.J., Ohloff, G., West, J.R., Young, J.C., 1980. Pheromonal attraction and allomonal interruption of *Ips pini* in California by the two enantiomers of ipsdienol. J. Chem. Ecol. 6, 703–717.
Birgersson, G., Bergström, G., 1989. Volatiles released from individual spruce bark beetle entrance holes: quantitative variations during the first week of attack. J. Chem. Ecol. 15, 2465–2483.

Birgersson, G., Schlyter, F., Löfqvist, J., Bergström, G., 1984. Quantitative variation of pheromone components in the spruce bark beetle *Ips typographus* from different attack phases. J. Chem. Ecol. 10, 1029–1055.

Birgersson, G., Schlyter, F., Bergström, G., Löfqvist, J., 1988. Individual variation in aggregation pheromone content of the bark beetle, *Ips typographus*. J. Chem. Ecol. 14, 1737–1761.

Blanche, C., Hodges, J., Nebeker, T., 1985. Changes in bark beetle susceptibility indicators in a lightning-struck loblolly pine. Can. J. For. Res. 15, 397–399.

Blažytė-Čereškienė, L., Apšegaitė, V., Radžiutė, S., Mozūraitis, R., Būda, V., Pečiulytė, D., 2015. Electrophysiological and behavioural responses of *Ips typographus* (L.) to *trans*-4-thujanol—a host tree volatile compound. Ann. For. Sci. http://dx.doi.org/10.1007/s13595-015-0494-5. http://link.springer.com/article/10.1007/s13595-015-0494-5.

Bleiker, K., Six, D., 2007. Dietary benefits of fungal associates to an eruptive herbivore: potential implications of multiple associates on host population dynamics. Environ. Entomol. 36, 1384–1396.

Blomquist, G.J., Figueroa-Teran, R., Aw, M., Song, M., Gorzalski, A., Abbott, N.L., Chang, E., Tittiger, C., 2010. Pheromone production in bark beetles. Insect Biochem. Mol. Biol. 40, 699–712.

Bohlmann, J., Gershenzon, J., 2009. Old substrates for new enzymes of terpenoid biosynthesis. Proc. Natl. Acad. Sci. U.S.A. 106, 10402–10403.

Boone, C.K., Six, D.L., Raffa, K.F., 2008. The enemy of my enemy is still my enemy: competitors add to predator load of a tree-killing bark beetle. Agric. For. Entomol. 10, 411–421.

Boone, C.K., Aukema, B.H., Bohlmann, J., Carroll, A.L., Raffa, K.F., 2011. Efficacy of tree defense physiology varies with bark beetle population density: a basis for positive feedback in eruptive species. Can. J. For. Res. 41, 1174–1188.

Borg, T.K., Norris, D.M., 1969. Feeding responses by *Hylurgopinus rufipes* to combined chemical and physical stimuli. Ann. Entomol. Soc. Am. 62, 730–733.

Breshears, D.D., Myers, O.B., Meyer, C.W., Barnes, F.J., Zou, C.B., Allen, C.D., McDowell, N.G., Pockman, W.T., 2009. Tree die-off in response to global change-type drought: mortality insights from a decade of plant water potential measurements. Front. Ecol. Environ. 7, 185–189.

Bruce, T.J.A., Wadhams, L.J., Woodcock, C.M., 2005. Insect host location: a volatile situation. Trends Plant Sci. 10, 269–274.

Bunt, W., Coster, J., Johnson, P., 1980. Behavior of the southern pine beetle on the bark of host trees during mass attack. Ann. Entomol. Soc. Am. 73, 647–652.

Burbano, E., Wright, M., Bright, D.E., Vega, F.E., 2011. New record for the coffee berry borer, *Hypothenemus hampei*, in Hawaii. J. Insect Sci. 11, 117.

Byers, J.A., 1983a. Influence of sex, maturity and host substances on pheromones in the guts of the bark beetles, *Ips paraconfusus* and *Dendroctonus brevicomis*. J. Insect Physiol. 29, 5–13.

Byers, J.A., 1983b. Sex-specific responses to aggregation pheromone: regulation of colonization density in the bark beetle *Ips paraconfusus*. J. Chem. Ecol. 9, 129–142.

Byers, J.A., 1993. Avoidance of competition by spruce bark beetles, *Ips typographus* and *Pityogenes chalcographus*. Experientia 49, 272–275.

Byers, J.A., 1996. An encounter rate model of bark beetle populations searching at random for susceptible host trees. Ecol. Model. 91, 57–66.

Byers, J.A., 2004. Chemical ecology of bark beetles in a complex olfactory landscape. In: Lieutier, F., Day, K.R., Battisti, A., Grégoire, J.-C., Evans, H.F. (Eds.), Bark and Wood Boring Insects in Living Trees in Europe, a Synthesis. Kluwer Academic Publishers, Dordrecht, pp. 89–134.

Byers, J.A., Zhang, Q., 2011. Chemical ecology of bark beetles in regard to search and selection of host trees. In: Liu, T., Kang, L. (Eds.), Recent Advances in Entomological Research. Springer, Berlin, Heidelberg, pp. 150–190.

Cale, J.A., Taft, S., Najar, A., Klutsch, J.G., Hughes, C.C., Sweeney, J.D., Erbilgin, N., 2015. Mountain pine beetle (*Dendroctonus ponderosae*) can produce its aggregation pheromone and complete brood development in naïve red pine (*Pinus resinosa*) under laboratory conditions. Can. J. For. Res. 45, 1873–1877.

Campbell, S.A., Borden, J.H., 2006a. Close-range, in-flight integration of olfactory and visual information by a host-seeking bark beetle. Entomol. Exp. Appl. 120, 91–98.

Campbell, S.A., Borden, J.H., 2006b. Integration of visual and olfactory cues of hosts and spi1;non-hosts by three bark beetles (Coleoptera: Scolytidae). Ecol. Entomol. 31, 437–449.

Chen, S., Luetje, C.W., 2012. Identification of new agonists and antagonists of the insect odorant receptor co-receptor subunit. PLoS One 7, e36784.

Chen, S., Luetje, C.W., 2013. Phenylthiophenecarboxamide antagonists of the olfactory receptor co-receptor subunit from a mosquito. PLoS One 8, e84575.

Chen, H.-B., Zhang, Z., Wang, H.-B., Kong, X.-B., 2010. Antennal morphology and sensilla ultrastructure of *Dendroctonus valens* LeConte (Coleoptera: Curculionidae, Scolytinae), an invasive forest pest in China. Micron 41, 735–741.

Clark, E.L., Carroll, A.L., Huber, D.P., 2010. Differences in the constitutive terpene profile of lodgepole pine across a geographical range in British Columbia, and correlation with historical attack by mountain pine beetle. Can. Entomol. 142, 557–573.

Claudianos, C., Lim, T., Young, M., Yan, S., Cristino, A.S., Newcomb, R.D., Gunasekaran, N., Reinhard, J., 2014. Odor memories regulate olfactory receptor expression in the periphery. Eur. J. Neurosci. 39, 1642–1654.

Conn, J., Borden, J., Hunt, D., Holman, J., Whitney, H., Spanier, O., Pierce Jr., H., Oehlschlager, A., 1984. Pheromone production by axenically reared *Dendroctonus ponderosae* and *Ips paraconfusus* (Coleoptera: Scolytidae). J. Chem. Ecol. 10, 281–290.

Cosanza, R., d'Arge, R., de Groot, R., Farber, S., Grasso, M., Hannon, B., Limburg, K., Naeem, S., O'Neill, R.V., Paruelo, J., Raskin, R.G., Sutton, P., van den Belt, M., 1997. The value of the world's ecosystem services and natural capital. Nature 387, 253–260.

Coulson, R.N., 1979. Population dynamics of bark beetles. Annu. Rev. Entomol. 24, 417–447.

Coulson, R.N., Hennier, P., Flamm, R., Rykiel, E., Hu, L., Payne, T., 1983. The role of lightning in the epidemiology of the Southern Pine Beetle. Z. Angew. Entomol. 96, 182–193.

Creeden, E.P., Hicke, J.A., Buotte, P.C., 2014. Climate, weather, and recent mountain pine beetle outbreaks in the western United States. For. Ecol. Manag. 312, 239–251.

Croset, V., Rytz, R., Cummins, S.F., Budd, A., Brawand, D., Kaessmann, H., Gibson, T.J., Benton, R., 2010. Ancient protostome origin of chemosensory ionotropic glutamate receptors and the evolution of insect taste and olfaction. PLoS Genet. 6, e1001064.

Cudmore, T.J., Björklund, N., Carroll, A.L., Lindgren, B.S., 2010. Climate change and range expansion of an aggressive bark beetle: evidence of higher beetle reproduction in naive host tree populations. J. Appl. Ecol. 47, 1036–1043.

Cullingham, C.I., Cooke, J.E., Dang, S., Davis, C.S., Cooke, B.J., Coltman, D.W., 2011. Mountain pine beetle host-range expansion threatens the boreal forest. Mol. Ecol. 20, 2157–2171.

Davis, R.S., Hood, S., Bentz, B.J., 2012. Fire-injured ponderosa pine provide a pulsed resource for bark beetles. Can. J. For. Res. 42, 2022–2036.

Delvas, N., Bauce, É., Labbé, C., Ollevier, T., Bélanger, R., 2011. Phenolic compounds that confer resistance to spruce budworm. Entomol. Exp. Appl. 141, 35–44.

Dickens, J.C., Payne, T.L., 1977. Bark beetle olfaction: pheromone receptor system in *Dendroctonus frontalis*. J. Insect Physiol. 23, 481–489.

Dickens, J.C., Payne, T.L., Ryker, L.C., Rudinsky, J.A., 1984. Single cell responses of the Douglas-fir beetle *Dendroctonus pseudotsugae* Hopkins (Coleoptera: Scolytidae) to pheromones and host odors. J. Chem. Ecol. 10, 583–600.

Dickens, J.C., Payne, T.L., Ryker, L.C., Rudinsky, J.A., 1985. Multiple acceptors for pheromonal enantiomers on single olfactory cells in the Douglas-fir beetle *Dendroctonus pseudotsugae* Hopk. (Coleoptera: Scolytidae). J. Chem. Ecol. 11, 1359–1370.

Dulaurent, A.M., Porte, A.J., van Halder, I., Vetillard, F., Menassieu, P., Jactel, H., 2012. Hide and seek in forests: colonization by the pine processionary moth is impeded by the presence of nonhost trees. Agric. For. Entomol. 14, 19–27.

Elkinton, J., Wood, D., Browne, L., 1981. Feeding and boring behavior of the bark beetle, Ips paraconfusus, in extracts of ponderosa pine phloem. J. Chem. Ecol. 7, 209–220.

El-Sayed, A., Suckling, D., Wearing, C., Byers, J., 2006. Potential of mass trapping for long-term pest management and eradication of invasive species. J. Econ. Entomol. 99, 1550–1564.

Embrey, S., Remais, J.V., Hess, J., 2012. Climate change and ecosystem disruption: the health impacts of the North American rocky mountain pine beetle infestation. Am. J. Public Health 102, 818–827.

Engsontia, P., Sanderson, A.P., Cobb, M., Walden, K.K.O., Robertson, H.M., Brown, S., 2008. The red flour beetle's large nose: an expanded odorant receptor gene family in *Tribolium castaneum*. Insect Biochem. Mol. Biol. 38, 387–397.

Erbilgin, N., Krokene, P., Kvamme, T., Christiansen, E., 2007. A host monoterpene influences *Ips typographus* (Coleoptera: Curculionidae, Scolytinae) responses to its aggregation pheromone. Agric. For. Entomol. 9, 135–140.

Evenden, J.C., Bedard, W.E., Struble, G.R., 1943. The mountain pine beetle, an important enemy of western pines. USDA Circular No. 664, pp. 1–25.

Faccoli, M., Schlyter, F., 2007. Conifer phenolic resistance markers are bark beetle antifeedant semiochemicals. Agric. For. Entomol. 9, 237–245.

Fadamiro, H.Y., Cossé, A.A., Baker, T.C., 1999. Fine-scale resolution of closely spaced pheromone and antagonist filaments by flying male *Helicoverpa zea*. J. Comp. Physiol. A. 185, 131–141.

Fares, Y., Sharpe, P.J.H., Magnuson, C.E., 1980. Pheromone dispersion in forests. J. Theor. Biol. 84, 335–359.

Ferrenberg, S., Mitton, J.B., 2014. Smooth bark surfaces can defend trees against insect attack: resurrecting a 'slippery' hypothesis. Funct. Ecol. 28, 837–845.

Fettig, C.J., Gibson, K.E., Munson, A.S., Negrón, J.F., 2014. Cultural practices for prevention and mitigation of mountain pine beetle infestations. For. Sci. 60, 450–463.

Flecke, C., Stengl, M., 2009. Octopamine and tyramine modulate pheromone-sensitive olfactory sensilla of the hawkmoth *Manduca sexta* in a time-dependent manner. J. Comp. Physiol. A. 195, 529–545.

Fox, A.N., Pitts, R.J., Robertson, H.M., Carlson, J.R., Zwiebel, L.J., 2001. Candidate odorant receptors from the malaria vector mosquito *Anopheles gambiae* and evidence of down-regulation in response to blood feeding. Proc. Natl. Acad. Sci. U.S.A. 98, 14693–14697.

Franceschi, V.R., Krokene, P., Christiansen, E., Krekling, T., 2005. Anatomical and chemical defenses of conifer bark against bark beetles and other pests. New Phytol. 167, 353–375.

Furniss, M., Schenk, J., 1969. Sustained natural infestation by the mountain pine beetle in seven new *Pinus* and *Picea* hosts. J. Econ. Entomol. 62, 518–519.

Gauthier, S., Bernier, P., Kuuluvainen, T., Shvidenko, A.Z., Schepaschenko, D.G., 2015. Boreal forest health and global change. Science 349, 819–822.

Gaylord, M.L., Kolb, T.E., Pockman, W.T., Plaut, J.A., Yepez, E.A., Macalady, A.K., Pangle, R.E., McDowell, N.G., 2013. Drought predisposes piñon–juniper woodlands to insect attacks and mortality. New Phytol. 198, 567–578.

Ghimire, B., Williams, C.A., Collatz, G.J., Vanderhoof, M., Rogan, J., Kulakowski, D., Masek, J.G., 2015. Large carbon release legacy from bark beetle outbreaks across Western United States. Glob. Change Biol. 21, 3087–3101.

Graf, M., Reid, M., Aukema, B., Lindgren, B., 2012. Association of tree diameter with body size and lipid content of mountain pine beetles. Can. Entomol. 144, 467–477.

Graham, K., 1959. Release by flight exercise of a chemotropic response from photopositive domination in a scolytid beetle. Nature 184, 283–284.

Gray, C.A., Runyon, J.B., Jenkins, M.J., Giunta, A.D., 2015. Mountain pine beetles use volatile cues to locate host limber pine and avoid non-host Great Basin bristlecone pine. PLoS One 10, e0135752.

Grégoire, J., Raffa, K., Lindgren, B., 2015. Economics and politics of bark beetles. In: Vega, F.E., Hofstetter, R.W. (Eds.), Bark Beetles: Biology and Ecology of Native and Invasive Species. Academic Press, Boston, MA, pp. 585–614.

Griffin, J.M., Turner, M.G., Simard, M., 2011. Nitrogen cycling following mountain pine beetle disturbance in lodgepole pine forests of Greater Yellowstone. For. Ecol. Manag. 261, 1077–1089.

Grodzki, W., Jakuš, R., Lajzová, E., Sitková, Z., Maczka, T., Škvarenina, J., 2006. Effects of intensive versus no management strategies during an outbreak of the bark beetle *Ips typographus* (L.) (Col.: Curculionidae, Scolytinae) in the Tatra Mts. in Poland and Slovakia. Ann. For. Sci. 63, 55–61.

Gu, X.-C., Zhang, Y.-N., Kang, K., Dong, S.-L., Zhang, L.-W., 2015. Antennal transcriptome analysis of odorant reception genes in the Red Turpentine Beetle (RTB), *Dendroctonus valens*. PLoS One 10, e0125159.

Guerrero, A., Feixas, J., Pajares, J., Wadhams, L.J., Pickett, J.A., Woodcock, C.M., 1997. Semiochemically induced inhibition of behaviour of *Tomicus destruens* (Woll.) (Coleoptera: Scolytidae). Naturwissenschaften 84, 155–157.

Hain, F.P., Duehl, A.J., Gardner, M.J., Payne, T.L., 2011. Natural history of the southern pine beetle. In: Coulson, R.N., Klepzig, K.D. (Eds.), Southern Pine Beetle II. GTR-SRS-140, Gen. Tech. Rep. SRS-140 ed. U.S. Department of Agriculture Forest Service, Southern Research Station, Asheville, NC, pp. 13–24.

Hallberg, E., 1982. Sensory organs in *Ips typographus* (Insecta: Coleoptera)—fine structure of antennal sensilla. Protoplasma 111, 206–214.

Hansson, B.S., Löfstedt, C., Löfqvist, J., Hallberg, E., 1986. Spatial arrangement of different types of pheromone-sensitive sensilla in a male moth. Nature 73, 269–270.

Hansson, B.S., Larsson, M.C., Leal, W.S., 1999. Green leaf volatile-detecting olfactory receptor neurones display very high sensitivity and specificity in a scarab beetle. Physiol. Entomol. 24, 121–126.

Harvey, B.J., Donato, D.C., Romme, W.H., Turner, M.G., 2013. Influence of recent bark beetle outbreak on fire severity and postfire tree regeneration in montane Douglas-fir forests. Ecology 94, 2475–2486.

Hayes, J.L., Strom, B.L., 1994. 4-Allylanisole as an inhibitor of bark beetle (Coleoptera: Scolytidae) aggregation. J. Econ. Entomol. 87, 1586–1594.

Hayes, J.L., Strom, B.L., Roton, L.M., Ingram Jr., L.L., 1994. Repellent properties of the host compound 4-allylanisole to the southern pine beetle. J. Chem. Ecol. 20, 1595–1615.

Haynes, K.J., Liebhold, A.M., Johnson, D.M., 2009. Spatial analysis of harmonic oscillation of gypsy moth outbreak intensity. Oecologia 159, 249–256.

Hood, S., Bentz, B., 2007. Predicting postfire Douglas-fir beetle attacks and tree mortality in the northern Rocky Mountains. Can. J. For. Res. 37, 1058–1069.

Huber, D.P., Gries, R., Borden, J.H., Pierce Jr., H.D., 2000. A survey of antennal responses by five species of coniferophagous bark beetles (Coleoptera: Scolytidae) to bark volatiles of six species of angiosperm trees. Chemoecology 10, 103–113.

Huber, D.P., Aukema, B.H., Hodgkinson, R.S., Lindgren, B.S., 2009. Successful colonization, reproduction, and new generation emergence in live interior hybrid spruce *Picea engelmannii* × *glauca* by mountain pine beetle *Dendroctonus ponderosae*. Agric. For. Entomol. 11, 83–89.

Hulcr, J., Pollet, M., Ubik, K., Vrkoč, J., 2005. Exploitation of kairomones and synomones by Medetera spp. (Diptera: Dolichopodidae), predators of spruce bark beetles. Eur. J. Entomol. 102, 655–662.

Hulcr, J., Ubik, K., Vrkoc, J., 2006. The role of semiochemicals in tritrophic interactions between the spruce bark beetle *Ips typographus*, its predators and infested spruce. J. Appl. Entomol. 130, 275–283.

Hulcr, J., Atkinson, T., Cognato, A., Jordal, B., McKenna, D., 2015. Morphology, taxonomy and phylogenetics of bark beetles. In: Vega, F.E., Hofstetter, R.W. (Eds.), Bark Beetles: Biology and Ecology of Native and Invasive Species. Academic Press, Boston, MA, pp. 41–84.

Hunt, D.W.A., Borden, J.H., Lindgren, B.S., Gries, G., 1989. The role of autoxidation of α-pinene in the production of pheromones of *Dendroctonus ponderosae* (Coleoptera: Scolytidae). Can. J. For. Res. 19, 1275–1282.

Hunter, A.F., Lechowicz, M.J., 1992. Foliage quality changes during canopy development of some northern hardwood trees. Oecologia 89, 316–323.

Hwang, S.-Y., Lindroth, R.L., 1997. Clonal variation in foliar chemistry of aspen: effects on gypsy moths and forest tent caterpillars. Oecologia 111, 99–108.

Hynum, B.G., Berryman, A.A., 1980. *Dendroctonus ponderosae* (Coleoptera: Scolytidae): pre-aggregation landing and gallery initiation on lodgepole pine. Can. Entomol. 112, 185–191.

Ignell, R., Root, C.M., Birse, R.T., Wang, J.W., Nässel, D.R., Winther, Å.M.E., 2009. Presynaptic peptidergic modulation of olfactory receptor neurons in *Drosophila*. Proc. Natl. Acad. Sci. U.S.A. 106, 13070–13075.

Jactel, H., Brockerhoff, E.G., 2007. Tree diversity reduces herbivory by forest insects. Ecol. Lett. 10, 835–848.

Jactel, H., Birgersson, G., Andersson, S., Schlyter, F., 2011. Non-host volatiles mediate associational resistance to the pine processionary moth. Oecologia 166, 703–711.

Jönsson, A.M., Appelberg, G., Harding, S., Bärring, L., 2009. Spatio-temporal impact of climate change on the activity and voltinism of the spruce bark beetle, *Ips typographus*. Glob. Change Biol. 15, 486–499.

Jönsson, A.M., Harding, S., Krokene, P., Lange, H., Lindelöw, Å., Økland, B., Ravn, H.P., Schroeder, L.M., 2011. Modelling the potential impact of global warming on *Ips typographus* voltinism and reproductive diapause. Clim. Chang. 109, 695–718.

Jost, R.W., Rice, A.V., Langor, D.W., Boluk, Y., 2008. Monoterpene emissions from Lodgepole and Jack pine bark inoculated with Mountain pine beetle-associated fungi. J. Wood Chem. Technol. 28, 37–46.

Kalinová, B., Břízová, R., Knížek, M., Turčáni, M., Hoskovec, M., 2014. Volatiles from spruce trap-trees detected by *Ips typographus* bark beetles: chemical and electrophysiological analyses. Arthropod Plant Inte. 8, 305–316.

Kärvemo, S., 2015. Outbreak dynamics of the spruce bark beetle *Ips typographus* in time and space. Acta Universitatis Agriculturae Sueciae (1652-6880), vol. 27. Department of Ecology. SLU, Swedish University of Agricultural Sciences, Uppsala. 44 pp. http://pub.epsilon.slu.se/12015/.

Kärvemo, S., Schroeder, L.M., 2010. A comparison of outbreak dynamics of the spruce bark beetle in Sweden and the mountain pine beetle in Canada (Curculionidae: Scolytinae).

Entomol. Tidskr. 131, 215–224. http://pub.epsilon.slu.se/5604/5601/karvemo_s_etal_110127.pdf.

Kärvemo, S., Rogell, B., Schroeder, M., 2014. Dynamics of spruce bark beetle infestation spots: importance of local population size and landscape characteristics after a storm disturbance. For. Ecol. Manag. 334, 232–240.

Kausrud, K.L., Grégoire, J.-C., Skarpaas, O., Erbilgin, N., Gilbert, M., Økland, B., Stenseth, N.C., 2011. Trees wanted-dead or alive! Host selection and population dynamics in tree-killing bark beetles. PLoS One 6, e18274.

Kausrud, K.L., Økland, B., Skarpaas, O., Grégoire, J.C., Erbilgin, N., Stenseth, N.C., 2012. Population dynamics in changing environments: the case of an eruptive forest pest species. Biol. Rev. 87, 34–51.

Keeling, C.I., Yuen, M.M.S., Liao, N.Y., Docking, T.R., Chan, S.K., Taylor, G.A., Palmquist, D.L., Jackman, S.D., Nguyen, A., Li, M., 2013. Draft genome of the mountain pine beetle, *Dendroctonus ponderosae* Hopkins, a major forest pest. Genome Biol. 14, R27.

Keen, F.P., 1938. Insect enemies of Western forest. Misc. Publ. No. 273, United States Department of Agriculture, Washington. 280 pp.

Kinn, D., Perry, T., Guinn, F., Strom, B., Woodring, J., 1994. Energy reserves of individual southern pine beetles (Coleoptera: Scolytidae) as determined by a modified phosphovanillin spectrophotometric method. J. Entomol. Sci. 29, 152–163.

Kirkendall, L.P., Biedermann, P.H.W., Jordal, B.H., 2015. Evolution and diversity of bark and ambrosia beetles. In: Vega, F.E., Hofstetter, R.W. (Eds.), Bark Beetles: Biology and Ecology of Native and Invasive Species. Elsevier, New York, NY, pp. 85–156.

Klepzig, K.D., Raffa, K., Smalley, E., 1991. Association of an insect-fungal complex with red pine decline in Wisconsin. For. Sci. 37, 1119–1139.

Klepzig, K.D., Smalley, E.B., Raffa, K.F., 1996. Combined chemical defenses against an insect-fungal complex. J. Chem. Ecol. 22, 1367–1388.

Koutroumpa, F.A., Kárpáti, Z., Monsempes, C., Hill, S.R., Hansson, B.S., Jacquin-Joly, E., Krieger, J., Dekker, T., 2014. Shifts in sensory neuron identity parallel differences in pheromone preference in the European corn borer. Front. Ecol. Evol. 2, 65.

Krokene, P., 2015. Conifer defense and resistance to bark beetles. In: Vega, F.E., Hofstetter, R.W. (Eds.), Bark Beetles: Biology and Ecology of Native and Invasive Species. Elsevier, New York, NY, pp. 177–207.

Kromann, S.H., Saveer, A.M., Binyameen, M., Bengtsson, M., Birgersson, G., Hansson, B.S., Schlyter, F., Witzgall, P., Ignell, R., Becher, P.G., 2015. Concurrent modulation of neuronal and behavioural olfactory responses to sex and host plant cues in a male moth. Proc. R. Soc. Lond. B Biol. Sci. 282, 20141884.

Kurz, W.A., Dymond, C.C., Stinson, G., Rampley, G.J., Neilson, E.T., Carroll, A.L., Ebata, T., Safranyik, L., 2008. Mountain pine beetle and forest carbon feedback to climate change. Nature 452, 987–990.

Kwon, J.Y., Dahanukar, A., Weiss, L.A., Carlson, J.R., 2007. The molecular basis of CO2 reception in *Drosophila*. Proc. Natl. Acad. Sci. U.S.A. 104, 3574–3578.

Lanier, G., 1983. Integration of visual stimuli, host odorants, and pheromones by bark beetles and weevils in locating and colonizing host trees. In: Ahmad, S. (Ed.), Herbivorous Insects: Host-Seeking Behavior and Mechanisms. Academic Press, New York, pp. 161–171.

Larsson, M.C., Leal, W.S., Hansson, B.S., 2001. Olfactory receptor neurons detecting plant odours and male volatiles in *Anomala cuprea* beetles (Coleoptera: Scarabaeidae). J. Insect Physiol. 47, 1065–1076.

Lawrence, R., Haack, R., 1995. Susceptibility of selected species of North American pines to shoot-feeding by an Old World scolytid: Tomicus piniperda. In: Hain, F.P., Salo, S.M.,

Ravlin, W.F., Payne, T.L., Raffa, K.F. (Eds.), Behavior, Population Dynamics, and Control of Forest Insects. Joint IUFRO Conference on Working Parties S2.07-05 and S2.07-06, 6–11 February 1994. Ohio State University, Wooster OH, Maui, HI, pp. 536–546.

Leal, W.S., 2012. Odorant reception in insects: roles of receptors, binding proteins, and degrading enzymes. Annu. Rev. Entomol. 58, 373–391.

Leufvén, A., Birgersson, G., 1987. Quantitative variation of different monoterpenes around galleries of *Ips typographus* (Coleoptera: Scolytidae) attacking Norway spruce. Can. J. Bot. 65, 1038–1044.

Leufvén, A., Bergström, G., Falsen, E., 1984. Interconversion of verbenols and verbenone by identified yeasts isolated from the spruce bark beetle *Ips typographus*. J. Chem. Ecol. 10, 1349–1361.

Lewis, K.J., Lindgren, B.S., 2002. Relationship between spruce beetle and tomentosus root disease: two natural disturbance agents of spruce. Can. J. For. Res. 32, 31–37.

Li, Z., Ni, J.D., Huang, J., Montell, C., 2014. Requirement for *Drosophila* SNMP1 for rapid activation and termination of pheromone-induced activity. PLoS Genet. 10, e1004600.

Lindgren, B., Raffa, K., 2013. Evolution of tree killing in bark beetles (Coleoptera: Curculionidae): trade-offs between the maddening crowds and a sticky situation. Can. Entomol. 145, 471–495.

Logan, J.A., Bentz, B.J., 1999. Model analysis of mountain pine beetle (Coleoptera: Scolytidae) seasonality. Environ. Entomol. 28, 924–934.

Logan, J.A., Powell, J.A., 2009. Ecological consequences of climate change altered forest insect disturbance regimes. In: Wagner, F.H. (Ed.), Climate Change in Western North America: Evidence and Environmental Effects. University of Utah Press, Salt Lake City, UT, pp. 98–109.

Logan, J.A., Macfarlane, W.W., Willcox, L., 2010. Whitebark pine vulnerability to climate-driven mountain pine beetle disturbance in the Greater Yellowstone Ecosystem. Ecol. Appl. 20, 895–902.

Lombardero, M.J., Ayres, M.P., Ayres, B.D., Reeve, J.D., 2000. Cold tolerance of four species of bark beetle (Coleoptera: Scolytidae) in North America. Environ. Entomol. 29, 421–432.

Lu, M., Wingfield, M.J., Gillette, N.E., Mori, S.R., Sun, J.H., 2010. Complex interactions among host pines and fungi vectored by an invasive bark beetle. New Phytol. 187, 859–866.

Lu, M., Wingfield, M.J., Gillette, N., Sun, J.-H., 2011. Do novel genotypes drive the success of an invasive bark beetle-fungus complex? Implications for potential reinvasion. Ecology 92, 2013–2019.

Lugo, A.E., 2015. Forestry in the Anthropocene. Science 349, 771.

Lusebrink, I., Erbilgin, N., Evenden, M.L., 2013. The lodgepole × jack pine hybrid zone in Alberta, Canada: a stepping stone for the mountain pine beetle on its journey East across the boreal forest? J. Chem. Ecol. 39, 1209–1220.

Manning, C.G., Reid, M.L., 2013. Sub-lethal effects of monoterpenes on reproduction by mountain pine beetles. Agric. For. Entomol. 15, 262–271.

Marini, L., Ayres, M.P., Battisti, A., Faccoli, M., 2012. Climate affects severity and altitudinal distribution of outbreaks in an eruptive bark beetle. Clim. Chang. 115, 327–341.

Marini, L., Lindelöw, Å., Jönsson, A.M., Wulff, S., Schroeder, L.M., 2013. Population dynamics of the spruce bark beetle: a long-term study. Oikos 122, 1768–1776.

Martemyanov, V.V., Pavlushin, S.V., Dubovskiy, I.M., Belousova, I.A., Yushkova, Y.V., Morosov, S.V., Chernyak, E.I., Glupov, V.V., 2015. Leaf surface lipophilic compounds as one of the factors of silver birch chemical defense against larvae of gypsy moth. PLoS One 10, e0121917.

Martin, D.M., Gershenzon, J., Bohlmann, J., 2003. Induction of volatile terpene biosynthesis and diurnal emission by methyl jasmonate in foliage of Norway spruce. Plant Physiol. 132, 1586–1599.

Matson, P., Hain, F., 1983. Host conifer defense strategies: a hypothesis. In: Safranyik, L. (Ed.), The Role of the Host in the Population Dynamics of Forest Insects: Joint Meeting of IUFRO Working Parties S2-07-05 and S2-07-06. September 4–7. Canadian Forestry Service, Pacific Forest Research Centre, Victoria, BC, Co-publisher USDA Forest Service, Banff, Alberta, Canada, pp. 33–42.

Mccullough, D.G., Sadof, C.S., 1998. Evaluation of an integrated management and compliance program for *Tomicus piniperda* (Coleoptera: Scolytidae) in pine Christmas tree fields. J. Econ. Entomol. 91, 785–795.

McKee, F.R., Huber, D.P., Aukema, B.H., 2013. Comparisons of mountain pine beetle (*Dendroctonus ponderosae* Hopkins) reproduction within a novel and traditional host: effects of insect natal history, colonized host species and competitors. Agric. For. Entomol. 15, 310–320.

McKee, F.R., Huber, D.P., Lindgren, B.S., Hodgkinson, R.S., Aukema, B.H., 2015. Effect of natal and colonised host species on female host acceptance and male joining behaviour of the mountain pine beetle (Coleoptera: Curculionidae) using pine and spruce. Can. Entomol. 147, 39–45.

Mezei, P., Grodzki, W., Blaženec, M., Škvarenina, J., Brandýsová, V., Jakuš, R., 2014. Host and site factors affecting tree mortality caused by the spruce bark beetle (*Ips typographus*) in mountainous conditions. For. Ecol. Manag. 331, 196–207.

Miller, D.R., Borden, J.H., 1990. β-Phellandrene: kairomone for pine engraver, *Ips pini* (Say) (Coleoptera: Scolytidae). J. Chem. Ecol. 16, 2519–2531.

Moeck, H.A., Wood, D.L., Lindahl, K.Q., 1981. Host selection behavior of bark beetles (Coleoptera: Scolytidae) attacking *Pinus ponderosa*, with special emphasis on the western pine beetle, *Dendroctonus brevicomis*. J. Chem. Ecol. 7, 49–83.

Morgan, R.E., Groot, P.d., Smith, S.M., 2004. Susceptibility of pine plantations to attack by the pine shoot beetle (*Tomicus piniperda*) in southern Ontario. Can. J. For. Res. 34, 2528–2540.

Müller, J., Bußler, H., Goßner, M., Rettelbach, T., Duelli, P., 2008. The European spruce bark beetle *Ips typographus* in a national park: from pest to keystone species. Biodivers. Conserv. 17, 2979–3001.

Mustaparta, H., Angst, M.E., Lanier, G.N., 1977. Responses of single receptor cells in the pine engraver beetle, *Ips pini* (Say) (Coleoptera: Scolytidae) to its aggregation pheromone, ipsdienol, and the aggregation inhibitor, ipsenol. J. Comp. Physiol. A. 121, 343–347.

Mustaparta, H., Angst, M.E., Lanier, G.N., 1979. Specialization of olfactory cells to insect and host produced volatiles in the bark beetle *Ips pini* (Say). J. Chem. Ecol. 5, 109–123.

Mustaparta, H., Angst, M.E., Lanier, G.N., 1980. Receptor discrimination of enantiomers of the aggregation pheromone ipsdienol in two species of *Ips*. J. Chem. Ecol. 6, 689–702.

Mustaparta, H., Tømmerås, B.Å., Baeckström, P., Bakke, J.M., Ohloff, G., 1984. Ipsdienol-specific receptor cells in bark beetles: structure–activity relationships of various analogues and of deuterium-labelled ipsdienol. J. Comp. Physiol. A 154, 591–596.

Nakagawa, S., Cuthill, I.C., 2007. Effect size, confidence interval and statistical significance: a practical guide for biologists. Biol. Rev. 82, 591–605.

Netherer, S., Matthews, B., Katzensteiner, K., Blackwell, E., Henschke, P., Hietz, P., Pennerstorfer, J., Rosner, S., Kikuta, S., Schume, H., 2015. Do water-limiting conditions predispose Norway spruce to bark beetle attack? New Phytol. 205, 1128–1141.

Niemiec, R.M., Lutz, D.A., Howarth, R.B., 2014. Incorporating carbon storage into the optimal management of forest insect pests: a case study of the Southern Pine Beetle (*Dendroctonus frontalis* Zimmerman) in the New Jersey Pinelands. Environ. Manage. 54, 875–887.

Økland, B., Bjørnstad, O.N., 2003. Synchrony and geographical variation of the spruce bark beetle (*Ips typographus*) during a non-epidemic period. Popul. Ecol. 45, 213–219.

Payne, T.L., Richerson, J.V., Dickens, J.C., West, J.R., Mori, K., Berisford, C.W., Hedden, R.L., Vité, J.P., Blum, M.S., 1982. Southern pine beetle: olfactory receptor and behavior discrimination of enantiomers of the attractant pheromone frontalin. J. Chem. Ecol. 8, 873–882.

Paynter, Q.E., Anderbrant, O., Schlyter, F., 1990. Behavior of male and female spruce bark beetles, *Ips typographus*, on the bark of host trees during mass attack. J. Insect Behav. 3, 529–543.

Pelosi, P., Zhou, J.J., Ban, L.P., Calvello, M., 2006. Soluble proteins in insect chemical communication. Cell. Mol. Life Sci. 63, 1658–1676.

Perkins, D.L., Jorgensen, C.L., Rinella, M.J., 2015. Verbenone decreases whitebark pine mortality throughout a mountain pine beetle outbreak. For. Sci. 61, 747–752.

Pettersson, E.M., Boland, W., 2003. Potential parasitoid attractants, volatile composition throughout a bark beetle attack. Chemoecology 13, 27–37.

Poland, T.M., Borden, J., 1998a. Disruption of secondary attraction of the spruce beetle, *Dendroctonus rufipennis*, by pheromones of two sympatric species. J. Chem. Ecol. 24, 151–166.

Poland, T.M., Borden, J.H., 1998b. Competitive exclusion of *Dendroctonus rufipennis* induced by pheromones of *Ips tridens* and *Dryocoetes affaber* (Coleoptera: Scolytidae). J. Econ. Entomol. 91, 1150–1161.

Pope, D., Coulson, R., Fargo, W., Gagne, J., Kelly, C., 1980. The allocation process and between-tree survival probabilities in *Dendroctonus frontalis* infestations. Res. Popul. Ecol. 22, 197–210.

Powell, J.A., Bentz, B.J., 2009. Connecting phenological predictions with population growth rates for mountain pine beetle, an outbreak insect. Landscape Ecol. 24, 657–672.

Powell, E.N., Raffa, K.F., 2011. Fire injury reduces inducible defenses of lodgepole pine against mountain pine beetle. J. Chem. Ecol. 37, 1184–1192.

Powell, E.N., Townsend, P.A., Raffa, K.F., 2012. Wildfire provides refuge from local extinction but is an unlikely driver of outbreaks by mountain pine beetle. Ecol. Monogr. 82, 69–84.

Pregitzer, P., Greschista, M., Breer, H., Krieger, J., 2014. The sensory neurone membrane protein SNMP1 contributes to the sensitivity of a pheromone detection system. Insect Mol. Biol. 23, 733–742.

Pureswaran, D.S., Sullivan, B.T., Ayres, M.P., 2006. Fitness consequences of pheromone production and host selection strategies in a tree-killing bark beetle (Coleoptera: Curculionidae: Scolytinae). Oecologia 148, 720–728.

Raffa, K.F., 1988. Host orientation behavior of *Dendroctonus ponderosae*: integration of token stimuli host and defenses. In: Mattson, W.J., Levieux, J., Bernard-Dagan, C. (Eds.), Mechanisms of Woody Plant Defenses Against Insects. Springer-Verlag, New York, pp. 369–390.

Raffa, K.F., 1991. Temporal and spatial disparities among bark beetles, predators, and associates responding to synthetic bark beetle pheromones: *Ips pini* (Coleoptera: Scolytidae) in Wisconsin. Environ. Entomol. 20, 1665–1679.

Raffa, K.F., 2001. Mixed messages across multiple trophic levels: the ecology of bark beetle chemical communication systems. Chemoecology 11, 49–65.

Raffa, K.F., Berryman, A.A., 1980. Flight responses and host selection by bark beetles. In: Berryman, A., Safranyik, L. (Eds.), Proceedings of the Second IUFRO Conference

on Dispersal of Forest Insects: Evaluation, Theory, and Management Implications. Conference Office, Cooperative Extension Service, Washington State University, Pullman, WA, pp. 213–233.
Raffa, K., Berryman, A., 1982. Gustatory cues in the orientation of *Dendroctonus ponderosae* (Coleoptera: Scolytidae) to host trees. Can. Entomol. 114, 97–104.
Raffa, K., Berryman, A., 1983. The role of host plant resistance in the colonization behavior and ecology of bark beetles (Coleoptera: Scolytidae). Ecol. Monogr. 53, 27–49.
Raffa, K.F., Berryman, A.A., 1987. Interacting selective pressures in conifer-bark beetle systems: a basis for reciprocal adaptations? Am. Nat. 129, 234–262.
Raffa, K.F., Aukema, B.H., Erbilgin, N., Klepzig, K.D., Wallin, K.F., 2005. Interactions among conifer terpenoids and bark beetles across multiple levels of scale: an attempt to understand links between population patterns and physiological processes. Recent Adv. Phytochem. 39, 79–118.
Raffa, K.F., Hobson, K.R., LaFontaine, S., Aukema, B.H., 2007. Can chemical communication be cryptic? Adaptations by herbivores to natural enemies exploiting prey semiochemistry. Oecologia 153, 1009–1019.
Raffa, K.F., Aukema, B.H., Bentz, B.J., Carroll, A.L., Hicke, J.A., Turner, M.G., Romme, W.H., 2008. Cross-scale drivers of natural disturbances prone to anthropogenic amplification: the dynamics of bark beetle eruptions. Bioscience 58, 501–517.
Raffa, K.F., Powell, E.N., Townsend, P.A., 2013. Temperature-driven range expansion of an irruptive insect heightened by weakly coevolved plant defenses. Proc. Natl. Acad. Sci. U.S.A. 110, 2193–2198.
Raffa, K.F., Aukema, B.H., Bentz, B.J., Carroll, A.L., Hicke, J.A., Kolb, T.E., 2015. Responses of tree-killing bark beetles to a changing climate. In: Björkman, C., Niemelä, P. (Eds.), Climate Change and Insect Pests. CABI, Oxfordshire, pp. 173–201.
Rankin, L.J., Borden, J., 1991. Competitive interactions between the mountain pine beetle and the pine engraver in lodgepole pine. Can. J. For. Res. 21, 1029–1036.
Reid, R., Whitney, H., Watson, J., 1967. Reactions of lodgepole pine to attack by *Dendroctonus ponderosae* Hopkins and blue stain fungi. Can. J. Bot. 45, 1115–1126.
Renwick, J.A.A., Vité, J.P., 1969. Bark beetle attractants: mechanism of colonization by *Dendroctonus frontalis*. Nature 224, 1222–1223.
Reynolds, K., Holsten, E., 1994. Relative importance of risk factors for spruce beetle outbreaks. Can. J. For. Res. 24, 2089–2095.
Risch, S.J., Andow, D., Altieri, M.A., 1983. Agroecosystem diversity and pest control: data, tentative conclusions, and new research directions. Environ. Entomol. 12, 625–629.
Robertson, H.M., Kent, L.B., 2009. Evolution of the gene lineage encoding the carbon dioxide receptor in insects. J. Insect Sci. 9, 1–14.
Robertson, H.M., Warr, C.G., Carlson, J.R., 2003. Molecular evolution of the insect chemoreceptor gene superfamily in *Drosophila melanogaster*. Proc. Natl. Acad. Sci. U.S.A. 100, 14537–14542.
Root, R.B., 1973. Organization of a plant-arthropod association in simple and diverse habitats: the fauna of collards (*Brassica oleracea*). Ecol. Monogr. 43, 95–124.
Root, C.M., Ko, K.I., Jafari, A., Wang, J.W., 2011. Presynaptic facilitation by neuropeptide signaling mediates odor-driven food search. Cell 145, 133–144.
Rudinsky, J.A., 1962. Ecology of Scolytidae. Annu. Rev. Entomol. 7, 327–348.
Rudinsky, J.A., Furniss, M.M., Kline, L.N., Schmitz, R.F., 1972. Attraction and repression of *Dendroctonus pseudotsugae* (Coleoptera: Scolytidae) by three synthetic pheromones in traps in Oregon and Idaho. Can. Entomol. 104, 815–822.
Ryall, K.L., Fahrig, L., 2005. Habitat loss decreases predator–prey ratios in a pine-bark beetle system. Oikos 110, 265–270.
Safranyik, L., Linton, D., 1983. Brood production by three spp. of *Dendroctonus* (Coleoptera: Scolytidae) in bolts from host and non-host trees. J. Entomol. Soc. B.C. 80, 10–13.

Safranyik, L., Shrimpton, D., Whitney, H., 1975. An interpretation of the interaction between lodgepole pine, the mountain pine beetle, and its associated blue stain fungi in western Canada. In: Baumgartner, D.M. (Ed.), Management of Lodgepole Pine Ecosystems. Washington State University Cooperative Extension Service, Pullman, WA, pp. 406–428.
Safranyik, L., Linton, D.A., Shore, T.L., 2000. Temporal and vertical distribution of bark beetles (Coleoptera: Scolytidae) captured in barrier traps at baited and unbaited lodgepole pines the year following attack by the mountain pine beetle. Can. Entomol. 132, 799–810.
Safranyik, L., Carroll, A., Régnière, J., Langor, D., Riel, W., Shore, T., Peter, B., Cooke, B., Nealis, V., Taylor, S., 2010. Potential for range expansion of mountain pine beetle into the boreal forest of North America. Can. Entomol. 142, 415–442.
Saint-Germain, M., Buddle, C.M., Drapeau, P., 2007. Primary attraction and random landing in host-selection by wood-feeding insects: a matter of scale? Agric. For. Entomol. 9, 227–235.
Sánchez-Gracia, A., Vieira, F.G., Rozas, J., 2009. Molecular evolution of the major chemosensory gene families in insects. Heredity 103, 208–216.
Sandoz, J.C., 2012. Olfactory plasticity. In: Binder, M.D., Hirokawa, N., Windhorst, U. (Eds.), Encyclopedia of Neuroscience. Springer Verlag, Berlin, pp. 1–5.
Sato, K., Touhara, K., 2009. Insect olfaction: receptors, signal transduction, and behavior. Results Probl. Cell Differ. 47, 121–138.
Saveer, A.M., Kromann, S.H., Birgersson, G., Bengtsson, M., Lindblom, T., Balkenius, A., Hansson, B.S., Witzgall, P., Becher, P.G., Ignell, R., 2012. Floral to green: mating switches moth olfactory coding and preference. Proc. R. Soc. Lond B Biol. Sci. 279, 2314–2322.
Schiebe, C., 2012. Attraction and resistance in the *Picea abies*—*Ips typographus* system. Acta Universitatis agriculturae Sueciae (1652–6880), 91 Department of Plant Protection Biology, Swedish University of Agricultural Sciences, Alnarp. 57 pp. http://pub.epsilon.slu.se/9211/.
Schiebe, C., Blaženec, M., Jakuš, R., Unelius, C.R., Schlyter, F., 2011. Semiochemical diversity diverts bark beetle attacks from Norway spruce edges. J. Appl. Entomol. 135, 726–737.
Schiebe, C., Hammerbacher, A., Birgersson, G., Witzell, J., Brodelius, P.E., Gershenzon, J., Hansson, B.S., Krokene, P., Schlyter, F., 2012. Inducibility of chemical defenses in Norway spruce bark is correlated with unsuccessful mass attacks by the spruce bark beetle. Oecologia 170, 183–198.
Schlyter, F., 2012. Semiochemical diversity in practice: anti-attractant semiochemicals reduces bark beetle attacks on standing trees—a first meta-analysis. Psyche J. Entomol. 2012. Article ID 268621.
Schlyter, F., Birgersson, G., 1999. Forest beetles. In: Hardie, J., Minks, A.K. (Eds.), Pheromones of Non-Lepidopteran Insects Associated with Agricultural Plants. CAB International, Oxford, pp. 113–148.
Schlyter, F., Johansson, A., 2010. Detection dog recognize pheromone from spruce bark beetle and follows it to source: a new tool from chemical ecology to forest protection. In: 26th ISCE Annual Meeting, ISCE 2010 Book of Abstracts, July 31–August 4. The International Society of Chemical Ecology, Inc, Tours, France. 384 pp, Poster P183.
Schlyter, F., Birgersson, G., Leufvén, A., 1989. Inhibition of attraction to aggregation pheromone by verbenone and ipsenol: density regulation mechanisms in bark beetle *Ips typographus*. J. Chem. Ecol. 15, 2263–2277.
Schwerdtfeger, F., 1955. Pathogenese der Borkenkäfer-Epidemie 1946–1950 in Nordwestdeutschland, Schriftenreihe Forstl. Fakult. Göttingen, p. 135.

Seybold, S., Quilici, D., Tillman, J., Vanderwel, D., Wood, D., Blomquist, G., 1995. *De novo* biosynthesis of the aggregation pheromone components ipsenol and ipsdienol by the pine bark beetles *Ips paraconfusus* Lanier and *Ips pini* (Say) (Coleoptera: Scolytidae). Proc. Natl. Acad. Sci. U.S.A. 92, 8393–8397.

Shepherd, R.F., 1966. Factors influencing the orientation and rates of activity of *Dendroctonus ponderosae* Hopkins (Coleoptera: Scolytidae). Can. Entomol. 98, 507–518.

Sherriff, R.L., Berg, E.E., Miller, A.E., 2011. Climate variability and spruce beetle (*Dendroctonus rufipennis*) outbreaks in south-central and southwest Alaska. Ecology 92, 1459–1470.

Siegert, N.W., McCullough, D.G., 2001. Preference of *Tomicus piniperda* (Coleoptera: Scolytidae) parent adults and shoot-feeding progeny adults for three pine species. Can. Entomol. 133, 343–353.

Siegert, N.W., McCullough, D.G., 2003. Colonization of Scots, red, and jack pine logs by *Tomicus piniperda* (Coleoptera: Scolytidae) in Michigan pine stands. Can. J. For. Res. 33, 2238–2244.

Simard, M., Powell, E.N., Raffa, K.F., Turner, M.G., 2012. What explains landscape patterns of tree mortality caused by bark beetle outbreaks in Greater Yellowstone? Glob. Ecol. Biogeogr. 21, 556–567.

Stoyenoff, J., Witter, J., Montgomery, M., 1994. Nutritional indices in the gypsy moth (*Lymantria dispar* (L.)) under field conditions and host switching situations. Oecologia 97, 158–170.

Strom, B.L., Roton, L.M., Goyer, R.A., Meeker, J.R., 1999. Visual and semiochemical disruption of host finding in the southern pine beetle. Ecol. Appl. 9, 1028–1038.

Su, C.-Y., Menuz, K., Reisert, J., Carlson, J.R., 2012. Non-synaptic inhibition between grouped neurons in an olfactory circuit. Nature 492, 66–71.

Sun, J., Lu, M., Gillette, N.E., Wingfield, M.J., 2013. Red turpentine beetle: innocuous native becomes invasive tree killer in China. Annu. Rev. Entomol. 58, 293–311.

Taft, S., Najar, A., Erbilgin, N., 2015. Pheromone production by an invasive bark beetle varies with monoterpene composition of its naïve host. J. Chem. Ecol. 41, 540–549.

Tauxe, G.M., MacWilliam, D., Boyle, S.M., Guda, T., Ray, A., 2013. Targeting a dual detector of skin and CO_2 to modify mosquito host seeking. Cell 155, 1365–1379.

Thalenhorst, W., 1958. Grundzüge der Populationsdynamik des grossen Fichtenborkenkäfers *Ips typographus* L., Schriftenreihe Forstl. Fakult. Göttingen, p. 126.

Tømmerås, B.Å., 1985. Specialization of the olfactory receptor cells in the bark beetle *Ips typographus* and its predator *Thanasimus formicarius* to bark beetle pheromones and host tree volatiles. J. Comp. Physiol. A 157, 335–342.

Tømmerås, B.Å., Mustaparta, H., 1987. Chemoreception of host volatiles in the bark beetle *Ips typographus*. J. Comp. Physiol. A 161, 705–710.

Tømmerås, B.Å., Mustaparta, H., 1989. Single cell responses to pheromones, host and non-host volatiles in the ambrosia beetle *Trypodendron lineatum*. Entomol. Exp. Appl. 52, 141–148.

Tømmerås, B.Å., Mustaparta, H., Gregoire, J.-C., 1984. Receptor cells in *Ips typographus* and *Dendroctonus micans* specific to pheromones of the reciprocal genus. J. Chem. Ecol. 10, 759–770.

Touhara, K., Vosshall, L., 2009. Sensing odorants and pheromones with chemosensory receptors. Annu. Rev. Physiol. 71, 307–332.

Unelius, R.C., Schiebe, C., Bohman, B., Andersson, M.N., Schlyter, F., 2014. Non-host volatile blend optimization for forest protection against the european spruce bark beetle, *Ips typographus*. PLoS One 9, e85381.

Ungerer, M.J., Ayres, M.P., Lombardero, M.J., 1999. Climate and the northern distribution limits of *Dendroctonus frontalis* Zimmermann (Coleoptera: Scolytidae). J. Biogeogr. 26, 1133–1145.

Vermeulen, A., Rospars, J.-P., 2004. Why are insect olfactory receptor neurons grouped into sensilla? The teachings of a model investigating the effects of the electrical interaction between neurons on the transepithelial potential and the neuronal transmembrane potential. Eur. Biophys. J. 33, 633–643.

Vosshall, L.B., Stocker, R.F., 2007. Molecular architecture of smell and taste in *Drosophila*. Annu. Rev. Neurosci. 30, 505–533.

Wagner, T.L., Fargo, W.S., Flamm, R.O., Coulson, R.N., Pulley, P.E., 1987. Development and mortality of *Ips calligraphus* (Coleoptera: Scolytidae) at constant temperatures. Environ. Entomol. 16, 484–496.

Wagner, T.L., Hennier, P.B., Flamm, R.O., Coulson, R.N., 1988. Development and mortality of *Ips avulsus* (Coleoptera: Scolytidae) at constant temperatures. Environ. Entomol. 17, 181–191.

Wallin, K.F., Raffa, K., 2000. Influences of external chemical cues and internal physiological parameters on the multiple steps of post-landing host selection behavior of *Ips pini* (Coleoptera: Scolytidae). Environ. Entomol. 29, 442–453.

Wallin, K.F., Raffa, K.F., 2001. Effects of folivory on subcortical plant defenses: can defense theories predict interguild processes? Ecology 82, 1387–1400.

Wallin, K.F., Raffa, K.F., 2002. Prior encounters modulate subsequent choices in host acceptance behavior by the bark beetle *Ips pini*. Entomol. Exp. Appl. 103, 205–218.

Wallin, K.F., Raffa, K.F., 2004. Feedback between individual host selection behavior and population dynamics in an eruptive herbivore. Ecol. Monogr. 74, 101–116.

Wallin, K.F., Rutledge, J., Raffa, K.F., 2002. Heritability of host acceptance and gallery construction behaviors of the bark beetle *Ips pini* (Coleoptera: Scolytidae). Environ. Entomol. 31, 1276–1281.

Werner, R.A., Holsten, E.H., Matsuoka, S.M., Burnside, R.E., 2006. Spruce beetles and forest ecosystems in south-central Alaska: a review of 30 years of research. For. Ecol. Manag. 227, 195–206.

West, D.R., Briggs, J.S., Jacobi, W.R., Negrón, J.F., 2014. Mountain pine beetle-caused mortality over eight years in two pine hosts in mixed-conifer stands of the southern Rocky Mountains. For. Ecol. Manag. 334, 321–330.

Williams, W.I., Robertson, I.C., 2008. Using automated flight mills to manipulate fat reserves in Douglas-fir beetles (Coleoptera: Curculionidae). Environ. Entomol. 37, 850–856.

Wilson, D.A., Best, A.R., Sullivan, R.M., 2004. Plasticity in the olfactory system: lessons for the neurobiology of memory. Neuroscientist 10, 513–524.

Wood, D.L., 1972. Selection and colonization of ponderosa pine by bark beetles. In: van Emden, H.F. (Ed.), Insect–Plant Relationships. Wiley, New York, pp. 101–117.

Wood, D.L., 1982. The role of pheromones, kairomones, and allomones in the host selection and colonization behavior of bark beetles. Annu. Rev. Entomol. 27, 411–446.

Ye, H., Haack, R., Petrice, T., 2005. *Tomicus piniperda* (Coleoptera: Scolytidae) reproduction and development in Scots, jack, red and eastern white pine under laboratory conditions. Great Lakes Entomol. 38, 13–22.

Yousuf, F., Carnegie, A.J., Bashford, R., Bedding, R.A., Nicol, H.I., Gurr, G.M., 2014a. Bark beetle (*Ips grandicollis*) disruption of woodwasp (*Sirex noctilio*) biocontrol: direct and indirect mechanisms. For. Ecol. Manag. 323, 98–104.

Yousuf, F., Gurr, G.M., Carnegie, A.J., Bedding, R.A., Bashford, R., Gitau, C.W., 2014b. Biology of the bark beetle *Ips grandicollis* Eichhoff (Coleoptera: Scolytinae) and its arthropod, nematode and microbial associates: a review of management opportunities for Australia. Aust. Entomol. 53, 298–316.

Zhang, Q.-H., Schlyter, F., 2003. Redundancy, synergism, and active inhibitory range of non-host volatiles in reducing pheromone attraction in European spruce bark beetle *Ips typographus*. Oikos 101, 299–310.

Zhang, Q.-H., Schlyter, F., 2004. Olfactory recognition and behavioural avoidance of angiosperm nonhost volatiles by conifer-inhabiting bark beetles. Agric. For. Entomol. 6, 1–19.
Zhang, Q.-H., Tolasch, T., Schlyter, F., Francke, W., 2002. Enantiospecific antennal response of bark beetles to spiroacetal (*E*)-conophthorin. J. Chem. Ecol. 28, 1839–1852.
Zhang, Q.-H., Schlyter, F., Chen, G., Wang, Y., 2007. Electrophysiological and behavioral responses of *Ips subelongatus* to semiochemicals from its hosts, non-hosts, and conspecifics in China. J. Chem. Ecol. 33, 391–404.

CHAPTER TWO

The Role of Population Genetic Structure in Understanding and Managing Pine Beetles

J.K. Janes*, P.D. Batista†,‡

*School of Environmental and Rural Science, The University of New England, Armidale, NSW, Australia
†University of Alberta, Edmonton, AB, Canada
‡University of Northern British Columbia, Prince George, BC, Canada

Contents

Abstract

Population genetic structure is the study of genetic variation in time and space. Traditionally, population genetic structure assessments provide information on the dispersal of species, mating behaviours and the delimitation of species and population boundaries. As such, the study of population genetic structure in bark beetles has become an important step in the ongoing management of outbreak populations. In this chapter, we review the current state of population genetic structure studies as they relate to bark beetles, with a particular emphasis on pine beetles. Furthermore, we show how pine beetle research may lead the way in promoting a more comprehensive and methodologically inclusive approach to explaining population genetic structure patterns, how they arise and what they might mean in a broader evolutionary context.

Advances in Insect Physiology, Volume 50
ISSN 0065-2806
http://dx.doi.org/10.1016/bs.aiip.2016.01.001

1. INTRODUCTION

The complex biology of pine beetles can have a significant impact on the spatial and temporal distribution of beetles, which in turn effects the genetic structure of beetle populations. Each year, outbreaks of these beetles contribute to widespread losses of valuable pine forests (Raffa et al., 2015). As a result, the management of these species, through early detection and population control, has become increasingly important. However, management requires the identification of management units, which relies on the accurate delimitation of species and their populations in space and time. This chapter reviews our current understanding of population genetic structure in pine beetles, framing it within a population genetics and evolutionary biology context. We highlight the most common uses and methods associated with population genetic structure and suggest direction for future studies as genomic resources continue to advance.

1.1 What is Population Genetic Structure?

A population is often defined as a group of individuals of the same species occupying a particular area at the same time (Roughgarden et al., 1989). However, observing individuals in space and time can be difficult, and researchers may come to different conclusions regarding the size of a given population, the number of populations and their interrelationships. This is where the field of *population genetics* can be extremely helpful. Population genetics is essentially the study of allele frequencies and how they vary in time and space as a result of natural selection, mutation, recombination and genetic drift (Holsinger and Weir, 2009). Through the incorporation of additional aspects such as gene flow, environmental change and population structure, population genetics can provide insights into adaptation and speciation (Slatkin, 1985).

The geographic structure of populations is fundamental to ecology and evolution. Essentially, the concept combines *demography* with population genetics and provides novel insights into demographic and evolutionary processes (Roderick, 1996). Often, the terms population genetic structure, population structure and geographic structure are used interchangeably. We will use *population genetic structure* throughout this chapter. Studies of population genetic structure are performed using indirect methods, in which data from genetic markers are used to infer the movement of individuals.

1.2 Biological Influences on Population Genetic Structure

Subfamily Scolytinae is a large and diverse group of over 200 genera and an estimated 6000 species in the weevil family, Curculionidae (Haran et al., 2013; Raffa et al., 2015). The subfamily is composed of subcortical-feeding (pine beetles) and wood-boring (ambrosia beetles) insects and is best known for its irruptive forest pest members that are capable of widespread damage and subsequent economic losses. Some of the better-known species from the subfamily include *Dendroctonus ponderosae* (mountain pine beetle), *Hypothenemus hampei* (coffee berry borer), *Ips typographus* (spruce ips) and *Scolytus scolytus* (large elm pine beetle).

The biology of pine beetles is fascinating. The group is characterized by a protected life cycle that is completed in galleries constructed within wood or cones. A short period of flight as an adult is the only time that individuals experience life outside their host. In addition, each species of beetle has specific adaptations that allow it to exploit its host. Thus, species can be monophagous (feeding on one type of host) or polyphagous (multiple hosts) (Rudinsky, 1962). Most species prefer to feed on, and breed in, weak or dying trees, in which case they are referred to as *secondary* pests (Wallner, 1987). However, some species can become *primary* pests, capable of overcoming mature healthy trees, as a result of a population irruption (Wallner, 1987) when suitable feeding and breeding material is abundant.

Historically, several species of *Dendroctonus* have periodically become primary pests in North America and destroyed millions of hectares of healthy pine stands, causing significant ecological and economic damage (Dale et al., 2001; Safranyik and Wilson, 2006; Schrey et al., 2008). The most recent of these outbreaks occurred in western USA and Canada when the mountain pine beetle (*D. ponderosae*) reached outbreak population levels around 2004 (Bleiker et al., 2010). The outbreak resulted in a significant range expansion for the species and the loss of millions of hectares of predominantly lodgepole pine (*Pinus contorta*) forests (Safranyik et al., 2010). In Europe, *I. typographus* (European spruce bark beetle) is arguably one of the most destructive pine beetles (Panayotov et al., 2011), occupying a similar *irruptive forest pest* role to the mountain pine beetle. Interestingly, while some pine beetles are known to irrupt and become primary pests, the phenomenon is relatively rare within the scolytids. Many species develop 'outbreak' populations when preferred resources are available but will not invade and kill healthy trees under such population pressure (Rudinsky, 1962).

Those species that do become primary pests exhibit preferences in host species, age, size and bark thickness. For example, within the genus *Dendroctonus*, *D. brevicomis* prefers older trees (Moeck et al., 1981), *D. ponderosae* (synonym *D. monticcolae*) younger standing trees (Safranyik, 2003), while *D. pseudotsugae* prefers freshly wind-felled or cut trees (Gillette et al., 2009). Furthermore, regional differences in ecology have been reported for pine beetles. For example, *Dendroctonus valens* regularly becomes a primary insect in China where it was introduced, but in North America it is restricted to being a secondary insect (Cai et al., 2008). What is perhaps more interesting to evolutionary biologists is the reported change in behaviour once a previously secondary insect becomes primary. Endemic populations of *I. typographus*, for example, are incapable of invading healthy trees, and this limitation is linked to the abundance of host tree availability (Viktorinova et al., 2011). Typically, weakened trees are found in low numbers thereby preventing an increase in beetle population density as a result of competition for resources. As soon as the availability of susceptible trees increases, competition is lowered and the population can increase. Shifting population dynamics begin to meet the conditions for population irruption and the epidemic phase. With increasing numbers of susceptible hosts, the increasing population of beetles can begin to attack healthy trees, overcoming them by sheer numbers and the primary attack begins.

Irruptive outbreaks, feeding preferences, competitive interactions, mating and dispersal behaviours, for example, can all affect the movement of beetles, the size of the population and the exchange of genetic material. These changes in behaviour, the abiotic and biotic factors that contribute to the changes, and the potential ecological and economic impacts have resulted in numerous pine beetle studies around the world. Scientists, forestry operators, governments and conservationists alike have been actively researching pine beetles in an attempt to better understand their biology, and most importantly, determining ways to better control these irruptive forest pests.

2. THE AGE OF DNA-BASED METHODS

Numerous questions relating to pine beetle biology cannot be answered through field observations alone. The majority of the life cycle of pine beetles is cryptic, owing to their characteristic behaviour of living within wood or cones. Pine beetles are also quite small, most averaging the size of a grain of rice, and in endemic phases they are at reasonably low population densities, such that they are not always easily observed.

A number of earlier studies have relied on collections of pine beetles, either by using traps or by removing tree material, in order to create laboratory studies (eg, Bentz et al., 2011; Fraser et al., 2014; Reid, 1958; Viktorinova et al., 2011). Keeping pine beetles alive in the lab can be difficult because the temperature, humidity and food availability are often very different (Warren, 1958; Yearian and Wilkinson, 1967). Once successfully 'kept' in the lab there can still be problems when it comes to 'natural' behaviour and physiology. For example, observing mating, larval development, emergence and flight behaviour all require significant investments of time in order to optimize laboratory conditions to promote such behaviours. While there have been a number of successes and advances in our understanding of pine beetle biology as a result, there will still be the question of how 'natural' these behaviours are under such 'unnatural' conditions. DNA-based methods have, in some ways, simplified these types of studies because genetic information allows many aspects relating to demographic and evolutionary processes to be inferred indirectly.

DNA-based methods have advanced significantly in recent years, with the range of markers and their respective precision and power increasing greatly. Genetic testing methods have become widely accessible and feasible to perform even in small laboratories. Ten to twenty years ago, a large study would have comprised a hundred samples using roughly 10–20 genetic markers (eg, Langor and Spence, 1991; Mock et al., 2007; Stock et al., 1984; Sturgeon and Mitton, 1986; Zúñiga et al., 2006). These studies were limited, not only by cost and technology, but also by computational power. Current studies are more likely to rely on *genomic* resources, encompassing a few thousand to a few hundred thousand markers and several hundred individuals (eg, Janes et al., 2014; Li et al., 2008).

2.1 Molecular Marker Choice

Molecular markers and their methods come in a variety of flavours—some are considered dominant, others co-dominant; some are restriction enzyme based while others rely on more complex chemical reactions. Molecular markers have changed over time and the number of markers available is constantly increasing as technology advances. However, choosing which marker to use largely depends on the question being asked. Here we detail some of the more commonly used markers in population genetic structure studies and highlight the advantages and disadvantages of each.

Allozymes are common biological enzymes that display variant forms, represented by different alleles (Avise, 2004a). Visualizing and assessing allozymes is relatively fast, cheap and simple. Samples are homogenized to release enzymes from cells, and the homogenates are placed into a gel and separated using electrophoresis. After more than 30 years of use, allozymes continue to provide valuable population-level information on the relatedness of individuals and the genetic variability within and between populations (Lewontin, 1991). With respect to allozyme studies on pine beetles, there are a few limitations that are worthy of note: (1) tissue must be alive or frozen before use (Taylor et al., 1994), (2) some allozymes will show little variation (Edwards and Hoy, 1993), (3) co-migrating bands are assumed to be *homologous*, (4) some bands might not follow Mendelian inheritance rules, (5) allozymes reveal a subset of the total genetic variation present (Lewontin, 1991) and (6) the pattern of ancestry among alleles cannot be defined (DeSalle and Volger, 1994).

Microsatellites are a popular choice for genetic studies as they have the potential to provide information on contemporary migration patterns and relatedness of individuals. Microsatellites, simple sequence repeats and short tandem repeats are small sections, usually two to six nucleotides long, that are repeated throughout the nuclear genome at a relatively high frequency (Miah et al., 2013; Slatkin, 1995). Microsatellites, like all genetic tools, do have some drawbacks: (1) the mutational mechanisms behind microsatellites are complex and often have implications for estimates of allele frequencies; (2) size-based identification of alleles assumes that all distinct alleles will vary in length, but this does not account for alleles of the same size that are from a different lineage (*homoplasies*); and (3) scoring microsatellites is not easily repeated across labs or even users (Kumar et al., 2009). However, microsatellites are likely to continue to be a reasonably cheap and easy alternative to other genetic methods because of the high information content per locus (Gupta et al., 1999).

Dominant markers such as random-amplified polymorphic DNA (RAPDs), amplified fragment length polymorphisms (AFLPs) and restriction fragment length polymorphisms (RFLPs) make use of restriction enzymes to cut DNA at specific sites. These markers exhibit dominance; heterozygotes and homozygotes are indistinguishable, as both will produce a band when visualized (Avise, 2004a). Therefore, the markers are scored as the presence or absence of a band. These markers are capable of producing a large number of loci that can be useful in genetic mapping, relatedness and population structure analyses (Vos et al., 1995). However, the dominant nature can result in (1) a loss of information as there are few alleles per locus and (2) bands that are of similar size, but actually represent different sections of DNA (Shan et al., 1999).

Sanger sequencing is a method rather than a specific marker, but its use has been by far the most widely applied. Sanger sequencing can be applied in two ways—the random, or shotgun, approach in which random fragments are amplified (Seeb et al., 2011); or the more common primer walking approach in which direct sequencing of unknown DNA within known sequence sites occurs (Franca et al., 2002). Sequences can be derived from nuclear DNA (eg, ITS—the internal transcribed spacer) (Taylor and Harris, 2012) or the mitochondrial genome (mtDNA) (Emerson et al., 2010). The biggest advantage of sequencing is that DNA regions can be selected depending on the level of genetic resolution required. For example, nuclear genes, such as ITS, accumulate mutations faster than mtDNA and provide information on the movement of both males and females as it is biparentally inherited (Avise, 2004a). In contrast, mtDNA is uniparentally inherited, via the mother, and therefore accumulates mutations at a slower rate due to a lack of recombination (Avise, 2004a). However, any mutations in mtDNA are likely to become fixed in a population even with very little migration because of its haploid nature and thus, smaller effective population size (Barrow et al., 2015). Sequencing remains a valuable method in studies of population genetic structure, but it does require careful consideration of marker choice with respect to specific taxa and research questions.

SNPs or single-nucleotide polymorphisms represent one of the most recent markers for revealing the evolutionary history of populations. Their occurrence throughout the genome also makes them ideal for analyses of speciation, population genetic history and diversity. SNPs are quickly becoming the marker of choice for many studies because they are relatively cheap, reproducible and reliable (Slate et al., 2009). However, the biggest complication to SNP analysis is a bias towards analysing only the most variable loci (Luikart et al., 2003). This artefact can be easily corrected by increasing the number of individuals used initially to screen for polymorphisms (Brumfield et al., 2003).

Next-generation sequencing (*NGS*) has become increasing popular as it offers the ability to sequence much larger data sets, far more rapidly than Sanger sequencing and at a greatly reduced cost per sample (Durmaz et al., 2015). NGS methods attach synthesized segments of known DNA sequence (adaptors) to the ends of target DNA. Thus, the adaptors allow multiple different target sequences to be read using the same sequencing primers. NGS methods are also miniaturized and highly parallel, allowing for high throughput in small reaction chambers or flow cells. Numerous platforms exist for NGS, each one differing slightly in terms of chemistry and/or sequence output. While NGS methods have made molecular studies more feasible in many aspects, the problem faced by researchers now relates to data handling, analysis and storage (Ekblom and Galindo, 2010). High-performance computing centres and a basic knowledge of bioinformatics are becoming essential in the pursuit of NGS-based studies.

3. COMMON METHODS OF ASSESSMENT

3.1 *F*-Statistics: The Original Method

The most commonly used measures of population genetic structure are Wright's *F*-statistics. Typically, species comprise several populations over their geographic range. These populations are usually divided into several subpopulations as a result of geographic isolation, ecological heterogeneity or differences in behaviour. Over time, these populations become increasingly different from one another as a result of *genetic drift*. Introduced in the 1930s, *F*-statistics provide a measure of genetic structure by describing the genetic diversity partitioned within and among populations (Wright, 1951). Small populations with little migration, or gene flow, tend to be more differentiated from one another than large populations with higher rates of migration. Thus, Wright (1951) developed a means of predicting the relationships among evolutionary processes, such as migration, mutation and drift, which contribute to genetic differentiation within a species.

As noted by Holsinger and Weir (2009), F_{ST} is much more than a simple measure of genetic differentiation. F_{ST} directly infers the variance in allele frequencies among populations, and on the other hand, the degree of similarity among individuals within a population (Leng and Zhang, 2011). At the population level, large values of F_{ST} indicate that allele frequencies among populations are very different. At the locus level, large values of F_{ST} indicate that one allele is under selection rather than drift. Several related statistics have since been described since Wright's F_{ST}. For example, R_{ST} is commonly used for microsatellite data, Φ_{ST} for sequence data and G_{ST} when genetic drift contributions to among-population differences are not of interest (Holsinger and Weir, 2009).

3.1.1 *Case Study:* F*-Statistics Assess the Influence of Increasing Host Specialization*

Population genetic structure, measured as genetic differentiation, can provide insights into modes of speciation. The more genetically structured and differentiated populations of a given species become, the higher the likelihood that the populations will be considered different species in time. The mechanisms by which populations of pine beetles become sufficiently differentiated have been widely studied, but still remain highly debated. For example, some believe that host plant influences dictate insect genetic differentiation (Kelley and Farrell, 1998; Kelley et al., 2000).

Let us consider how host plants, or increasing specialization, would influence population structure and perhaps lead to speciation. Given that some pine beetles specialize in feeding on particular tree species while others are generalists, the consequences of diet, in relation to host species, have posed some interesting evolutionary questions. For example, it has been hypothesized that *sympatric speciation* of pine beetles may occur through the isolation of populations as a result of increasing dietary specialization (Bush, 1975). This process of speciation would likely present as increasing genetic differentiation among populations, and thus local adaptation of the beetle to specific hosts (as per Van Zandt and Mopper, 1998). Alternatively, diet may lead to *allopatric speciation* in which the patchy distribution of the host tree species reduces the potential for gene flow among the insect populations through increasing isolation (Barton and Charlesworth, 1984). Thus, the effects of genetic drift should be greater on a specialist feeder because they have a limited capacity to find suitable alternate food sources in a patchy landscape (Peterson and Denno, 1998; Thompson, 2004).

Kelley et al. (2000) assessed the relationship between host specialization and patterns of genetic differentiation of sister pine beetles: *Dendroctonus jeffreyi* and *D. ponderosae*. *D. jeffreyi* is a specialist feeder, thus its distribution should rely on the distribution of *Pinus jeffreyi*, and all populations of *D. jeffreyi* should be equally adapted to its host. In contrast, *D. ponderosae* is a generalist capable of feeding on several pine species (Kelley and Farrell, 1998), hence the beetle should have a wider, and less patchy distribution that facilitates greater gene flow. However, some populations of *D. ponderosae* may exhibit higher levels of genetic differentiation and structure, if they have been exposed to different hosts because they must locally adapt to that particular host species. Using estimates of F_{ST}, Kelley et al. (2000) assessed the genetic differentiation and inferred the population genetic structure of *D. jeffreyi* and *D. ponderosae* within and among two sites.

They found that genetic differentiation, and therefore population genetic structure, of the generalist was greater among beetles using different hosts within the same population than that of the specialist using one host over similar spatial scales (Kelley et al., 2000). However, they found this pattern to be inconsistent among populations and loci used.

3.1.2 Case Study: Assessing the Effects of Geographic Isolation with F-Statistics

Just as some support the idea of increasing specialization resulting in speciation, others consider increasing geographic isolation of populations the primary driving factor. With respect to geographic isolation, the model of isolation by distance (IBD) proposed by Wright (1943) is still one of the most commonly used. The IBD model describes the process of increasing genetic differentiation correlated with increasing geographic distance. One study by Ruiz et al. (2009) more specifically assessed the effects of geographic isolation on genetic structure on pine beetles. In contrast to the Kelley et al. (2000) study, Ruiz et al. (2009) removed any influence of host specialization limiting the research to one pine beetle (*D. pseudotsugae*) found on one host (*Pseudotsuga menziesii* var. *glauca*). *P. menziesii* var. *glauca* is an interior variant of Douglas-fir found throughout the Rocky Mountains in North America (Hermann and Lavender, 1990). Ruiz et al. (2009) found strong genetic differentiation, and thus structure, between geographically separated populations of *D. pseudotsugae* making use of the same host. The estimates of F_{ST} obtained by Ruiz et al. (2009) led the authors to conclude that dispersal and geographic distance have greater effects on population genetic structure and differentiation than host specialization would. Interestingly, the estimates of genetic differentiation observed between *D. pseudotsugae* populations across the range of *P. menziesii* var. *glauca* were sufficient to support two subspecies of beetle: *D. p. pseudotsugae* in North America and *D. p. barragani* in Mexico.

3.1.3 Case Study: F-Statistics and the Shift Towards Landscape Community Genetics

A novel extension of the use of *F*-statistics in population structure studies is the incorporation of regression methods to assess co-evolutionary relationships. Pine beetles share close evolutionary relationships with their hosts and numerous species also co-evolve with fungal symbionts. These fungal symbionts often assist the beetle in overcoming the host defence system, and the relationship can range from positive to negative (Six and Wingfield, 2011).

In mountain pine beetles (*D. ponderosae*), the relationship is commonly believed to be a positive mutualism. The beetle transports fungus in mycangia to the new host, the fungus assists in overcoming the host tree defences by colonizing the phloem of the tree and draining valuable nutrients and resources (Solheim, 1995). In return, the beetle larvae feed on fungal contaminated phloem, which results in larger, faster developing beetles (Six and Wingfield, 2011). While population structure has been assessed extensively in individual taxa within pine beetle systems, very few have tried to compare structure patterns among taxa.

A recent paper by James et al. (2011) used estimates of F_{ST} to assess the contemporary population genetic structure among *D. ponderosae* and its primary fungal symbiont *Grosmannia clavigera*. Ordinations of pairwise F_{ST} estimates were used to assess the population genetic structure of both species. Both beetles and fungi exhibited similarly weak genetic structure ($F_{ST} < 0.04$) across the study region. Comparisons of principle coordinates analyses using pairwise F_{ST} estimates revealed a strong pattern of population structure within *D. ponderosae* that correlated with latitude—a pattern that was not found in the fungal symbiont (James et al., 2011).

The authors expanded on the traditional uses of F_{ST} by incorporating a landscape genetic component in which measures of spatial connectivity, such as climate suitability and pine volume, were assessed using redundancy analyses (James et al., 2011). These methods showed that pine volume and climate suitability best predicted beetle genetic variation and structure, whereas no spatial connectivity variables were identified that predicted fungal structure (James et al., 2011). Thus, a novel twist on a traditional method revealed that in spite of a close symbiotic relationship among beetle and fungus, the two taxa did not always display similar patterns of population structure.

3.2 Relatedness: The Extreme Similarity Assessment

Pine beetles exhibit a range of mating and dispersal systems. For example, *Xylosandrus germanus*, an ambrosia beetle, exhibits *haplodiploidy* (in which males are haploid), the females are the sole dispersers and mating was presumed to occur prior to female dispersal using sibling males (Peer and Taborsky, 2005). *Dendroctonus micans*, the great European spruce bark beetle, also uses a system in which females were presumed to mate with sibling males, however both males and females can disperse (Fraser et al., 2014). In contrast, *D. ponderosae* was reported to rely more heavily on monogamous, but non-sibling matings (Bleiker et al., 2013; Reid, 1958). Information on the level of relatedness among individuals within and among populations can provide additional information to studies of population genetic structure (Wang and Santure, 2009). For example, the proportion of highly genetically related individuals in a population provides indirect evidence for the predominant mating system being used (Berger-Wolf et al., 2007). This information can be used to infer pedigree structure, which contributes to our understanding of the spatial genetic structure patterns present (Jacquemyn et al., 2006).

Common means of assessing relationships among individuals include the use of relatedness estimators and paternity analyses. Relatedness estimators are often used in natural populations when a true pedigree is not feasible (Wang, 2002). These estimators are derived using either a likelihood approach to compare allele frequencies among pairs of individuals (ie, Goodnight and Queller, 1999) or a moment estimators approach identifying the probability of identity-by-decent (ie, Queller and Goodnight, 1989). Essentially, these methods measure the biological relationships among individuals. A value close to one indicates a high level of inbreeding or relatedness, whereas a value close to zero suggests little inbreeding. Alternatively, paternity analyses offer more detailed estimates of paternity and sibship. Several paternity assignment programmes exist (eg, Marshall et al., 1998; Wang and Santure, 2009), and each has its own strengths and weaknesses.

3.2.1 Case Study: Using Paternity and Relatedness to Infer Population Genetic Structure

Our understanding of evolutionary processes suggests that a species with limited gene flow should exhibit strong population genetic structure. For example, *X. germanus* has very limited male dispersal coupled with local mating, thus the potential for gene flow among populations is limited (Peer and Taborsky, 2005). In contrast, both sexes of *D. micans* are capable of dispersing even though local mating occurs. A recent study by Fraser et al. (2014) specifically assessed the extent of local mating in *D. micans*. Instead of using a more rigorous relatedness estimator or paternity analysis, the authors inferred sibship, and therefore paternity, by assessing linear regressions of pooled offspring calibrated microsatellite allele peaks with those from adult beetles derived from the same population (Fraser et al., 2014). While this approach was unconventional, it did show support for multiple matings within *D. micans*. The authors concluded that males will almost always mate with a female sibling, but that males will also disperse and opportunistically mate with an unrelated female later (Fraser et al., 2014). This information suggests that population genetic structure between populations of *D. micans* is not as strong as previously thought because there is more potential for migration and therefore gene flow.

As previously mentioned, *D. ponderosae* was believed to have a largely monogamous mating system (Reid, 1956) with a reasonable proportion (3–12%) of these matings occurring prior to emergence (Bleiker et al., 2013). However, Janes et al. (2016) applied relatedness estimators and paternity analyses to SNP data to reveal a different story. Their study showed a number of galleries contained half-siblings, while others contained high numbers of unrelated larvae (Janes et al., 2016). This information is changing the perception of population genetic structure of mountain pine beetle, at least under epidemic conditions. For example, assuming a monogamous mating system with some level of sibling mating, populations would be expected to show a signal of population genetic structure due to a higher level of inbreeding compared to an outcrossing species. However, a confirmed polygamous mating system coupled with dispersal of both sexes results in an absence of population genetic structure as a result of increased gene flow, migration and genetic recombination events.

3.3 Bayesian Clustering Methods: One of These Is Not Like the Others

One of the most popular ways of assessing population genetic structure requires the use of Bayesian clustering methods to assign individuals from a sample into populations based on genetic similarity. This type of approach is particularly useful for learning about evolutionary relationships, and assigning individuals of unknown origin to discrete populations without having to subjectively pre-assign individuals first. While a number of software programs use a Bayesian clustering algorithm of some kind (eg, Corander et al., 2003; Francois et al., 2006), arguably the most widely used is that of Pritchard et al. (2000).

The approach used by Pritchard et al. (2000) and employed in the program STRUCTURE assumes a model in which there are *K* populations, each population characterized by a set of allele frequencies at each locus. This method then assigns individuals to populations based on their genotypes. Simultaneously the method estimates the allele frequencies for each population, thereby providing an estimate of the number of genetic populations sampled and the relative assignment of each individual to those populations.

3.3.1 Case Study: Beware the Scale of Population Genetic Structure

A study conducted by Schrey et al. (2008) set out to define the fine-scale population structure of *Dendroctonus frontalis* from six sites in Mississippi using microsatellite markers. The authors assessed the data using the admixture model within STRUCTURE (Pritchard et al., 2000), which allows for a proportion of an individual's genotype to have originated from another 'population'. The optimal number of clusters was allowed to run from 1 to 6 with the likelihood of each being estimated by the highest probability log likelihood (Schrey et al., 2008). Interestingly, the clustering method indicated that the six sites actually comprised a single genetic population—there was no population genetic structure present. The authors stated that dispersal, and hence gene flow, was sufficient enough within the state of Mississippi to prevent any signal of population genetic structure at that spatial scale (Schrey et al., 2008). While the research did not necessarily have the outcome expected by the authors, it did provide sufficient genetic evidence to suggest that previous reported dispersal distances were underestimated for *D. frontalis*.

A few years later the authors expanded on their work. Schrey et al. (2011) sampled 19 sites across southeastern USA, resulting in a much larger and geographically wider study. In addition, the authors chose to compare three of the most popular Bayesian clustering methods, implemented through the programs BAPS (Corander et al., 2003), STRUCTURE (Pritchard et al., 2000) and TESS (Francois et al., 2006). Interestingly, the authors chose to assess the likelihood of different numbers of K among the programs. For example, $K=1$–27 was assessed in BAPS, $K=2$–10 in TESS and $K=1$–5 in STRUCTURE (Schrey et al., 2011). The results did vary among the programs. Both BAPS and TESS confirmed two genetic populations across the expanded study range; whereas STRUCTURE consistently identified a single genetic population (Schrey et al., 2011). The authors concluded that, at a larger spatial scale, population genetic structure is present within the *D. frontalis* range. Effectively, sites from eastern USA were genetically differentiated from sites in western USA (Schrey et al., 2011).

The different patterns of population genetic structure observed at different spatial scales, in this instance, are reflective of different processes. At a small spatial scale (ie, one state, as per Schrey et al. (2008)), the population genetic structure was influenced more by demographic processes such as population size and mating system. At a larger spatial scale (ie, several states, as per Schrey et al., 2011), the pattern was more strongly influenced by evolutionary processes such as geography and migration. The identification of two genetic populations across southeastern USA corresponded with a geographical barrier (the Appalachian Mountains) and the likely historical dispersal routes taken by the beetle. Thus, the spatial scale at which population genetic structure is assessed can influence results.

3.4 Phylogeography: Travelling Back in Time

Unlike population genetic analysis, which essentially examine allele frequencies between populations, the field of *phylogeography* examines the phylogenetic relationships of populations in respect to their geographic distribution and *biogeographic* history. With respect to animal-based studies, the creation of this field is largely a result of the increased use of mtDNA sequencing techniques opening up the use of phylogenetic methods in examining the relationship between the geography and genealogy of a species (Avise, 2004b). Phylogeographic analyses can be viewed as a historical population genetic structure approach in that they provide insights into geographical origins, range expansions and contractions and the effects of environmental and climatic changes on dispersal events (Drummond and Bouckaert, 2015).

Historically, phylogeography relied on *maximum parsimony* methods to infer ancestral locations on a phylogenetic tree (Drummond and Bouckaert, 2015). However, model-based inference approaches are more commonly used today as they allow for some level of probability surrounding the uncertainty associated with the phylogenetic tree to be estimated (Drummond and Bouckaert, 2015). Some of the more popular methods include structured coalescent models (eg, Beerli and Paclczewski, 2010) and Bayesian inferences using various models such as structured birth–death (eg, Mr Bayes, BEAST).

3.4.1 Case Study: Phylogeography Explains Differential Management of European Pine Beetles

Populations of *I. typographus*, the eight-spined spruce bark beetle, across Europe were used to assess phylogeographic questions relating to gene flow, migration routes and population genetic structure (Stauffer et al., 1999). Throughout Europe *I. typographus* causes significant losses of *Picea abies* (Norway spruce) when populations outbreak. In order to control losses of this ecologically and economically important spruce, synthetic aggregation pheromones have been produced (Stauffer et al., 1999). These pheromones are used to lure individuals into traps, thereby controlling the population size, to some degree. However, different populations of *I. typographus* react differently to the lures (Stauffer et al., 1999).

Genotyping 21 populations across a combination of isozymes and mtDNA sequences, Stauffer et al. (1999) applied maximum parsimony phylogeny methods to infer phylogeography. They found that gene flow among European populations was high and that the population genetic structure of *I. typographus* had been influenced by post-glaciation events (Stauffer et al., 1999). In particular, the *haplotype* diversity suggested that beetle populations had survived the last glacial maximum in refuges to the south and north of Moscow, around Croatia and Italy—a refugium pattern that is also seen within the host tree. Stauffer et al. (1999) inferred that the beetle had expanded into Europe from the refugia in the Apennine and Dinaric Alps from a northeast direction, before dispersing further north into Scandinavia. This pattern indicated that the expansion of the beetle did not follow the expansion of its host tree from the refuge in Kostroma, north of Moscow, into Western Europe and Scandinavia, suggesting a potential barrier was present in those areas (Fig. 1). The lower level of haplotype diversity within Scandinavia, owing to the more recent evolutionary history, correlated well with previous evidence showing that Scandinavian populations are more easily managed using synthetic pheromone lures than Central European

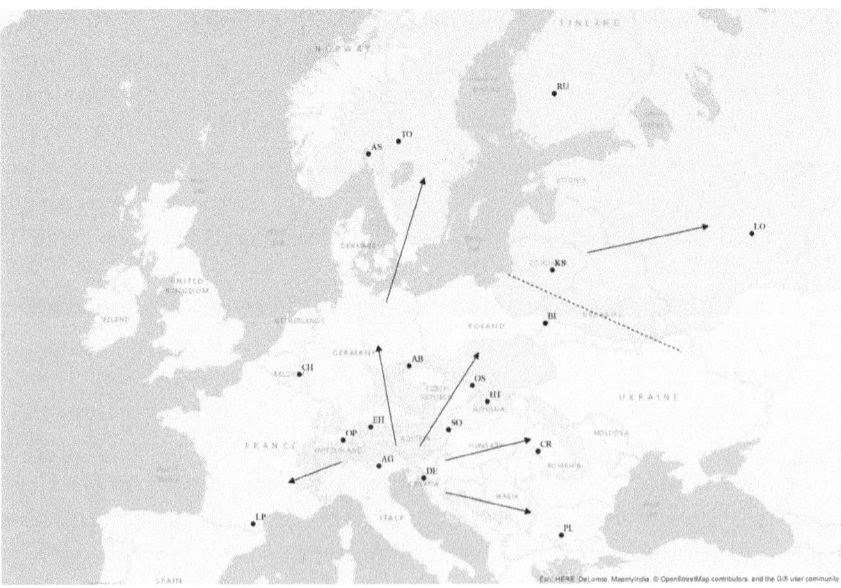

Fig. 1 Map of the phylogeographic dispersal of the eight-spined spruce bark beetle throughout Europe. *After Stauffer, C., Lakatos, F., Hewitt, G.M., 1999. Phylogeography and postglacial colonization routes of* Ips typographus *L. (Coleoptera, Scolytidae). Mol. Ecol. 8, 763–773.* (See the colour plate.)

populations within higher haplotype diversity. A similar phylogeographic pattern was observed by Sallé et al. (2007). Thus, population genetic structure effects on behavioural traits can be inferred through phylogeography.

3.4.2 Case Study: Is Gene Flow Historical or Contemporary?

Typically, when explaining the geographic distribution of genetic variation, we discuss relatively recent demographic events (Maroja et al., 2007). However, the origins of that population genetic structure may be the result of more ancient *vicariance* and isolation events. A paper by Maroja et al. (2007) investigated the effects of historical and contemporary gene flow within populations of *D. rufipennis*, the spruce beetle, using phylogeographic methods. The spruce beetle occurs throughout northern USA, ranging from Alaska to Newfoundland and extending south along the western coast to California, Arizona and New Mexico (Maroja et al., 2007). Within its range, white spruce (*Picea glauca*) and Engelmann spruce (*Picea englemannii*) are the most favoured hosts (Kelley and Farrell, 1998). Within British Columbia, the two spruce species hybridize (Maroja et al., 2007) and beetles within this hybrid zone have been shown to have conflicting mtDNA and nuclear signatures, in spite of the two markers being concordant across the rest of the range (Maroja et al., 2007) (Fig. 2).

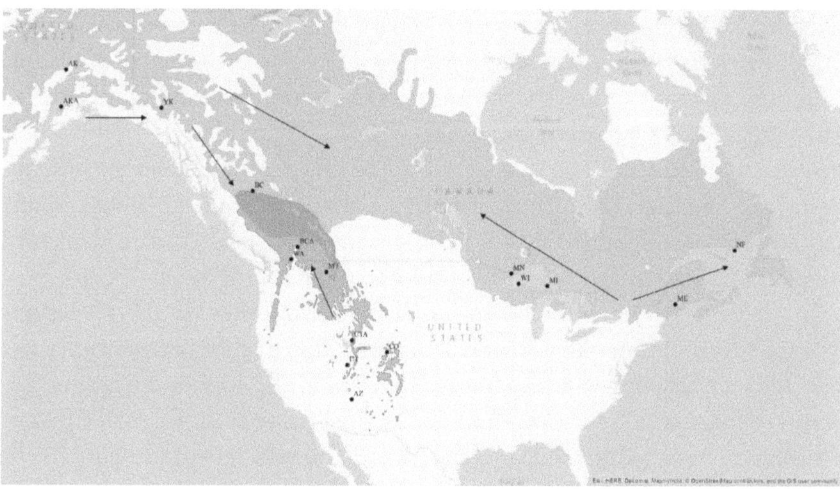

Fig. 2 Historical movement of the spruce beetle throughout North America. *After Maroja, L.S., Bogdanowicz, S.M., Wallin, K.F., Raffa, K.F., Harrison, R.G., 2007. Phylogeography of spruce beetles (*Dendroctonus rufipennis *Kirby) (Curculionidae: Scolytinae) in North America. Mol. Ecol. 16, 2560.* (See the colour plate.)

Both mtDNA and nuclear microsatellites revealed three groupings of *D. rufipennis*, two distinct northern groups and one group that represented populations running the length of the Rocky Mountains. However, the two markers differed in attributing beetles from the spruce hybrid zones to the three groupings. These results suggested that there were at least three glacial refugia for *D. rufipennis*. The authors also reported that populations within the Pacific Northwest, including British Columbia, represented a transition zone in which beetles from multiple refugia had come together in secondary contact. Maroja et al. (2007) were able to show that there had been historical isolation of populations, in the form of three glacial refugia, which had resulted in significant genetic divergence, and hence strong population genetic structure, of these groups. In contrast, the phylogeographic methods also revealed more recent signatures of gene flow in the Pacific Northwest where populations appeared to be maintaining an area of secondary contact.

3.5 Old Becomes New: Structure Meets Function

Early studies of the genetic structure of pine beetles were largely based on isozyme or allozyme variation. These early studies, like others using alternative markers, showed that population genetic structure patterns were largely the result of geographic isolation and host tree specialization (eg, Langor and Spence, 1991; Stock et al., 1978; Sturgeon and Mitton, 1986). What was particularly interesting about these early studies was the use of putatively non-neutral markers to elucidate these patterns. The vast majority of molecular studies use *neutral markers*, this means that the genetic variation present tells us nothing about adaptation and fitness within the population. For example, it makes no difference if an individual in the population is a *heterozygote* or a *homozygote* if that particular locus is not under selection. Thus, neutral markers provide valuable information on gene flow, migration and dispersal independent of adaptive selection (Holderegger and Wagner, 2006).

In contrast, *adaptive markers* are subject to natural selection. For example, if a population of pine beetles was distributed along an altitudinal cline, they may still be exchanging genes but the beetles at the higher elevation were subject to intense selection on their ability to tolerate cold. Effectively, some of the genes controlling cold tolerance would be different among the beetles in higher and lower elevations as a result of selection, in spite of the continued gene flow. If adaptive markers at cold tolerance genes were used to assess population genetic structure, the result may be different to that obtained using neutral markers. Whether or not the genetic variation at adaptive genes should, or can, be used to infer population genetic structure, or be correlated with neutral genetic variation is still under debate.

3.5.1 Case Study: Does Adaptive Variation Measure Up with Neutral?

Several studies have looked at genetic differentiation among *D. ponderosae* populations in response to host species using allozymes. For example, Stock et al. (1984) used six polymorphic allozymes across aspartate aminotransferase, acid phosphatase, esterase, peptidase, leucine aminopeptidase 2 and phosphoglucose isomerase. Sturgeon and Mitton (1986) used five polymorphic loci, and Langor and Spence (1991) used 14 loci. Kelley et al. (2000) identified three polymorphic enzymes from esterase, peptidase and phosphoglucose isomerase protein groups. All of these studies identified a pattern of population genetic structure showing that there was significant genetic differentiation among populations inhabiting different host tree species. However, a study by Mock et al. (2007) using AFLP markers was unable to detect the same pattern, suggesting that *D. ponderosae* population genetic structure was not affected by host tree. While adaptive and neutral markers provided different population genetic structure evidence, it does not mean that one choice was necessarily more accurate or meaningful. Both Langor and Spence (1991) and Mock et al. (2007) concluded that the adaptive markers were reflecting selective pressures on *D. ponderosae* from the host trees. These selective pressures were unlikely to be detected by the AFLP markers, serving as a reminder that marker choice can influence the results of a study.

3.5.2 Case Study: Gene Function Can Help Explain Population Genetic Structure Patterns

The use of high-throughput markers such as SNPs has been providing molecular studies with a greater ability to understand the evolutionary processes driving population genetic structure patterns. The next step is to begin to understand the direct effects of selection on genes, and how adaptation can influence population genetic structure. For example, how does standing genetic variation respond to selective pressure on particular genes, how will population genetic structure reflect these changes and how can an understanding of functional genetics help us interpret the patterns we see? A recent paper by Janes et al. (2014) was the first study within a pine beetle system to incorporate aspects of population genetic structure and selective sweeps to provide a detailed view of how the mountain pine beetle (*D. ponderosae*) breached the Canadian Rocky Mountains.

The mountain pine beetle underwent rapid range expansion in Canada during its last population irruption. The beetle became firmly established in northern British Columbia and Alberta (Safranyik et al., 2010) with anecdotal reports that the beetle had also spread to the Northwest Territories and Saskatchewan. While the spread of the beetle was worrisome, it was more concerning that *D. ponderosae* had been able to switch host. Cullingham et al. (2011) confirmed that *D. ponderosae* had begun using jack pine (*Pinus banksiana*) as a host and appeared to be in the process of locally adapting to this novel host species courtesy of a jack pine × lodgepole pine hybrid zone in Alberta. Not only had the mountain pine beetle done the unthinkable by breaching what had been considered a natural barrier—the Canadian Rocky Mountains; but now it appeared as though the beetle would be free to continue expanding through the entire Canadian boreal forest.

Janes et al. (2014) identified a number of genes that they believed would be subject to natural selection in the expanding populations. For example, genes associated with cold tolerance, detoxification, dispersal capacity, pheromone production and olfaction were considered candidate genes as they were likely to be affected by a novel host and environment. Genotyping over 500 beetles from 27 populations, the study assessed the population genetic structure and connectivity of populations across more than 1000 SNPs (Janes et al., 2014). The authors confirmed a previously identified north–south pattern of genetic differentiation in Canadian populations, suggesting that *D. ponderosae* was indeed exhibiting differential gene flow in relation to geography (Janes et al., 2014). The population genetic structure pattern observed indicated that *D. ponderosae* had expanded from historical populations in southwestern British Columbia, and that the expansion had occurred in two directions—north and east through British Columbia, while simultaneously dispersing east and then north in Alberta and British Columbia (Janes et al., 2014).

These results provided some insight into the dispersal patterns and the level of genetic connectivity among subpopulations within Canada. However, the authors incorporated selective sweeps in order to further explain the evolutionary processes at work. *Selective sweeps*, or outlier detection, assess the level of genetic differentiation (F_{ST} based) among loci and populations to identify statistically significant aberrant loci in relation to a 'neutral' mean value. Using a conservative approach, Janes et al. (2014) identified several SNPs under *directional selection* that had a functional association with cholesterol synthesis, ion transport and muscle contraction. The genes identified had significant adaptive potential in facilitating the continued expansion of mountain pine beetle within Canadian boreal environments.

4. CAN PINE BEETLE POPULATION GENETIC STRUCTURE LEAD THE WAY?

Examining the population genetic structure of pine beetles has helped us to understand the environmental and evolutionary forces that have shaped these species over time and space. Here, we have provided a short history of progress that such methods and studies have made over the years. However, traditional population genetic methods, such as population genetic structure, can only tell us so much. These methods are largely limited to helping us explain *what* has happened; they cannot tell us *how* it happened. Studies of the mountain pine beetle system, because of its incredible ecological and economic significance, have been dominating the pine beetle literature in recent years, providing an exceptional basis for further understanding complex insect pest systems. The recent study by Janes et al. (2014) was the first pine beetle study to more fully address how population genetic structure patterns develop. The paper has provided a strong foundation for future studies, highlighting the need for pine beetle research in the area of population genetic structure studies to work towards more methodologically inclusive research. For example, although Janes et al. (2014) identified several functional regions under directional selection that could be used to explain mountain pine beetle range expansion, further work in the area of functional genetics is required to fully understand how these genes provide a selective advantage, and what their tolerances are. In the age of genome-wide data, population genetics and pine beetle biology have the opportunity to advance significantly by incorporating functional genetics. Given that pine beetles are such significant forest pests, they have reasonably fast generation times and are becoming increasingly easier to trap and maintain in laboratories; there appears to be an open niche for pine beetle studies to fill with respect to melding population genetics and functional genetics.

REFERENCES

Avise, J.C., 2004a. Molecular techniques. In: Molecular Markers, Natural History and Evolution, second ed. Sinauer Associates, Inc., MA, USA, pp. 55–113.

Avise, J.C., 2004b. Species phylogenetics and macroevolution. In: Molecular Markers, Natural History and Evolution, second ed. Sinauer Associates, Inc., MA, USA, pp. 401–473.

Barrow, L.N., Bigelow, A.T., Phillips, C.A., Lemmon, E.M., 2015. Phylogeographic inference using Bayesian model comparison across a fragments chorus frog species complex. Mol. Ecol. 24, 4739–4758.

Barton, N.H., Charlesworth, B., 1984. Genetic revolutions, founder effects and speciation. Annu. Rev. Ecol. Syst. 15, 133–164.

Beerli, P., Paclczewski, M., 2010. Unified framework to evaluate panmixia and migration direction among multiple sampling locations. Genetics 185, 313–326.

Bentz, B., Bracewell, R., Mock, K.E., Pfrender, M.E., 2011. Genetic architecture and phenotypic plasticity of thermally-regulated traits in an eruptive species, *Dendroctonus ponderosae*. Evol. Ecol. 25, 1269–1288.

Berger-Wolf, T.Y., Sheikh, S.I., DasGupta, B., Ashley, M.V., Caballero, I.C., Chaovalitwongse, W., Putrevu, S.L., 2007. Reconstructing sibling relationships in wild populations. Bioinformatics 23, i49–i56.

Bleiker, K.P., Carroll, A.L., Smith, G.D., 2010. Mountain pine beetle range expansion: assessing the threat to Canada's boreal forest by evaluating the endemic niche. Mountain Pine Beetle Working Paper 2010-02, Natural Resources Canada, British Columbia. 23 pp.

Bleiker, K.P., Heron, R.J., Braithwaite, E.C., Smith, G.D., 2013. Preemergence mating in the mass-attacking bark beetle, *Dendroctonus ponderosae* (Coleoptera: Curculionidae). Can. Entomol. 145, 12–19.

Brumfield, R.T., Beerli, P., Nickerson, D.A., Edwards, S.V., 2003. The utility of single nucleotide polymorphisms in inferences of population history. Trends Ecol. Evol. 18, 249–256.

Bush, G.L., 1975. Modes of animal speciation. Annu. Rev. Ecol. Syst. 6, 339–364.

Cai, Y.-W., Cheng, X.-Y., Xu, R.-M., Duan, D.-H., Kirkendall, L.R., 2008. Genetic diversity and biogeography of red turpentine beetle *Dendroctonus valens* in its native and invasive regions. Insect Sci. 15, 291–301.

Corander, J.C., Waldmann, P., Silanpaa, M.J., 2003. Bayesian analysis of genetic differentiation between populations. Genetics 163, 367–374.

Cullingham, C.I., Cooke, J.E., Dang, S., Davis, C.S., Cooke, B.J., Coltman, D.W., 2011. Mountain pine beetle host-range expansion threatens boreal forest. Mol. Ecol. 20, 2157–2171.

Dale, V.H., Joyce, L.A., McNulty, S., Neilson, R.P., Ayres, M.P., Flannigan, M.D., Hanson, P.J., Irland, L.C., Lugo, A.E., Peterson, C.J., Simberloff, D., Swanson, F.J., Stocks, B.J., Wotton, B.M., 2001. Climate change and forest disturbances. Bioscience 51, 723–734.

DeSalle, R., Volger, A., 1994. Phylogenetic analysis on the edge: the application of cladistic techniques at the population level. In: Golding, B. (Ed.), Non-Neutral Evolution: Theories and Molecular Data. Chapman and Hall, New York, pp. 154–174.

Drummond, A.J., Bouckaert, R.R., 2015. Bayesian Evolutionary Analysis with BEAST. Cambridge University Press, Cambridge.

Durmaz, A.A., Karaca, E., Demkow, U., Toruner, G., Schoumans, J., Cogulu, O., 2015. Evolution of genetic techniques: past, present, and beyond. BioMed. Res. Int. 2015, 461524. http://dx.doi.org/10.1155/2015/461524. Epub 2015 Mar 22.

Edwards, O.R., Hoy, M.A., 1993. Polymorphism in two parasitoids detected using random amplified polymorphic DNA polymerase chain reaction. Biol. Control 3, 243.

Ekblom, R., Galindo, J., 2010. Applications of next generation sequencing in molecular ecology of non-model organisms. Heredity 107, 1–15.

Emerson, K.J., Merz, C.R., Catchen, J.M., Hohenlohe, P.A., Cresko, W.A., Bradshaw, W.E., Holzapfel, C.M., 2010. Resolving postglacial phylogeography using high-throughput sequencing. Proc. Natl. Acad. Sci. 17, 16196–16200.

Franca, L.T.C., Carriho, E., Kist, T.B.L., 2002. A review of DNA sequencing techniques. Quart. Rev. Biophys. 35, 169–200.

Francois, O., Ancelet, S., Guillot, G., 2006. Bayesian clustering using hidden Markov random fields in spatial population genetics. Genetics 174, 805–816.

Fraser, C.I., Brahy, O., Mardulyn, P., Dohet, L., Mayer, F., Grégoire, J.C., 2014. Flying the nest: male dispersal and multiple paternity enables extrafamilial matings for the invasive bark beetle *Dendroctonus micans*. Heredity 113, 327–333.

Gillette, N.E., Mehmel, C.J., Webster, J.N., Mori, S.R., Erbilgin, N., Wood, D.L., Stein, J.D., 2009. Aerially applied methylcyclohexenone-releasing flakes protect *Pseudotsuga menziesii* stands from attack by *Dendroctonus pseudotsugae*. For. Ecol. Manag. 257, 1231–1236.

Goodnight, K.F., Queller, D.C., 1999. Computer software for performing likelihood tests of pedigree relationship using genetic markers. Mol. Ecol. 8, 1231–1234.

Gupta, P.K., Varshney, R.K., Sharma, P.C., Ramesh, B., 1999. Molecular markers and their applications in wheat breeding. Plant Breed. 118, 369–390.

Haran, J., Timmermans, M.J.T.N., Vogler, A.P., 2013. Mitogenome sequences stabilize the phylogenetics of weevils (Curculionoidea) and establish the monophyly of larval ectophagy. Mol. Phylogenet. Evol. 67, 156–166.

Hermann, R.K., Lavender, D.P., 1990. *Pseudotsuga menziesii* (Mirb.) Franco Douglas fir. In: Burns, R.M., Honkala, B.H. (Eds.), Silvics of North America, vol. 1. Conifers. Agriculture Handbook 654, USDA Forest Service, Washington, DC, pp. 527–540.

Holderegger, R., Wagner, H.H., 2006. A brief guide to Landscape Genetics. Landsc. Ecol. 21, 793–796.

Holsinger, K.E., Weir, B.S., 2009. Genetics in geographically structured populations: defining, estimating and interpreting FST. Nat. Rev. Genet. 10, 639–650.

Jacquemyn, H., Brys, R., Vandepitte, K., Honnay, O., Roldan-Ruiz, I., 2006. Fine-scale genetic structure of life history stages in the food-deceptive orchid *Orchis purpurea*. Mol. Ecol. 15, 2801–2808.

James, P.M.A., Coltman, D.W., Murray, B.W., Hamelin, R.C., Sperling, F.A.H., 2011. Spatial genetic structure of a symbiotic beetle-fungal system: toward multi-taxa integrated landscape genetics. In: Joly, S. (Ed.), PLoS One 6, e25359.

Janes, J.K., Li, Y., Keeling, C.I., Yuen, M.M., Boone, C.K., Cooke, J.E., Bohlmann, J., Huber, D.P., Murray, B.W., Coltman, D.W., Sperling, F.A., 2014. How the mountain pine beetle (*Dendroctonus ponderosae*) breached the Canadian Rocky mountains. Mol. Biol. Evol. 31, 1803–1815.

Janes, J.K., Roe, A.D., Rice, A.V., Gorrell, J.C., Coltman, D.W., Langor, D.W., Sperling, F.A., 2016. Polygamy and an absence of fine-scale structure in *Dendroctonus ponderosae* (Hopk.) (Coleoptera: Curcilionidae) confirmed using molecular markers. Heredity 116 (1), 68–74. http://dx.doi.org/10.1038/hdy.2015.71. Epub 2015 Aug 19.

Kelley, S.T., Farrell, B.D., 1998. Is specialization a dead end? The phylogeny of host use in *Dendroctonus* bark beetles (Scolytidae). Evolution 52, 1731–1743.

Kelley, A.T., Farrell, B.D., Mitton, J.B., 2000. Effects of specialization on genetic differentiation in sister species of bark beetles. Heredity 84, 218–227.

Kumar, P., Gupta, V.K., Misra, A.K., Modi, D.R., Pandey, B.K., 2009. Potential of molecular markers in plant biotechnology. Plant Omics J. 2, 141–162.

Langor, D.W., Spence, J.R., 1991. Host effects on allozyme and morphological variation of the mountain pine beetle *Dendroctonus ponderosae* Hopkins (Coleoptera: Scolytidae). Can. Entomol. 123, 395–410.

Leng, L., Zhang, D., 2011. Measuring population differentiation using GST or D? A simulation study with microsatellite DNA markers under a finite island model and nonequilibrium conditions. Mol. Ecol. 20, 2494–2509.

Lewontin, R.C., 1991. Electrophoresis in the development of evolutionary genetics: milestone or millstone. Genetics 128, 657–662.

Li, J.Z., Absher, D.M., Tang, H., Southwick, A.M., Casto, A.M., Ramachandran, S., Cann, H.M., Barsh, G.S., Feldman, M., Cavalli-Sforza, L.L., Myers, R.M., 2008. Worldwide human relationships inferred from genome-wide patterns of variation. Science 319, 1100–1104.

Luikart, G., England, P.R., Tallmon, D., Jordan, S., Taberlet, P., 2003. The power and promise of population genomics: from genotyping to genome typing. Nat. Rev. Genet. 4, 981–994.

Maroja, L.S., Bogdanowicz, S.M., Wallin, K.F., Raffa, K.F., Harrison, R.G., 2007. Phylogeography of spruce beetles (*Dendroctonus rufipennis* Kirby) (Curculionidae: Scolytinae) in North America. Mol. Ecol. 16, 2560.

Marshall, T.C., Slate, J., Kruuk, L.E.B., Pemberton, J.M., 1998. Statistical confidence for likelihood-based paternity inference in natural populations. Mol. Ecol. 7, 639–655.

Miah, G., Rafii, M.Y., Ismail, M.R., Puteh, A.B., Rahim, H.A., Islam, Kh.N., Latif, M.A., 2013. A review of microsatellite markers and their applications in rice breeding programs to improve blast disease resistance. Int. J. Mol. Sci. 14, 22499–22528.

Mock, K.E., Bentz, B.J., Oneill, E.M., Chong, J.P., Orwin, J., Pfrender, M.E., 2007. Landscape-scale genetic variation in a forest outbreak species, the mountain pine beetle (*Dendroctonus ponderosae*). Mol. Ecol. 16, 553–568.

Moeck, H.A., Wood, D.L., Lindahl, K.Q., 1981. Host selection behavior of bark beetles (Coleoptera: Scolytidae) attacking *Pinus ponderosa*, with special emphasis on the western pine beetle, *Dendroctonus brevicomis*. J. Chem. Ecol. 7, 49–83.

Panayotov, M., Kulakowski, D., Laranjeiro Dos Santos, L., Bebi, P., 2011. Wind disturbances shape old Norway spruce-dominated forest in Bulgaria. For. Ecol. Manag. 262, 470–481.

Peer, K., Taborsky, M., 2005. Outbreeding depression, but no inbreeding depression in haplodiploid ambrosia beetles with regular sibling mating. Evolution 59, 317–323.

Peterson, M.A., Denno, R.F., 1998. The influence of dispersal and diet breadth on patterns of genetic isolation by distance in phytophagus insects. Am. Nat. 152, 428–446.

Pritchard, J.K., Stephens, M., Donnelly, P., 2000. Inference of population structure using multilocus genotype data. Genetics 155, 945–959.

Queller, D.C., Goodnight, K.F., 1989. Estimating relatedness using genetic markers. Evolution 43, 258–275.

Raffa, K.F., Gregoire, J., Lindgren, B.S., 2015. Natural history and ecology of pine beetles. In: Pine Beetles. Biology and Ecology of Native and Invasive Species. Elsevier Inc., London, UK, pp. 1–40.

Reid, R.W., 1956. Studies on the biology of the mountain pine beetle, *Dendroctonus Monticolae* Hopk.: interim report 1955–1: Forest research report. Government of Canada, Department of Northern Affairs and National Resources, Forestry Branch, Forest Biology Laboratory, Calgary, AB. 48 pp.

Reid, R.W., 1958. The behaviour of the Mountain Pine Beetle, *Dendroctonus monticolae* Hopk., during mating, egg laying and gallery construction. Can. Entomol. 90, 505–509.

Roderick, G.K., 1996. Geographic structure of insect populations: gene flow, phylogeography and their uses. Annu. Rev. Entomol. 41, 325–352.

Roughgarden, J., May, R.M., Levin, S.A. (Eds.), 1989. Perspectives in Ecological Theory. Princeton University Press, Princeton, NJ.

Rudinsky, J.A., 1962. The ecology of the Scolytidae. Annu. Rev. Entomol. 7, 327–348.

Ruiz, E.A., Rinehart, J.E., Hayes, J.L., Zúñiga, G., 2009. Effect of geographic isolation on genetic differentiation in *Dendroctonus pseudotsugae* (Coleoptera: Curculionidae). Hereditas 146, 79–92.

Safranyik, L., 2003. Mountain Pine Beetle Epidemiology in Lodgepole Pine. Natural Resources Canada, Victoria, BC.

Safranyik, L., Wilson, B. (Eds.), 2006. The Mountain Pine Beetle: A Synthesis of Biology, Management, and Impacts on Lodgepole Pine. Natural Resources Canada, Pacific Forestry Center, Victoria, BC.
Safranyik, L., Carroll, A.L., Régnière, J., Langor, D.W., Riel, W.G., Shore, T.L., Peter, B., Cooke, B.J., Nealis, V.G., Taylor, S.W., 2010. Potential for range expansion of mountain pine beetle into the boreal forest of North America. Can. Entomol. 142, 415–441.
Sallé, A., Arthofer, W., Lieutier, F., Stauffer, C., Kerdelhué, C., 2007. Phylogeography of a host-specific insect: genetic structure of *Ips typographus* in Europe does not reflect the past fragmentation of its host. Biol. J. Linn. Soc. 90, 239–246.
Schrey, N.M., Schrey, A.W., Heist, E.J., Reeve, J.D., 2008. Fine-scale genetic population structure of southern pine beetle (Coleoptera: Curculionidae) in Mississippi forests. Environ. Entomol. 37, 271–276.
Schrey, N.M., Schrey, A.W., Heist, E.J., Reeve, J.D., 2011. Genetic heterogeneity in a cyclical Forest pest, the Southern Pine Beetle, *Dendroctonus frontalis*, is differentiated into east and west groups in the southeastern United States. J. Insect Sci. 11, 1–10.
Seeb, J.E., Carvalho, G.R., Hauser, L., Naish, K., Roberts, S., Seeb, L.W., 2011. Single-nucleotide polymorphism (SNP) discovery and applications of SNP genotyping in non-model organisms. Mol. Ecol. Resour. 11, 1.
Shan, X., Blake, T.K., Talbert, L.E., 1999. Conversion of AFLP markers to sequence-specific PCR markers in barley and wheat. Theor. Appl. Genet. 98, 1072–1078.
Six, D.L., Wingfield, M.J., 2011. The role of phytopathogenicity in bark beetle–fungus symbioses: a challenge to the classic paradigm. Annu. Rev. Entomol. 56, 255–272.
Slate, J., Gratten, J., Beraldi, D., Stapley, J., Hale, M., Pemberton, J.M., 2009. Gene mapping in the wild with SNPs: guidelines and future directions. Genetica 136, 97–107.
Slatkin, M., 1985. Gene flow in natural populations. Annu. Rev. Ecol. Syst. 16, 393–430.
Slatkin, M., 1995. Measure of population subdivision based on microsatellite allele frequencies. Genetics 139, 457–462.
Solheim, H., 1995. Early stages of blue-stain fungus invasion of lodgepole pine sapwood following mountain pine beetle attack. Can. J. Bot. 73, 70–74.
Stauffer, C., Lakatos, F., Hewitt, G.M., 1999. Phylogeography and postglacial colonization routes of *Ips typographus* L. (Coleoptera, Scolytidae). Mol. Ecol. 8, 763–773.
Stock, M.W., Guenther, J.D., Pitman, G.B., 1978. Implications of genetic differences between mountain pine beetle populations to integrated pest management. In: Berryman, A.A., Amman, G.D., Stark, R.W., Kibbee, D.L. (Eds.), Theory and practice of mountain pine beetle management in lodgepole pine forests. Symposium, 25–27 April, Pullman, Washington, University of Idaho, College of Forest Resources, pp. 197–204.
Stock, M.W., Amman, G.D., Higby, P.K., 1984. Genetic variation among mountain pine beetle. Ann. Entomol. Soc. Am. 77, 760–764.
Sturgeon, K.B., Mitton, J.B., 1986. Allozyme and morphological differentiation of mountain pine beetles *Dendroctonus ponderosae* Hopkins (Coleoptera: Scolytidae) associated with host tree. Evolution 40, 290–302.
Taylor, H.R., Harris, W.E., 2012. An emergent science on the brink of irrelevance: a review of the past 8 years of DNA barcoding. Mol. Ecol. Resour. 12, 377–388.
Taylor, D.J., Finston, T.L., Hebert, P.D.N., 1994. The 15% solution for preservation. Trends Evol. Ecol. 9, 230.
Thompson, N., 2004. The Coevolutionary Process. Chicago University Press, Chicago, IL.
Van Zandt, P.A., Mopper, S., 1998. A meta-analysis of adaptive deme formation in phytophagus insect populations. Am. Nat. 152, 595–604.
Viktorinova, I., Kucerova, L., Bohmova, M., Henry, I., Jindra, M., Dolezal, P., Zurovcova, M., Zurovec, M., 2011. Characterization of two closely related α-amylase paralogs in the bark beetle, *Ips typographus* (L.). Arch. Insect Biochem. Physiol. 77, 179–198.

Vos, P., Hogers, R., Bleeker, M., Reijans, M., Van de Lee, T., Hornes, M., Frijters, A., Pot, J., Peleman, J., Kuiper, M., Zabeau, M., 1995. AFLP: a new technique for DNA fingerprinting. Nucleic Acids Res. 23, 4407–4414.

Wallner, W.E., 1987. Factors affecting insect population dynamics: differences between outbreak and non-outbreak species. Annu. Rev. Entomol. 32, 317–340.

Wang, J., 2002. An estimator for pairwise relatedness using molecular markers. Genetics 160, 1203–1315.

Wang, J., Santure, A.W., 2009. Parentage and sibship inference from multilocus genotype data under polygamy. Genetics 181, 1579–1594.

Warren, G.L., 1958. A method of rearing bark- and cambium-feeding beetles with particular reference to *Hylobius warreni* Wood (Coleoptera: Curculionidae). Can. Entomol. 90 (7), 425–428.

Wright, S., 1943. Isolation by distance. Genetics 28, 114–138.

Wright, S., 1951. The genetic structure of populations. Ann. Eugen. 15, 323–354.

Yearian, W.C., Wilkinson, R.C., 1967. Development of three *Ips* bark beetles on a phloem-based rearing medium. Fla. Entomol. 50, 43–45.

Zúñiga, G., Cisneros, R., Salinas-moreno, Y., Hayes, J.L., Rinehart, J.E., 2006. Genetic structure of *Dendroctonus mexicanus* (Coleoptera: Curculionidae: Scolytinae) in the trans-Mexican volcanic belt. Ann. Entomol. Soc. Am. 99, 945–958.

CHAPTER THREE

The Proteomics and Transcriptomics of Early Host Colonization and Overwintering Physiology in the Mountain Pine Beetle, *Dendroctonus ponderosae* Hopkins (Coleoptera: Curculionidae)

D.P.W. Huber, J.A. Robert
Natural Resources and Environmental Studies Institute, University of Northern British Columbia, Prince George, BC, Canada

Contents

Abstract

Two important phases in the life of the mountain pine beetle, *Dendroctonus ponderosae* Hopkins (Coleoptera: Curculionidae), are the periods of host colonization, where they must overcome host defences to successfully reproduce, and larval overwintering, during which time the insects under the bark of the tree must endure extended periods of deep cold. Proteomic and transcriptomic research into these portions of the mountain pine beetle life cycle have shed light on physiological processes that were already known to play important roles (detoxification of host secondary metabolites, reproductive physiology, and biosynthesis of cryoprotectants) and have revealed new territory for research (immune response, stress physiology). We discuss our overall findings in the context of an insect that is spreading to a new geographic range and encountering new hosts and a changing climate.

Advances in Insect Physiology, Volume 50
ISSN 0065-2806
http://dx.doi.org/10.1016/bs.aiip.2015.12.001

1. INTRODUCTION

The mountain pine beetle is mainly a semelparous organism (Safranyik and Carroll, 2006), which, like many other organisms including some vertebrates such as salmon, reproduce once in their lifetime (although there are variations, eg, Reid, 1958 and references therein). After maturing during the course of a year under the bark of a host pine tree, new adult mountain pine beetles leave the relative protection of the host tree in which they matured and fly to find a new host in which to reproduce. As with salmon, this migration of sorts—more accurately a dispersal (Safranyik et al., 1992)—from a depleted natal host to a fresh, new host is fraught with danger. The insect does not feed during its dispersal flight (Reid, 1958). The adult life can be abruptly cut short due to the vicissitudes of weather: short rain squalls, summer wind storms, extreme heat, and unexpected cold. Predators, adapted to hone in on mountain pine beetle pheromone signals, stalk the boles of trees on which it lands. And in a mixed forest of nonhost conifers (Gray et al., 2015; Pureswaran and Borden, 2005; Pureswaran et al., 2004) and angiosperms (Huber and Borden, 2001, 2003; Huber et al., 2000), it needs to find one of the often few trees in the stand that will be suitable and susceptible to attack (Raffa and Berryman, 1983).

Mountain pine beetles, like many other bark beetles, have evolved powerful aggregation and antiaggregation pheromone systems (covered elsewhere in this volume) that guide many individuals to the correct hosts and that provide for coordination of mass attacks on host trees. This provides the insects with some level of advantage in the battle against the tree. In combination with detection of, and response to, volatile compounds emanating from host (Conn et al., 1983) and nonhost trees (Huber et al., 2000), as well as conspecific-produced pheromones, many mountain pine beetles often converge on one or a few trees in a stand (Hynum and Berryman, 1980; Lindgren and Borden, 1993; Pitman and Vité, 1969). Together, they stand a better chance of overcoming the tree's copious defences and successfully reproducing (Raffa and Berryman, 1983).

But even when the insect finds a host tree, the danger has not passed. A tree, like any other organism, has a vested interest in its survival. Because trees do not move after they germinate from a seed, they have necessarily evolved strong physical and chemical defences against all manner of enemies. And pines are well adapted to cope with bark beetle attacks, releasing both constitutive and induced resin-based defences against the invading insects and their often pathogenic symbionts (Franceschi et al, 2005; Raffa et al., 2005; Seybold et al., 2006; covered elsewhere in this volume). Tree defences are often classified as either 'constitutive' (Wittstock and Gershenzon, 2002) or 'induced' (Karban and Myers, 1989). Constitutive defences are those that are continually in place, as opposed to induced defences that are stimulated by insect attack or pathogen infection and are detectable in the plant only after such an assault. Many secondary metabolites that are present in the resin of a tree prior to attack by an insect are constitutive defences, as are physical barriers like the outer bark and other structural material that an invading insect must contend with. As mountain pine beetles attack, the tree releases resin stored in ducts (Franceschi et al, 2005; Reid and Watson, 1966). The resin is both a physical barrier to entry by the beetles and a mixture of chemicals, some of them toxic to the insect or its symbionts (Seybold et al., 2006). During this host colonization phase—a phase which we will examine in detail in terms of gene expression later in this chapter—invading mountain pine beetles need to cope with chemical defences, produce aggregation and antiaggregation pheromones, and complete their reproductive activity. Thus a mountain pine beetle, again somewhat like a salmon, enters an inhospitable environment and must survive long enough to reproduce while surrounded by competitors rushing to do the same thing. The process of dispersal and host colonization, while as short as mere hours or days at the most (Safranyik et al., 1992), is a crucial moment in the life of a mountain pine beetle. Insects that fail at this, either through maladaptation or simple bad luck, do not pass their genes along to the next generation.

If a pair of beetles is successful in host colonization, the female will lay eggs along the walls of a gallery constructed in the phloem (Cerezke, 1995). Larvae will soon hatch and will begin to feed on bark and phloem tissues that are replete with the secondary metabolites left behind from the tree's battle with the larvae's parents and their compatriots (Clark et al., 2012). While the larvae undoubtedly receive some nutrients from the maternal provisioning of the eggs, they need to ingest substantial amounts of toxin-laden phloem tissue in order to begin development and to get ready for the impending winter. The deep cold of winter, lasting up to a half year or more in some locations within the insects' geographical range, is often a major mortality factor for larvae (Cole, 1981). As with the dispersal and host colonization process, the ability, or lack thereof, of immature mountain pine beetles to survive the winter is both a major factor in natural selection (covered elsewhere in this volume) and in the growth dynamics of populations (Hicke et al., 2006;

Régnière and Bentz, 2007; Safranyik et al., 2010). A second part of this chapter will look in closer detail at larval overwintering physiology.

Because adult host colonization and larval overwintering are such vital points both for individual insects and in terms of the ecological and economic impact of expanding populations, recent research in our laboratory has targeted these processes. We have used new transcriptomics and proteomics methods to examine which genes are being expressed in insects during these times, and how that expression results in the accumulation of physiological relevant proteins.

Transcriptomic methods are powerful tools in advancing our understanding of organismal biology that allow us a window into the genes that are being expressed in an organism or tissue under the influence of various endogenous or exogenous factors. With these methods, we can assess an instant 'photograph' of expressed genes and compare that photograph to others at different times. However, just because we can determine which mRNAs are being produced in an organism or one of its tissues at a certain time, we do not know how (or even whether) those transcripts are being processed into their enzymes and other protein products. For the most part, it is the complement of proteins in a cell determines how that cell functions in the organism.

Parallel proteomic investigations, therefore, have the potential to help better interpret transcriptomic results. However, proteomics studies are also not without their limitations. For instance, a transcriptomic analysis may show increases in a particular mRNA, but a subsequent proteomic analysis on the same tissue may show no significant change in the level of the protein product. If a protein, vital to that organism at that time, is being rapidly produced and degraded while providing its metabolic function to the cells where it is expressed, then the proteomic result would not reflect the importance of that protein in the organism, but the transcriptomic result would. In addition, limitations of the sequencing methods and databases available for protein identifications from samples result in far fewer identified proteins in comparison to the gene transcript identifications. In each sample, we were able to annotate over 800 proteins based on the sequences obtained and on the available databases. In comparison, we were able to align the sequenced transcripts (many with an annotated gene function or name) of over 13,000 gene models from the mountain pine beetle genome Keeling et al., 2012, 2013a,b,c). Thus because of the sequencing technology as well as the available databases, our transcript data sets contain over an order of magnitude more information than we could glean from the proteomic data. But, by using both methods, in parallel when possible, we can better elucidate physiological function.

In recent work, we have used transcriptomic and proteomic methods on male and female adults attacking host lodgepole pine (compared with starved males and females), and larvae over the course of a winter (comparing larvae in early autumn with late autumn, and early spring with late spring). In each case, we used the same sample of insects for both proteomics and transcriptomics analyses; we were comparing apples to apples. So we feel that our four data sets represent a substantially complete picture of the physiological workings of these insects during these two life phases.

Specifically, we examined whole-body transcriptomes and proteomes in the two ecophysiological contexts. In the host colonization experiments, male and female mountain pine beetles that were collected in their natal bolts in southern British Columbia were emerged from their expended host material in our laboratory. Control males and females were then starved for 24 h prior to being extracted for transcriptomic and proteomic analysis. Treatment females were placed into predrilled holes in fresh lodgepole pine bolts and were partnered with a male. The pairs were then screened within the bolt and were removed 24 h later. Insects that showed evidence of feeding in the phloem were used in the transcriptomic and proteomic analyses. Statistical comparisons were made between starved and attacking individuals and between males and females. All treatments (male/female, starved/colonizing) were replicated four times.

In the larval overwintering work, we collected larvae from whole trees fitted with temperature dataloggers in the eastern central interior of British Columbia, near the Alberta border in the autumn of 2008 and spring of 2009. Some insects from the larger sample pool were used in specific transcriptomic work on the glycerol biosynthetic pathway (Fraser, 2011), and results from that work, along with knowledge of the temperature regime at the site that year, led us to choose four groups of insects from the sample pool for more extensive transcriptomic and proteomic studies. The larvae used in the study were collected in September and November (autumn) and March and May (spring). In general, our statistical analyses compared shifts in gene expression and protein levels across the autumn and across the spring. All treatments (dates) were replicated four times.

As with studies of this type that produce copious amounts of data (significant and otherwise), choosing which stories to tell and which to omit depends both on the patterns in the data and, to an extent, on the metaphorical lenses donned by the researchers. Others, doubtless, would look at the same data and see other complementary stories, and we certainly invite interested researchers to do that. In addition, we had to decide whether to extract RNA and proteins from whole insect bodies or to dissect out particular tissues for examination. While investigation of specific tissues would provide additional detail, we opted for whole-body extractions in order to gather a full, but perhaps less detailed, picture of physiological events in the insects. Given what we have found from this broader approach, we are now able to design more specific follow-up studies.

In general, we observed, and describe here, evidence of stress physiology, detoxification processes, pheromone biosynthesis, immune function, and rapid shifts in reproductive physiology during host colonization; along with detoxification, stress physiology, cold hardening, and developmental processes during larval overwintering. Over the course of this chapter, we will detail some of our findings that are particularly relevant to these general categories in the two life phases. In all cases, unless otherwise noted, where we refer to genes or gene products expressed in our experiments, we are citing Bonnett et al. (2012) (overwintering proteomics), Robert et al. (2013) (host colonization transcriptomics), Pitt et al. (2014) (host colonization proteomics), and J.A. Robert et al. unpublished data (overwintering transcriptomics).

2. HOST COLONIZATION

Trees maintain a variety of formidable defences and vigorously resist attacks by bark beetles and their associated microorganisms (Franceschi et al, 2005; Raffa et al., 2005; Seybold et al., 2006). The best-studied chemical defences, in terms of their effects on bark beetles and their symbiotic fungi, are the terpenoids (for review, see Bohlmann, 2012). The defensive aspects of conifer terpenoids are made more complex by the fact that bark beetles often detect them alone or in conjunction with aggregation pheromone components emitted by other colonizing conspecifics to locate host trees (Seybold et al., 2006). In addition, many bark beetle pheromone components are similar to host terpenoids.

Upon encountering a host tree—found through a combination of visual cues (Campbell and Borden, 2006; Shepherd, 1966), avoidance of nonhosts (Huber and Borden, 2001, 2003; Huber et al., 2000; Pureswaran and Borden, 2005; Pureswaran et al., 2004), detection of host volatiles (Billings et al., 1976; Wood, 1982), detection of con- or heterospecific pheromone emissions (Byers, 1989), and direct testing of candidate trees (Raffa and Berryman, 1982a,b)—an attacking bark beetle first encounters the outer bark of the tree. Bark beetles maintain a number of wood-degrading enzymes

(Pauchet et al., 2010) and we detected some of these present in the bodies of colonizing insects. We found that some are upregulated during the early attack phase. Plant cell wall-degrading enzymes are a group of enzymes defined by their ability to break down the polysaccharide (eg, cellulose) and glycoprotein components of plant cell walls (Pauchet et al., 2010). These components are part of the physical defences of the host tree and one of the barriers that mountain pine beetle must contend with in order to successfully reproduce in a selected host tree.

In our study of the physiological changes in adult mountain pine beetle during early host colonization, we saw the upregulation of transcripts coding for a variety of plant cell wall degrading enzymes (eg, glucanases, endopolygalacturonases, pectin methylesterases), and this upregulation is different between males and females. Females are the first to colonize and chew into a suitable host tree, and it is in the females that we see the largest change and variety in the expression of plant cell wall degrading enzyme transcripts, several of them increasing over 50-fold above the levels measured in the control females. In our proteomics analysis, both males and females had detectable plant cell wall degrading enzymes, but the relative values of the proteins did not show the dynamic range that was apparent in the transcript expression data. The presence of these enzymes during early host colonization indicates some level of constitutive expression. In other words, bark beetles are primed to begin to digest tree tissue prior to their arrival at the tree. And indeed, our control beetles (ie, those not yet exposed to host plant tissue) show detectable expression of transcripts annotated as plant cell wall degrading enzymes. This is consistent with the attacking bark beetles needing to feed in their natal host and then bore through the outer bark to emerge for dispersal. The same enzymes that would aid in those predispersal tasks would be used during host colonization.

In the mountain pine beetle, the female is the pioneering sex (Latty and Reid, 2009) and is the first to arrive at a host tree to begin the process of excavating through the outer bark and phloem. As the first arrivers, they are more susceptible to harm from early constitutive defences (Latty and Reid, 2009). In addition to producing the cell wall degrading enzymes that allow them to overcome the host tree's physical barriers to attack, they also produce the initial components of the overall pheromone bouquet that will emanate from a tree under attack (Libbey et al., 1985; Pureswaran et al., 2000). Males and other females detect these emissions within the mix of the host volatiles released by female excavation and within the bark and phloem. Upon joining the aggregation, they release pheromone components, with the male bouquet differing from that of the female in composition and timing (Pureswaran et al., 2000). At some point in the host colonization process, when the insects detect that the tree has lost the ability to effectively defend itself, or when an optimum number of conspecifics are present in the aggregation, an antiaggregation pheromone—in this case verbenone—is produced (Hunt and Borden, 1990; Hunt et al., 1989; Pitman et al., 1969; Rudinsky et al., 1974; Ryker and Yandell, 1983). The dynamics of the physiological and biosynthetic processes of pheromone biosynthesis by bark beetles is outlined elsewhere in this volume. In our studies, transcript changes reflecting the shift in pheromone production within the first 24 h of beetle colonization was primarily evident in male beetles. We observed a male-specific increase in the expression of transcripts encoding two keys enzymes in the mevalonate pathway—3-hydroxy-3-methylglutaryl-CoA synthase and 3-hydroxy-3-methylglutaryl-CoA reductase. This suggests a physiological shift from the production of *exo*-brevicomin (Song et al., 2014a,b), a pheromone that attracts females (Conn et al., 1983; Rudinsky et al., 1974), to frontalin, an antiaggregation pheromone produced from isoprenyl precursors made by the mevalonate pathway (Keeling et al., 2013a,b,c). Thus a significant rise in the key enzymes in the mevalonate pathway supports the shift in male pheromone production during the first 24 h of host colonization from the production of chemicals to attract females to one that deters competition.

It should be noted that our work only covered the very first hours of host colonization, and that in a somewhat contrived situation in which females were forced into a particular location on the bark in cut bolts rather than standing trees, and where neither sex was able to exercise any mate choice (Pureswaran and Borden, 2005). In addition, all of the insects that we assayed for comparison against controls were in the presence of the opposite sex. Thus further detailed work could be done on long-term trends in gene regulation during the full course of a typical colonization event—for example, late colonization and the release of antiaggregation pheromone components—ideally in a more natural setting, along with gene regulation of the two sexes not in association with each other. As detailed in other chapters in this volume and elsewhere (eg, Keeling et al., 2013a,b,c; Song et al., 2014a,b), a great deal of this information is already known for genes specifically related to the main pheromone biosynthetic pathways, thanks to much previous and ongoing excellent work in this field, but there doubtless still many aspects that remain to be uncovered in this complex and dynamic phenomenon.

Upon breaching the barrier of the outer bark, females, and later the males who join them, encounter phloem saturated with secondary metabolites (Seybold et al., 2006). Phloem tissue contains resin ducts (Franceschi et al., 2005; Reid and Watson, 1966), some of which the insects may sever, often resulting in insects doused with resin and literally swimming in the sticky and toxic fluid (Keeling et al., 2006; Raffa and Berryman, 1982a,b). Again, because our experiments were limited to the first day of host colonization and because we were using freshly cut bolts (and hence likely only constitutive defences were present as cut trees could not respond to attacks), the effects that we saw should be seen as limited to the insects dealing with constitutive chemical defences. Many of the same enzymes may be operative in detoxification of both constitutive and induced secondary metabolites. First, while there are qualitative differences between plant constitutive and induced defences, including pine (Ott et al., 2011), many of the differences are qualitative in nature. Thus the insects deal with many of the same secondary metabolites—albeit in differing amounts—in the early constitutive phase as in the later induced phase. Second, some detoxification enzymes are known to function on a variety of related substrates (Feyereisen, 1999). A classic example of this is some of the cytochromes P450 in parsnip webworms that have been shown to each detoxify several of the furanocoumarins in their wild parsnip hosts (Li et al., 2004). Other examples of this sort exist—including in taxa as disparate from mountain pine beetles as humans (Danielson, 2002)—so it is likely that at least a subset of the proteins and unregulated genes that we detected in the early colonization phase are also part of the detoxification machinery in later colonization. One antennal-specific cytochrome P450 is known to use pine terpenoids as a substrate (Keeling et al., 2013a,b,c), so such activity is not unexpected elsewhere in the insect's body.

Mountain pine beetle adults do not often have second chances at mating once they have entered a host tree. There are likely exceptions to this, for instance D.H. and others have observed an early emergence of previous year adults that, particularly in large infestations, have the potential to colonize and reproduce in a second tree (and see Reid, 1958 and associated references). In addition, in other bark beetle species, such as *Ips pini* (Reid and Roitberg, 1995), males are known to abandon galleries after mating to search for more mating opportunities in the same year. In any case, it is likely that host colonizing mountain pine beetles are investing as much of their energy as possible in their reproductive opportunity and therefore any energy invested in the metabolism of toxins reduces reproductive output. Thus there is likely a balancing act between detoxification and excretion of byproducts, and overall tolerance of the host defence compounds.

Evidence for the tolerance of tree-produced toxins may be seen in the proliferation of a variety of detoxification and stress physiology-related gene regulation and protein products. In our work, 24 h after exposure to host tissue, the transcripts for cytochromes P450, transferases, esterases, ABC transporters, and alcohol dehydrogenases showed large and significant changes. Each of these could play a role in the detoxification of host tree defence compounds and enable the attacking beetles to complete their reproductive cycle. Cytochromes P450 are very common enzymes that are often implicated in the detoxification of a wide range of substrates in insects (Feyereisen, 1999; Li et al., 2004); several have been identified that play a role in the detoxification of terpenoids and in pheromone production in bark beetles (Aw et al., 2010; Keeling et al., 2006; Tittiger et al., 2005). In our study of mountain pine beetle adults during early host colonization, the gene transcripts encoding for cytochromes P450 showed a significant and varied change for both male and female beetles. Cytochromes P450 are notoriously difficult to assign a function based on sequence information (Feyereisen, 1999), but it is likely that several of the cytochromes P450 that were observed to increase following feeding, especially those associated with the midgut and fat body, play a role in the detoxification of host defence compounds. Similarly, glutathione-*S*-transferases and glucosyl/glucuronosyl transferases alter the solubility of substrates to facilitate excretion from the insect body, and they have been shown to function in concert with cytochromes P450 to detoxify xenobiotics (Sheenan et al., 2001). Both male and female beetles showed significantly increased expression of several transferases within 24 h of exposure to host tissue. An identified glutathione-*S*-transferase protein was also elevated in males relative to controls in our proteome data. In addition, transcripts annotated as esterases—enzymes implicated in insect resistance to insecticides (Li et al., 2007)—showed increased expression in female beetles. An ABC transporter, similar to those generally known for their role in sequestration of toxic compounds (Labbé et al., 2010), also increased significantly in females after exposure to host tissue in our experiments. Because a portion of toxic pine resin is comprised of terpenoid alcohols, the observed upregulation of transcripts annotated as alcohol dehydrogenases in both males and females may also play a role in aiding the adult beetle to survive long enough to mate, excavate galleries, and lay eggs.

The presence of stress physiology-related proteins and upregulation of such genes indicates the possibility of some secondary metabolites affecting vital biochemical processes by direct effect on proteins. While the specific effects of diverse conifer secondary metabolites on protein or other targets in bark beetles requires more investigation, we do know that some plant secondary metabolites specifically target protein function (Acamovic and Brooker, 2005). Some of the stress physiology-related proteins—heat shock proteins (Feder and Hofmann, 1999) and calreticulin (Fink, 1999), for example—that were strongly upregulated shortly after exposure to the host are known to specifically act to maintain correct protein folding. Other upregulated genes and protein products are known to be involved in protection against cellular oxidative damage—eg, superoxide dismutase (Lopez-Martinez et al., 2008)—again pointing to potential effects of secondary metabolites on portions of the cellular machinery.

It is possible that an upregulated stress response is present in colonizing bark beetles, not only in the context of stress related to exposure to host toxins but also to help them to deal with the effects of substantial flight exercise during dispersal. Our experiments did not subject the insects to precolonization flight, so if this is the case, then perhaps some stress-related proteins are upregulated immediately following flight due to exogenous or endogenous signals related to the initiation of host colonization. That is, a stress response related to investment in repair and maintenance of tissues following the dispersal flight might only be initiated upon the completion of successful host finding when bark beetle flight muscles are known to be catabolized for energy (Bhakthan et al., 1970; Reid, 1958; Sahota, 1975).

In any case, future work targeting the effects specific resin chemicals that are likely to interact with proteins would contribute to helping us understand tree defence and would even potentially reveal new and useful pharmaceuticals for human use. The stress response in host-attacking mountain pine beetles is, to our knowledge, not discussed elsewhere in the literature and certainly deserves further work that will lead to a better understanding of host colonization success and failure in this and related species.

While the presence of an upregulated stress response in host colonizing adults was somewhat surprising to us, even more surprising was the existence of what seems to be a substantial immune response. For example, transcripts annotated as C-type lectins and scavenger proteins were upregulated in both male and female beetles. These groups of enzymes are responsible for the identification of foreign cells slated for destruction by the immune system (Tanji et al., 2006). Transcripts annotated as having a role in the identification of bacterial invaders or in an antiviral response showed significant upregulation. Almost no information exists for the role of these transcripts in insects, and so the impact of the beetle's immune response during host colonization is an area that is currently wide open for study.

Just as a variety of predators and parasitoids feed on bark beetles, it is also highly likely that pathogens, spread prior to emergence from the host tree or encountered on the bark of potential hosts or during mating encounters, opportunistically or specifically infect adult bark beetles. Thus in addition to the stress of dealing with a dispersal flight and host tree defences, colonizing bark beetles could also face pathogenic agents that could reduce their lifespan and fecundity in what is likely their sole mating attempt. In addition, depending on the pathogen, an infected adult may spread disease to offspring that either affect their growth and development or that persists with them until adulthood to infect the next generation.

Immunity in bark beetles remains understudied, or more to the point, mainly unstudied in the fullest sense. Bark beetles, including mountain pine beetles, do interact with a variety of microorganisms. Of particular note are the symbiotic and pathogenic fungi (Reid et al., 1967; Whitney, 1971) that are often carried in specialized structures called mycangia. In an early large-scale proteomic study of southern pine beetle (*Dendroctonus frontalis*) mycangia proteome, Pechanova et al. (2008) found a large number of proteins with a variety of annotated functions associated with those structures. One overall message from that study was that the insects are not merely receptacles and vehicles for fungal spores but also likely play some as yet to be determined role in the life cycle and physiology of the microorganisms (Adams et al., 2013) that they carry. That is, while their symbiotic fungi doubtless affect bark beetle physiology and behaviour, it seems that the tight connection between the insect and the fungi is a two-way street. Immunological functions could play a role in this partnership by, for instance, regulating the phenology and activity of associated symbiotic or other microorganisms that are known to impact reproductive success of adults or their larvae as well (Therrien et al., 2015).

The role of bacteria and viruses in the reproductive success of mountain pine beetle is particularly interesting considering recent discoveries about vitellogenin (Salmela et al., 2015). In our work, we noted that vitellogenin was upregulated by 1400-fold (transcriptome) and 2.3-fold (proteome) in female mountain pine beetles almost immediately upon encountering host material. Work by Salmela et al. (2015) in honeybees has shown that vitellogenin binds to pathogens and to other metabolic indicators of pathogen presence. Upon being deposited into eggs, the vitellogenin carries parts of these pathogens with them, and thus has the potential to be the source for intergenerational immune priming in insects. If this is the case in bark beetles as in honeybees, then the sudden onset of vitellogenin synthesis in females only after the dispersal flight makes sense in an immunological context as that is when the female is most likely to be encountering pathogens that may be detrimental to her young.

Having found a host tree, entered it, encountered the tree defences and associated pathogens, deposited and perhaps helped to activate symbiotic fungi, mountain pine beetles begin to excavate an ovipositional gallery and lay eggs. There has been a substantial amount of research on adult condition as it relates to flight exercise and later reproductive output. Our results, particularly the massive upregulation of vitellogenin and its precursors, indicate that females do not commit energy to producing reproductive products (ie, provisioned eggs) until they are sure that they do not require further energy for dispersal. Other evidence that points toward this temporal partitioning of resource use come from past work on degradation of wing muscles following initiation of host colonization in some bark beetles (Reid, 1958). We observed a differential production of protein products involved in cytoskeletal and anatomical maintenance and rearrangement such as an increase in calponin (an enzyme that plays a role in the regulation of muscle contraction) with a simultaneous decrease in actinin and kinesin (proteins that play a role in the structure of muscle tissue) in adult males after exposure to host tree tissue.

Having now successfully deposited fertilized eggs, a pair of bark beetles has completed that portion of their life cycle. However unless their soon-to-hatch larvae also survive to make the same difficult journey, the work will have been in vain.

3. LARVAL OVERWINTERING

Mountain pine beetle eggs are deposited in small niches on the sides of the ovipositional gallery and hatch shortly thereafter. The very young larvae begin to mine through the phloem tissue, moving perpendicularly from the vertical ovipositional gallery and feeding as they go. While much attention has been paid to the tree defences encountered by the attacking parents, work in our research program has recently shown that the larvae encounter higher levels of some secondary metabolites than their parents (Clark et al., 2012). This makes sense in the context of constitutive and induced defensive responses by attacked hosts. When the parental generation first encounters the tree, they encounter only the constitutive defences.

The tree, detecting the attack, begins to synthesize and excrete induced resin and, in the case in other conifers (Nagy et al., 2000), modify its anatomy to more effectively permeate its affected tissues with secondary metabolites (Reid and Watson, 1966; Reid et al., 1967). This response is rapid enough that even though in a large scale mass attack the tree will still die, but prior to death, the phloem becomes highly saturated with resin compounds. These compounds are variably volatile and do not necessarily quickly dissipate from the tree. Some of fungi that were inoculated into the tree by the attacking adults are known to be able to detoxify some conifer secondary metabolites (Hesse-Orce et al., 2010), but growth of the fungi takes some time. Thus, larvae are born into a potentially toxic wasteland that has been left behind after the chemical warfare conducted by the host tree against their parents.

This presents a substantial problem for the larvae. While they hatch in the heat of summer, winter is always fast approaching; summer can cool to autumn temperatures prior to the equinox, and as the days shorten, a sudden cold snap could arrive at any time. The early instar larvae need to develop overwintering defences against early cold snaps and the later deep and extended cold of winter in just a few short weeks of feeding in host phloem, and they need to do this in an environment and food source that is saturated with toxins left over from their parents' battle with the tree.

It is therefore likely that detoxification of host secondary metabolites is a vital process in the larvae as well as in the adults. While the adults must cope with constitutive defences and the early onslaught of induced defences, once they have successfully deposited eggs in the gallery, their task is completed. And while some adults survive winter and may reemerge early the following year (D. Huber, personal observation) or even in the same year (Reid, 1958), most do not. As such, we expected to find substantial evidence of detoxification metabolism in larvae.

In the autumn larvae, no less than 11 transcripts annotated as cytochromes P450 increase significantly in expression between September and November. All of these show no change over the winter, with the exception of two that show a threefold decrease between November and March, and all show either no change or a significant decrease in spring. In addition, two glutathione-*S*-transferases, enzymes involved in the transfer of a glutathione moiety to potentially toxic chemicals entering the insect body in order to increase hydrophilicity and excretion, changed significantly over the winter in both the gene expression and proteomics study. In the transcriptomics study, two putative glutathione-*S*-transferases increased significantly in expression between September and November, and the expression of both genes decreased in the period between March and May (but the decrease is only significant for one). This significant decrease was echoed in the proteomics data for a detected glutathione-*S*-transferase in the spring.

In general, we observed maintenance of constitutive levels of proteins but increased expression of genes involved in detoxification activity during the course of the autumn—cytochromes P450, glutathione-*S*-transferases, and esterases. In the spring, we observed—with a few exceptions—decreases in the levels of protein representatives of these gene families across the course of the season. This indicates that by the time that we assayed larvae (late-September) they had already built up their defences against host toxins. This is logical, of course, because by this time, the larvae would have been feeding in the phloem tissue for several weeks or even several months. Not having the ability to feed and remove toxins would have proven fatal. On the other hand, the maintenance or drop in detoxification-related expression and gene products in the spring coincides with the likely drop in the levels of phloem-based secondary metabolites due to natural chemical and physical processes and the detoxification activity of beetle-associated symbiotic fungi. If, in fact, levels of secondary metabolites remaining in host tissue are lower in the spring (something that we have yet to test), then the maintenance of high levels of detoxification enzymes would represent a metabolic expense at a time when completing development in a timely manner for subsequent dispersal and reproduction is important.

Much like the adults, the larvae seem to be coping with and preparing for physiological stress during the autumn. Many of the same stress physiology-related genes and protein products that we saw expressed in host colonizing adults, plus some others, were either constitutively expressed in autumn larvae or upregulated in autumn larvae and downregulated in spring larvae. These include heat shock proteins, proteins associated with stressors such as heat, cold, and desiccation as they help stabilize proteins within the cell (Denlinger and Lee, 2010). Expression of two transcripts annotated as heat shock proteins shows a pattern of increase in autumn, stability over the winter, and then a significant decrease in spring. In addition, the largest fold change in expression that we observed for larvae between September and November was again annotated as a galactose-specific C-type lectin, an enzyme associated with immune response in insects (Tanji et al., 2006). This gene transcript increased over 100-fold in the autumn. This suggests that, not only are the developing larvae exposed to the threat of toxic host compounds and abiotic stressors but also to microbial pathogens.

Constitutively expressed stress physiology genes are likely to be, at least in part, responsible for helping the larvae to cope with resin toxins as it feeds. The various proteins and genes that showed increased levels of expression between September and November are more likely to be involved in dealing with oxidative stress and the presence of free radicals brought about by cold temperatures and subsequent physiological processes (eg, dehydration) that might be involved in the cold-hardening response (Denlinger and Lee, 2010). At the very least, we can say that the stress response in autumn mountain pine beetle larvae is real and substantial, although at this point, we do not know the specifics of its function. The drop in stress-related gene expression and protein levels in the spring, like the drop in detoxification metabolism, also makes sense if levels of secondary metabolites and substantial risk of deep cold have dissipated at this time. In terms of an energy budget, spring is the time for the larvae to spend their energy and any metabolic surplus left over from the winter on growth and development in order to be ready for their dispersal flight and subsequent chance at reproduction in the coming summer.

An obvious physiological shift that should be required for autumn mountain pine beetle larvae is the development of cold hardening. Mountain pine beetles are not freeze tolerant; that is, they cannot allow their bodily fluids to freeze. Instead, as with other freeze resistant insects, they employ several strategies, including the accumulation of antifreeze compounds in their haemolymph to reduce their supercooling point (Bentz and Mullins, 1999). The supercooling point is the point, below the normal freezing temperature of water, where the haemolymph will freeze, and mountain pine beetles are known to be able to reduce this to around −35 °C (Safranyik, 1998), and perhaps colder in some cases.

One of the main cryoprotectant compounds that has been observed in overwintering mountain pine beetles is glycerol (Bentz and Mullins, 1999), and our results showed an increase in the level of glycerol biosynthetic proteins (and related gene expression) in the autumn as well as a subsequent drop in the same proteins and expression in the spring. In one of the pathways leading to the production of glycerol, the transcripts for two key enzymes in the catalytic steps immediately preceding the synthesis of glycerol (glycerol-3-phosphate dehydrogenase and glycerol kinase) were significantly upregulated in autumn larvae between September and November. These same enzymes show a large drop (decreasing 10- and 6-fold, respectively) between samples of larvae taken in March and May the following spring.

These results closely match those of other work in our laboratory (Fraser, 2011) in which real time quantitative PCR was employed to monitor gene expression specific to most of the steps in glycerol biosynthesis from glycolysis onwards. The insects used in the overwintering proteomic and transcriptomic studies were from the same population (geographically and temporally) as those used in the RT-qPCR work. Those insects, collected near to Valemount, British Columbia, also represent a seeming transitional population in terms of their population genetics (Gayathri Samarasekera et al., 2012), although this was not known at the time of collection. Thus while the glycerol results coincide well with our expectations (ie, increases and decreases of cold-hardening gene expression in autumn and spring, respectively), it is possible that this trend is not going to be identical for all populations across the much larger range of this insect. While we chose our samples for analysis based on the temperature regime measured at the trees—using samples taken prior to and after the first substantial cold—it is not yet known if exogenous signals (temperature, light levels filtering through the bark, etc.) or endogenous signals (developmental, circadian/circannual physiology, etc.) or a combination are responsible for the timing of these processes.

Other cryoprotectant metabolism also showed similar autumn/spring shifts in gene expression. For instance protein levels of 2-deoxyglucose-6-phosphate phosphatase, an enzyme that produces 2-deoxyglucose, increased substantially in the autumn and decreased very dramatically in the spring. 2-Deoxyglucose is known to regulate glycolysis and cell growth (Barban and Schulze, 1961; Wan et al., 2004)—upstream reactions from glycerol biosynthesis—and could itself also be a cryoprotectant. α,α,-Trehalose-phosphate synthase also showed significant increases in protein level in the autumn and significant decreases in the spring. While trehalose generally makes up a large portion of the haemolymph carbohydrates in insects, it is also a cryoprotectant (Baust and Lee, 1983). From these results it is likely that glycerol is not the only cryoprotectant employed by mountain pine beetle larvae during autumn cold hardening.

Larvae that are able to survive and adequately feed in toxin-laden phloem in the autumn, that endure the stress of those toxins and of overwintering, and that produce enough cryoprotectants to avoid freezing in the depths of the winter must then continue their growth and development in anticipation of summer dispersal and reproduction. Gene expression shifts noticeably from detoxification, immune response, and stress response during the autumn to growth and development in during the spring thaw.

Of the transcripts that show greater than 10-fold decrease in expression in the autumn, eight are annotated as insect cuticle proteins. Conversely, in the spring, one of the transcripts showing over 1000-fold increase is annotated as a digestive enzyme in beetles, cathepsin L-like proteinase (Murdock et al., 1987). We found several dozen growth and development-related proteins in autumn larvae, but only a caprin homologue (cytoplasmic activation/proliferation-associated protein) showed a shift in levels between September and November (a small but significant decrease). We also found several dozen (a few more than in the spring) proteins in the same category in the spring, and in this case, a dozen of them showed significant increases between March and May and two others showed substantial and significant decreases during the same time period. In the latter case, the two enzymes had annotations similar to juvenile hormone epoxide hydrolase, indicating a likely shift in juvenile hormone titres during later development. Many of the transcripts that show large and significant increases in the spring are unannotated—a great deal of work needs to be done in order to understand the physiological processes occurring in mountain pine beetle as they come out of overwintering and begin the process of pupation and emergence.

A good summary of mountain pine beetle overwintering might be to say that autumn is a time of intense and frantic preparation for winter done within the still-hostile environment of the dead host. In the spring, the surviving larvae must complete development with access to what appears to be a more benign food source, and they reallocate their metabolic expenses away from stress physiology, detoxification, and cold hardening and towards development. In essence, just as with adult host colonization, this phase is a dynamic and active one for the insect in terms of physiology, and we have a great deal more to learn about the numerous processes that have been partially uncovered by this recent research.

4. IMPLICATIONS

Due to the economic and ecological impacts of mountain pine beetle activity, the volume of past research on this species has possibly made it one of the better known insects in the world. Following the massive outbreaks in central and northern British Columbia and parts of the United States during the turn of the current century, the concentration of available research funding, and thus effort, has revealed more than we could have imagined in a variety of biological (and beyond) contexts. The research and ideas outlined in this chapter, elsewhere in this book, and throughout the recent literature, provide massive amounts of information for guiding policy and management, as well as many new questions to be answered. Substantial funding and subsequent research effort by many excellent scientists was no accident—it was in large part due to the anticipation of movement of the beetle from its historical redoubt west of the Rocky Mountains in British Columbia into the boreal forests of Alberta.

Around 2006, the first established populations of mountain pine beetles were observed in lodgepole pine forests in the northeast corner of British Columbia, a region that is on the eastern slope of the Rocky Mountains and that had not previously been exposed to the insect (Janes et al., 2014). During the next several years the beetle moved through the eastern population of lodgepole pine, into a region in west-central Alberta where lodgepole and jack pine substantially hybridize, and from there into pure jack pine stands in central and east-central Alberta (Cullingham et al., 2011). In addition to this movement of the outbreak into a new host and a new geographical region to the east, the insect is also currently expanding its range further north in British Columbia and Alberta towards to pine forests of Yukon Territory and the Northwest Territories.

These range shifts are unprecedented in recorded history and we have no reason to believe that they do not represent a completely unique phenomenon, at least since the previous ice age. To put it succinctly, we seem to be witnessing the movement of an important North American species into new hosts and into a new range where it will doubtless experience a novel climate. Mountain pine beetle is a native species to this continent, but it also represents an invasive species on a (large) regional scale. A number of factors likely at play in this movement include climate change affecting both temperature regime and tree vitality and defence; human activity such as the suppression of fire and creation of large even-aged stands through logging practices; and year-by-year meteorological conditions during the flight period. These factors, and others, are reviewed and assessed elsewhere (eg, Anderegg et al., 2015; Janes et al., 2014; Keane et al., 2015; Sambaraju et al., 2012; Tobin et al., 2014).

The question then becomes, 'what can be done about this?' The most likely answer is, frankly, 'not much' if what is meant is direct control and eradication of the insect in its new range. If climate is the major driver of the phenomenon, then without immediate and radical shifts in fossil fuel use, the factors that may reduce or stop the spread of this insect in its new range are novel tree defences in the new host, stand structure as it affects dispersal in the new geographic range, or ideally timed and severe cold snaps over large land areas. Even with radical curtailment of carbon and other greenhouse gas emissions, a great deal of the future warming and other climatic change, particularly at the higher latitudes at which this portion of the mountain pine beetle infestation is active, is already inevitable.

There are three foreseeable and logical outcomes to the current spread of mountain pine beetle into jack pine forests, none of which are under direct human control. First, perhaps there will be a substantial cold snap in the new range in a near future winter while the beetle's foray into its new range is still not too extensive. Perhaps it will be deep cold in the mid-winter, or substantial cold in the mid-autumn or early spring when the larvae have reduced levels of resistance to cold or stress. The progression of climate change, however, likely means fewer deep cold events in North America (Screen et al., 2015), so the odds of climate-based solutions worsen each year, even while the infestation in the new range potentially covers more territory.

Second, perhaps the new jack pine host maintains defences that slow beetle population growth substantially, or perhaps the nutrients in jack pines are insufficient for growing larvae to produce appropriate overwintering defences. In terms of host defence in jack pine at this time, from what we know (Clark et al., 2014; Lusebrink et al, 2011; Taft et al., 2015), the trees are likely at best equally, and at worst, less defended from mountain pine beetles than are lodgepole pines. Beyond that, mountain pine beetles are able, in some circumstances, to successfully aggregate, mate, and reproduce even in some nonpines (Huber et al., 2009) demonstrating a substantial level of flexibility in host use. Almost every pine species in their large geographical range—other than a few exceptions—is susceptible to them (Furniss and Carolin, 1977; Gray et al., 2015). There is no reason to expect that jack pine, which easily hybridizes with lodgepole pine, will not be a usable host. We do still need to learn more about the mountain pine beetle in its new range, but at this point, it is reasonable to assume that there is potential for an outbreak in jack pine stands similar to the one that occurred in lodgepole pine in British Columbia and elsewhere.

Third, mountain pine beetle in the jack pine boreal forest may be the new normal. If that is the case, then we will have to learn to manage it there, as silviculturists have done in lodgepole pine forests for generations. Development of models (Shore and Safranyik, 1992) of susceptibility, risk, and spread based on mountain pine beetle physiology as it relates to cold tolerance and host colonization will be vital to managers in this new region (Aukema et al., 2008; Safranyik et al., 2010). Our proteomics and transcriptomics results provide new hints of the important aspects of mountain pine beetle physiology to monitor and study on a deeper level, and give a research into management and the development of new models a further leg-up as the insect and humans both move into uncharted territory.

5. CONCLUSION

Our first proteomic- and transcriptomic-based look at mountain pine beetle physiology, following the development of the genomic data required for its analysis, has shown that adult insects colonizing trees put a premium on detoxification, resistance to stress, immune function, and conservation of as much energy as possible for reproductive output. Larval insects feed in highly toxic tissues and seasonally adjust their stress physiology, detoxification machinery, and developmental processes in response to host toxins and the onset and departure of winter cold. During both life phases and stages we observe very dynamic gene expression and, presumably, concomitant physiological shifts. Our work has provided a large number of target genes and proteins that are now available for study related to mountain pine beetle physiology in the context of the identity, quantity, and timing of defences in a new host; climate in a new geographical range (depth of cold, timing of cold, interaction of winter onset, and host quality); and the potential encounter of new pathogens, or new interactions, that might further affect immune and stress responses.

We look forward to further work with some of these processes in our research program, and to hearing about work by others, as we move forward in understanding and managing this important, yet also very interesting, insect.

REFERENCES

Acamovic, T., Brooker, J., 2005. Biochemistry of plant secondary metabolites and their effects in animals. Proc. Nutr. Soc. 64, 403–412. http://dx.doi.org/10.1079/pns2005449.

Adams, A., Aylward, F., Adams, S., Erbilgin, N., Aukema, B., Currie, C., Suen, G., Raffa, K., 2013. Mountain pine beetles colonizing historical and naive host trees are associated with a bacterial community highly enriched in genes contributing to terpene metabolism. Appl. Environ. Microbiol. 79, 3468–3475. http://dx.doi.org/10.1128/aem.00068-13.

Anderegg, W.R.L., Hicke, J.A., Fisher, R.A., Allen, C.D., Aukema, J., Bentz, B., Hood, S., et al., 2015. Tree mortality from drought, insects, and their interactions in a changing climate. New Phytol. 208 (3), 674–683 http://dx.doi.org/10.1111/nph.13477.

Aukema, B., Carroll, A., Zheng, Y., Zhu, J., Raffa, K., Moore, R., Stahl, K., Taylor, S., 2008. Movement of outbreak populations of mountain pine beetle: influences of spatiotemporal patterns and climate. Ecography 31, 348–358. http://dx.doi.org/10.1111/j.2007.0906-7590.05453.x.

Aw, T., Schlauch, K., Keeling, C., Young, S., Bearfield, J., Blomquist, G., Tittiger, C., 2010. Functional genomics of mountain pine beetle (*Dendroctonus ponderosae*) midguts and fat bodies. BMC Genomics 11, 215. http://dx.doi.org/10.1186/1471-2164-11-215.

Barban, S., Shulze, H., 1961. The effects of 2-deoxyglucose on the growth and metabolism of cultured human cells. J. Biol. Chem. 236, 1887–1890.

Baust, J., Lee, R., 1983. Population differences in antifreeze/cryoprotectant accumulation patterns in an antarctic insect. Oikos 40, 120–124.

Bentz, B., Mullins, D., 1999. Ecology of mountain pine beetle (Coleoptera: Scolytidae) cold hardening in the intermountain west. Environ. Entomol. 28, 577–587. http://dx.doi.org/10.1093/ee/28.4.577.

Bhakthan, N.M.G., Borden, J.H., Nair, K.K., 1970. Fine structure of degenerating and regenerating flight muscles in a bark beetle, *Ips confusus* I. Degeneration. J. Cell Sci. 6, 807–819.

Billings, R., Gara, R., Hrutfiord, B., 1976. Influence of ponderosa pine resin volatiles on the response of dendroctonus ponderosae to synthetic *trans*-verbenol. Environ. Entomol. 5, 171–179. http://dx.doi.org/10.1093/ee/5.1.171.

Bohlmann, J., 2012. Pine terpenoid defences in the mountain pine beetle epidemic and in other conifer pest interactions: specialized enemies are eating holes into a diverse, dynamic and durable defence system. Tree Physiol. 32, 943–945. http://dx.doi.org/10.1093/treephys/tps065.

Bonnett, T., Robert, J., Pitt, C., Fraser, J., Keeling, C., Bohlmann, J., Huber, D., 2012. Global and comparative proteomic profiling of overwintering and developing mountain pine beetle, *Dendroctonus ponderosae* (Coleoptera: Curculionidae), larvae. Insect Biochem. Mol. Biol. 42, 890–901. http://dx.doi.org/10.1016/j.ibmb.2012.08.003.

Byers, J., 1989. Chemical ecology of bark beetles. Experientia 45, 271–283. http://dx.doi.org/10.1007/bf01951813.

Campbell, S., Borden, J., 2006. Integration of visual and olfactory cues of hosts and non-hosts by three bark beetles (Coleoptera: Scolytidae). Ecol. Entomol. 31, 437–449. http://dx.doi.org/10.1111/j.1365-2311.2006.00809.x.

Cerezke, H., 1995. Egg gallery, brood production, and adult characteristics of mountain pine beetle, *Dendroctonus ponderosae* Hopkins (Coleoptera: Scolytidae), in three pine hosts. Can. Entomol. 127, 955–965. http://dx.doi.org/10.4039/ent127955-6.

Clark, E., Huber, D., Carroll, A., 2012. The legacy of attack: implications of high phloem resin monoterpene levels in lodgepole pines following mass attack by mountain pine beetle, *Dendroctonus ponderosae* Hopkins. Environ. Entomol. 41, 392–398.

Clark, E., Pitt, C., Carroll, A., Lindgren, B., Huber, D., 2014. Comparison of lodgepole and jack pine resin chemistry: implications for range expansion by the mountain pine beetle, *Dendroctonus ponderosae* (Coleoptera: Curculionidae). PeerJ 2, e240. http://dx.doi.org/10.7717/peerj.240.

Cole, W., 1981. Some risks and causes of mortality in mountain pine beetle populations: a long-term analysis. Res. Popul. Ecol. 23, 116–144. http://dx.doi.org/10.1007/bf02514096.

Conn, J., Borden, J., Scott, B., Friskie, L., Pierce Jr., H., Oehlschlager, A., 1983. Semiochemicals for the mountain pine beetle, *Dendroctonus ponderosae* (Coleoptera: Scolytidae) in British Columbia: field trapping studies. Can. J. Forest Res. 13, 320–324. http://dx.doi.org/10.1139/x83-045.

Cullingham, C., Cooke, J., Dang, S., Davis, C., Cooke, B., Coltman, D., 2011. Mountain pine beetle host-range expansion threatens the boreal forest. Mol. Ecol. 20, 2157–2171. http://dx.doi.org/10.1111/j.1365-294x.2011.05086.x.

Danielson, P., 2002. The Cytochrome P450 superfamily: biochemistry, evolution and drug metabolism in humans. Curr. Drug Metab. 3, 561–597. http://dx.doi.org/10.2174/1389200023337054.

Denlinger, D., Lee, R., 2010. Low Temperature Biology of Insects. Cambridge University Press, Cambridge.

Feder, M., Hofmann, G., 1999. Heat-shock proteins, molecular chaperones, and the stress response: evolutionary and ecological physiology. Annu. Rev. Physiol. 61, 243–282. http://dx.doi.org/10.1146/annurev.physiol.61.1.243.

Feyereisen, R., 1999. Insect P450 enzymes. Annu. Rev. Entomol. 44, 507–533. http://dx.doi.org/10.1146/annurev.ento.44.1.507.

Fink, A., 1999. Chaperone-mediated protein folding. Physiol. Rev. 79, 425–449.

Franceschi, V., Krokene, P., Christiansen, E., Krekling, T., 2005. Anatomical and chemical defenses of conifer bark against bark beetles and other pests. New Phytol. 167, 353–376. http://dx.doi.org/10.1111/j.1469-8137.2005.01436.x.

Fraser, J., 2011. Cold Tolerance and Seasonal Gene Expression in *Dendroctonus ponderosae*. M.Sc. thesis, University of Northern British Columbia.

Furniss, R., Carolin, V., Keen, F., 1977. Western Forest Insects. U.S. Dept. of Agriculture, Forest Service, Washington.

Gayathri Samarasekera, G., Bartell, N., Lindgren, B., Cooke, J., Davis, C., James, P., Coltman, D., Mock, K., Murray, B., 2012. Spatial genetic structure of the mountain pine beetle (*Dendroctonus ponderosae*) outbreak in western Canada: historical patterns and contemporary dispersal. Mol. Ecol. 21, 2931–2948. http://dx.doi.org/10.1111/j.1365-294x.2012.05587.x.

Gray, C., Runyon, J., Jenkins, M., Giunta, A., 2015. Mountain pine beetles use volatile cues to locate host limber pine and avoid non-host Great Basin bristlecone pine. PLoS One 10, e0135752. http://dx.doi.org/10.1371/journal.pone.0135752.

Hesse-Orce, U., Diguistini, S., Keeling, C., Wang, Y., Li, M., Henderson, H., Docking, R., Liao, N., Robertson, G., Holt, R., Jones, S., Bohlmann, J., Breuil, C., 2010. Gene discovery for the bark beetle-vectored fungal tree pathogen *Grosmannia clavigera*. BMC Genomics 11, 536. http://dx.doi.org/10.1186/1471-2164-11-536.

Hicke, J.A., Logan, J.A., Powell, J., Ojima, D.S., 2006. Changing temperatures influence suitability for modeled mountain pine beetle (*Dendroctonus ponderosae*) outbreaks in the western United States. J. Geophys. Res. 111, G02019. http://dx.doi.org/10.1029/2005jg000101.

Huber, D., Borden, J., 2001. Protection of lodgepole pines from mass attack by mountain pine beetle, *Dendroctonus ponderosae*, with nonhost angiosperm volatiles and verbenone. Entomol. Exp. Appl. 99, 131–141. http://dx.doi.org/10.1046/j.1570-7458.2001.00811.x.

Huber, D., Borden, J., 2003. Comparative behavioural responses of *Dryocoetes confusus* Swaine, *Dendroctonus rufipennis* (Kirby), and *Dendroctonus ponderosae* Hopkins (Coleoptera: Scolytidae) to angiosperm tree bark volatiles. Environ. Entomol. 32, 742–751. http://dx.doi.org/10.1603/0046-225x-32.4.742.

Huber, D., Gries, R., Borden, J., Pierce Jr., H., 2000. A survey of antennal responses by five species of coniferophagous bark beetles (Coleoptera: Scolytidae) to bark volatiles of six species of angiosperm trees. Chemoecology 10, 103–113. http://dx.doi.org/10.1007/pl00001811.

Huber, D., Aukema, B., Hodgkinson, R., Lindgren, B., 2009. Successful colonization, reproduction, and new generation emergence in live interior hybrid spruce *Picea engelmannii* × *glauca* by mountain pine beetle *Dendroctonus ponderosae*. Agric. For. Entomol. 11, 83–89. http://dx.doi.org/10.1111/j.1461-9563.2008.00411.x.

Hunt, D., Borden, J., 1990. Conversion of verbenols to verbenone by yeasts isolated from *Dendroctonus ponderosae* (Coleoptera: Scolytidae). J. Chem. Ecol. 16, 1385–1397. http://dx.doi.org/10.1007/bf01021034.

Hunt, D., Borden, J., Lindgren, B., Gries, G., 1989. The role of autoxidation of α-pinene in the production of pheromones of *Dendroctonus ponderosae* (Coleoptera: Scolytidae). Can. J. Forest Res. 19, 1275–1282. http://dx.doi.org/10.1139/x89-194.

Hynum, B., Berryman, A., 1980. *Dendroctonus ponderosae* (Coleoptera: Scolytidae): pre-aggregation landing and gallery initiation on lodgepole pine. Can. Entomol. 112, 185–191. http://dx.doi.org/10.4039/ent112185-2.

Janes, J., Li, Y., Keeling, C., Yuen, M., Boone, C., Cooke, J., Bohlmann, J., Huber, D., Murray, B., Coltman, D., Sperling, F., 2014. How the mountain pine beetle (*Dendroctonus ponderosae*) breached the Canadian rocky mountains. Mol. Biol. Evol. 31, 1803–1815. http://dx.doi.org/10.1093/molbev/msu135.

Karban, R., Myers, J.H., 1989. Induced plant responses to herbivory. Annu. Rev. Ecol. Syst. 331–348.

Keane, R.E., Loehman, R., Clark, J., Smithwick, E.A.H., Miller, C., 2015. Exploring interactions among multiple disturbance agents in forest landscapes: simulating effects of fire, beetles, and disease under climate change. In: Simulation Modeling of Forest Landscape Disturbances. Springer International Publishing, pp. 201–231.

Keeling, C., Bearfield, J., Young, S., Blomquist, G., Tittiger, C., 2006. Effects of juvenile hormone on gene expression in the pheromone-producing midgut of the pine engraver beetle, *Ips pini*. Insect Mol. Biol. 15, 207–216. http://dx.doi.org/10.1111/j.1365-2583.2006.00629.x.

Keeling, C., Henderson, H., Li, M., Yuen, M., Clark, E., Fraser, J., Huber, D., Liao, N., Roderick Docking, T., Birol, I., Chan, S., Taylor, G., Palmquist, D., Jones, S., Bohlmann, J., 2012. Transcriptome and full-length cDNA resources for the mountain pine beetle, *Dendroctonus ponderosae* Hopkins, a major insect pest of pine forests. Insect Biochem. Mol. Biol. 42, 525–536. http://dx.doi.org/10.1016/j.ibmb.2012.03.010.

Keeling, C., Chiu, C., Aw, T., Li, M., Henderson, H., Tittiger, C., Weng, H., Blomquist, G., Bohlmann, J., 2013a. Frontalin pheromone biosynthesis in the mountain pine beetle, *Dendroctonus ponderosae*, and the role of isoprenyl diphosphate synthases. Proc. Natl. Acad. Sci. 110, 18838–18843. http://dx.doi.org/10.1073/pnas.1316498110.

Keeling, C., Henderson, H., Li, M., Dullat, H., Ohnishi, T., Bohlmann, J., 2013b. CYP345E2, an antenna-specific cytochrome P450 from the mountain pine beetle, *Dendroctonus ponderosae* Hopkins, catalyses the oxidation of pine host monoterpene volatiles. Insect Biochem. Mol. Biol. 43, 1142–1151. http://dx.doi.org/10.1016/j.ibmb.2013.10.001.

Keeling, C., Yuen, M., Liao, N., Roderick Docking, T., Chan, S., Taylor, G., Palmquist, D., Jackman, S., Nguyen, A., Li, M., Henderson, H., Janes, J., Zhao, Y., Pandoh, P., Moore, R., Sperling, F., Huber, D., Birol, I., Jones, S., Bohlmann, J., 2013c. Draft genome of the mountain pine beetle, *Dendroctonus ponderosae* Hopkins, a major forest pest. Genome Biol. 14, R27. http://dx.doi.org/10.1186/gb-2013-14-3-r27.

Labbé, R., Caveney, S., Donly, C., 2010. Genetic analysis of the xenobiotic resistance-associated ABC gene subfamilies of the Lepidoptera. Insect Mol. Biol. 20, 243–256. http://dx.doi.org/10.1111/j.1365-2583.2010.01064.x.

Latty, T., Reid, M., 2009. First in line or first in time? Effects of settlement order and arrival date on reproduction in a group-living beetle *Dendroctonus ponderosae*. J. Anim. Ecol. 78, 549–555. http://dx.doi.org/10.1111/j.1365-2656.2009.01529.x.

Li, X., Baudry, J., Berenbaum, M., Schuler, M., 2004. Structural and functional divergence of insect CYP6B proteins: from specialist to generalist cytochrome P450. Proc. Natl. Acad. Sci. 101, 2939–2944. http://dx.doi.org/10.1073/pnas.0308691101.

Li, X., Schuler, M., Berenbaum, M., 2007. Molecular mechanisms of metabolic resistance to synthetic and natural xenobiotics. Annu. Rev. Entomol. 52, 231–253. http://dx.doi.org/10.1146/annurev.ento.51.110104.151104.

Libbey, L., Ryker, L., Yandell, K., 1985. Laboratory and field studies of volatiles released by *Dendroctonus ponderosae* Hopkins (Coleoptera, Scolytidae). Z. Angew. Entomol. 100, 381–392. http://dx.doi.org/10.1111/j.1439-0418.1985.tb02795.x.

Lindgren, B., Borden, J., 1993. Displacement and aggregation of mountain pine beetles, *Dendroctonus ponderosae* (Coleoptera: Scolytidae), in response to their antiaggregation and aggregation pheromones. Can. J. Forest Res. 23, 286–290. http://dx.doi.org/10.1139/x93-038.

Lopez-Martinez, G., Elnitsky, M., Benoit, J., Lee, R., Denlinger, D., 2008. High resistance to oxidative damage in the Antarctic midge *Belgica antarctica*, and developmentally linked expression of genes encoding superoxide dismutase, catalase and heat shock proteins. Insect Biochem. Mol. Biol. 38, 796–804. http://dx.doi.org/10.1016/j.ibmb.2008.05.006.

Lusebrink, I., Evenden, M., Blanchet, F., Cooke, J., Erbilgin, N., 2011. Effect of water stress and fungal inoculation on monoterpene emission from an historical and a new pine host of the mountain pine beetle. J. Chem. Ecol. 37, 1013–1026. http://dx.doi.org/10.1007/s10886-011-0008-3.

Murdock, L., Brookhart, G., Dunn, P., Foard, D., Kelley, S., Kitch, L., Shade, R., Shukle, R., Wolfson, J., 1987. Cysteine digestive proteinases in Coleoptera. Comp. Biochem. Physiol. B 87, 783–787. http://dx.doi.org/10.1016/0305-0491(87)90388-9.

Nagy, N., Franceschi, V., Solheim, H., Krekling, T., Christiansen, E., 2000. Wound-induced traumatic resin duct development in stems of norway spruce (Pinaceae): anatomy and cytochemical traits. Am. J. Bot. 87, 302. http://dx.doi.org/10.2307/2656626.

Ott, D., Yanchuk, A., Huber, D., Wallin, K., 2011. Genetic variation of lodgepole Pine, *Pinus contorta* var. *latifolia*, chemical and physical defenses that affect mountain pine beetle, *Dendroctonus ponderosae*, attack and tree mortality. J. Chem. Ecol. 37, 1002–1012. http://dx.doi.org/10.1007/s10886-011-0003-8.

Pauchet, Y., Wilkinson, P., Chauhan, R., ffrench-Constant, R., 2010. Diversity of beetle genes encoding novel plant cell wall degrading enzymes. PLoS One 5, e15635. http://dx.doi.org/10.1371/journal.pone.0015635.

Pechanova, O., Stone, W., Monroe, W., Nebeker, T., Klepzig, K., Yuceer, C., 2008. Global and comparative protein profiles of the pronotum of the southern pine beetle, *Dendroctonus frontalis*. Insect Mol. Biol. 17, 261–277. http://dx.doi.org/10.1111/j.1365-2583.2008.00801.x.

Pitman, G., Vité, J., 1969. Aggregation behavior of *Dendroctonus ponderosae* (Coleoptera: Scolytidae) in response to chemical messengers. Can. Entomol. 101, 143–149. http://dx.doi.org/10.4039/ent101143-2.

Pitman, G., Vité, J., Kinzer, G., Fentiman, A., 1969. Specificity of population-aggregating pheromones in *Dendroctonus*. J. Insect Physiol. 15, 363–366. http://dx.doi.org/10.1016/0022-1910(69)90282-0.

Pitt, C., Robert, J., Bonnett, T., Keeling, C., Bohlmann, J., Huber, D., 2014. Proteomics indicators of the rapidly shifting physiology from whole mountain pine beetle, *Dendroctonus ponderosae* (Coleoptera: Curculionidae), adults during early host colonization. PLoS One 9, e110673. http://dx.doi.org/10.1371/journal.pone.0110673.

Pureswaran, D., Borden, J., 2005. Primary attraction and kairomonal host discrimination in three species of *Dendroctonus* (Coleoptera: Scolytidae). Agric. For. Entomol. 7, 219–230. http://dx.doi.org/10.1111/j.1461-9555.2005.00264.x.

Pureswaran, D., Gries, R., Borden, J., Pierce Jr., H., 2000. Dynamics of pheromone production and communication in the mountain pine beetle, *Dendroctonus ponderosae* Hopkins, and the pine engraver, *Ips pini* (Say) (Coleoptera: Scolytidae). Chemoecology 10, 153–168. http://dx.doi.org/10.1007/pl00001818.

Pureswaran, D., Gries, R., Borden, J., 2004. Antennal responses of four species of tree-killing bark beetles (Coleoptera: Scolytidae) to volatiles collected from beetles, and their host and nonhost conifers. Chemoecology 14, 59–66. http://dx.doi.org/10.1007/s00049-003-0261-1.

Raffa, K., Berryman, A., 1982a. Gustatory cues in the orientation of *Dendroctonus ponderosae* (Coleoptera: Scolytidae) to host trees. Can. Entomol. 114, 97–104. http://dx.doi.org/10.4039/ent11497-2.

Raffa, K., Berryman, A., 1982b. Physiological differences between lodgepole pines resistant and susceptible to the mountain pine beetle and associated microorganisms. Environ. Entomol. 11, 486–492. http://dx.doi.org/10.1093/ee/11.2.486.

Raffa, K., Berryman, A., 1983. The role of host plant resistance in the colonization behavior and ecology of bark beetles (Coleoptera: Scolytidae). Ecol. Monogr. 53, 27. http://dx.doi.org/10.2307/1942586.

Raffa, K., Aukema, B., Erbilgin, N., Klepzig, K., Wallin, K., 2005. Interactions among conifer terpenoids and bark beetles across multiple levels of scale: an attempt to understand links between population patterns and physiological processes. Recent Adv. Phytochem. 39, 80–118.

Régnière, J., Bentz, B., 2007. Modeling cold tolerance in the mountain pine beetle, *Dendroctonus ponderosae*. J. Insect Physiol. 53, 559–572. http://dx.doi.org/10.1016/j.jinsphys.2007.02.007.

Reid, R., 1958. Internal changes in the female mountain pine beetle, *Dendroctonus monticolae* hopk., associated with egg laying and flight. Can. Entomol. 90, 464–468. http://dx.doi.org/10.4039/ent90464-8.

Reid, M., Roitberg, B., 1995. Effects of body size on investment in individual broods by male pine engravers (Coleoptera: Scolytidae). Can. J. Zool. 73, 1396–1401. http://dx.doi.org/10.1139/z95-164.

Reid, R., Watson, J., 1966. Sizes, distributions, and numbers of vertical resin ducts in lodgepole pine. Can. J. Bot. 44, 519–525. http://dx.doi.org/10.1139/b66-062.

Reid, R., Whitney, H., Watson, J., 1967. Reactions of lodgepole pine to attack by Dendroctonus ponderosae hopkins and blue stain fungi. Can. J. Bot. 45, 1115–1126. http://dx.doi.org/10.1139/b67-116.

Robert, J., Pitt, C., Bonnett, T., Yuen, M., Keeling, C., Bohlmann, J., Huber, D., 2013. Disentangling detoxification: gene expression analysis of feeding mountain pine beetle illuminates molecular-level host chemical defense detoxification mechanisms. PLoS One 8, e77777. http://dx.doi.org/10.1371/journal.pone.0077777.

Rudinsky, J., Morgan, M., Libbey, L., Putnam, T., 1974. Antiaggregative-rivalry pheromone of the mountain pine beetle, and a new arrestant of the southern pine beetle. Environ. Entomol. 3, 90–98. http://dx.doi.org/10.1093/ee/3.1.90.

Ryker, L., Yandell, K., 1983. Effect of verbenone on aggregation of *Dendroctonus ponderosae* Hopkins (Coleoptera, Scolytidae) to synthetic attractant. Z. Angew. Entomol. 96, 452–459. http://dx.doi.org/10.1111/j.1439-0418.1983.tb03698.x.

Safranyik, L., 1998. Mortality of mountain pine beetle larvae, *Dendroctonus ponderosae* (Coleoptera: Scolytidae) in logs of lodgepole pine (*Pinus contorta* var. *latifolia*) at constant low temperatures. J. Entomol. Soc. British Columbia 95, 81–88.

Safranyik, L., Carroll, A.L., 2006. The biology and epidemiology of the mountain pine beetle in lodgepole pine forests. In: Safranyik, L., Wilson, W.R. (Eds.), The Mountain Pine Beetle: A Synthesis of Biology, Management, and Impacts on Lodgepole Pine. Natural Resources Canada, Canadian Forest Service, Pacific Forestry Centre, Victoria, British Columbia, pp. 3–66 (Chapter 1), 304 p.

Safranyik, L., Linton, D., Silversides, R., McMullen, L., 1992. Dispersal of released mountain pine beetles under the canopy of a mature lodgepole pine stand. J. Appl. Entomol. 113, 441–450. http://dx.doi.org/10.1111/j.1439-0418.1992.tb00687.x.

Safranyik, L., Carroll, A., Régnière, J., Langor, D., Riel, W., Shore, T., Peter, B., Cooke, B., Nealis, V., Taylor, S., 2010. Potential for range expansion of mountain pine beetle into the boreal forest of North America. Can. Entomol. 142, 415–442. http://dx.doi.org/10.4039/n08-cpa01.

Sahota, T.S., 1975. Effect of juvenile hormone on acid phosphatases in the degenerating flight muscles of the Douglas-fir beetle, *Dendroctonus pseudotsugae*. J. Insect Physiol. 21, 471–478.
Salmela, H., Amdam, G., Freitak, D., 2015. Transfer of immunity from mother to offspring is mediated via egg-yolk protein vitellogenin. PLoS Pathog. 11, e1005015. http://dx.doi.org/10.1371/journal.ppat.1005015.
Sambaraju, K., Carroll, A., Zhu, J., Stahl, K., Moore, R., Aukema, B., 2012. Climate change could alter the distribution of mountain pine beetle outbreaks in western Canada. Ecography 35, 211–223. http://dx.doi.org/10.1111/j.1600-0587.2011.06847.x.
Screen, J., Deser, C., Sun, L., 2015. Reduced risk of North American cold extremes due to continued arctic sea ice loss. Bull. Am. Meteorol. Soc. 96, 1489–1503. http://dx.doi.org/10.1175/bams-d-14-00185.1.
Seybold, S., Huber, D., Lee, J., Graves, A., Bohlmann, J., 2006. Pine monoterpenes and pine bark beetles: a marriage of convenience for defense and chemical communication. Phytochem. Rev. 5, 143–178. http://dx.doi.org/10.1007/s11101-006-9002-8.
Sheehan, D., Meade, G., Foley, V., Dowd, C., 2001. Structure, function and evolution of glutathione transferases: implications for classification of non-mammalian members of an ancient enzyme superfamily. Biochem. J. 360, 1–16. http://dx.doi.org/10.1042/bj3600001.
Shepherd, R., 1966. Factors influencing the orientation and rates of activity of *Dendroctonus ponderosae* Hopkins (Coleoptera: Scolytidae). Can. Entomol. 98, 507–518. http://dx.doi.org/10.4039/ent98507-5.
Shore, T., Safranyik, L., 1992. Susceptibility and Risk Rating Systems for the Mountain Pine Beetle in Lodgepole Pine Stands. Forestry Canada, Pacific Forestry Centre, Victoria, BC.
Song, M., Delaplain, P., Nguyen, T., Liu, X., Wickenberg, L., Jeffrey, C., Blomquist, G., Tittiger, C., 2014a. *exo*-Brevicomin biosynthetic pathway enzymes from the mountain pine beetle, *Dendroctonus ponderosae*. Insect Biochem. Mol. Biol. 53, 73–80. http://dx.doi.org/10.1016/j.ibmb.2014.08.002.
Song, M., Gorzalski, A., Nguyen, T., Liu, X., Jeffrey, C., Blomquist, G., Tittiger, C., 2014b. *exo*-Brevicomin biosynthesis in the fat body of the mountain pine beetle, *Dendroctonus ponderosae*. J. Chem. Ecol. 40, 181–189. http://dx.doi.org/10.1007/s10886-014-0381-9.
Taft, S., Najar, A., Erbilgin, N., 2015. Pheromone production by an invasive bark beetle varies with monoterpene composition of its naïve host. J. Chem. Ecol. 41, 540–549. http://dx.doi.org/10.1007/s10886-015-0590-x.
Tanji, T., Ohashi-Kobayashi, A., Natori, S., 2006. Participation of a galactose-specific C-type lectin in *Drosophila* immunity. Biochem. J. 396, 127–138. http://dx.doi.org/10.1042/bj20051921.
Therrien, J., Mason, C., Cale, J., Adams, A., Aukema, B., Currie, C., Raffa, K., Erbilgin, N., 2015. Bacteria influence mountain pine beetle brood development through interactions with symbiotic and antagonistic fungi: implications for climate-driven host range expansion. Oecologia 179, 467–485. http://dx.doi.org/10.1007/s00442-015-3356-9.
Tittiger, C., Keeling, C., Blomquist, G., 2005. Some insights into the remarkable metabolism of the bark beetle midgut. Recent Adv. Phytochem. 39, 57–78.
Tobin, P., Parry, D., Aukema, B., 2014. The influence of climate change on insect invasions in temperate forest ecosystems. In: Fenning, T. (Ed.), Challenges and Opportunities for the World's Forests in the 21st Century. Springer, The Netherlands, pp. 267–293.
Wan, R., Camandola, S., Mattson, M., 2004. Dietary supplementation with 2-deoxy-D-glucose improves cardiovascular and neuroendocrine stress adaptation in rats. Am. J. Physiol. Heart Circ. Physiol. 287, H1186–H1193. http://dx.doi.org/10.1152/ajpheart.00932.2003.
Whitney, H., 1971. Association of *Dendroctonus ponderosae* (Coleoptera: Scolytidae) with blue stain fungi and yeasts during brood development in lodgepole pine. Can. Entomol. 103, 1495–1503. http://dx.doi.org/10.4039/ent1031495-11.

Wittstock, U., Gershenzon, J., 2002. Constitutive plant toxins and their role in defense against herbivores and pathogens. Curr. Opin. Plant Biol. 5, 300–307. http://dx.doi.org/10.1016/s1369-5266(02)00264-9.

Wood, D., 1982. The role of pheromones, kairomones, and allomones in the host selection and colonization behavior of bark beetles. Annu. Rev. Entomol. 27, 411–446. http://dx.doi.org/10.1146/annurev.en.27.010182.002211.

CHAPTER FOUR

Semiochemicals in the Natural History of Southern Pine Beetle *Dendroctonus frontalis* Zimmermann and Their Role in Pest Management

B.T. Sullivan
USDA Forest Service, Southern Research Station, Pineville, LA, United States

Contents

Advances in Insect Physiology, Volume 50
ISSN 0065-2806
http://dx.doi.org/10.1016/bs.aiip.2015.12.002

Abstract

The southern pine beetle, *Dendroctonus frontalis* Zimmermann is generally considered to be one of the most significant biotic mortality agents of pines within North America, with a range stretching from New England to eastern Texas and from Arizona south to Nicaragua. As with other aggressive pine beetles, it relies on semiochemicals for coordinating the mass attacks necessary for colonization of healthy pines. Over the past 50 years its chemical ecology has received extensive study aimed at development of effective and practical semiochemical-based management strategies which might replace the destructive and costly techniques in practice. I review the literature on the chemical ecology of this insect with particular attention to the functional categorizations assigned to different semiochemicals and the data underlying these assignments. Additionally, I attempt to identify conflicts and knowledge gaps within current understanding of the chemical ecology of this insect that might represent a significant hindrance to progress in development of effective semiochemical-based management strategies.

1. INTRODUCTION

1.1 Bark Beetle Population Dynamics

Pine beetles, as do all bark beetles, feed and reproduce within the bark (particularly the phloem) of their host trees and most species, including the minority tree-killing or "aggressive" species, generally occur at low densities in the environment. They infest trees already killed or seriously weakened by other factors, such as disease, fire, windthrow, harvesting activities, climatic stresses, and damage from other species of insects. Under such conditions, their influences are considered to be generally positive relative to human concerns (eg, initiating the processes of breakdown of coarse woody debris and associated nutrient cycles; removal of unthrifty trees). These beetles normally have no impact on healthy pines, since such trees are able to generate a sufficient resin defensive response to kill or eject small numbers of beetles that may penetrate the bark and breach the resin ducts of the living host tissue. However, in the case of the aggressive, tree-killing bark beetle species, populations may reach outbreak densities sufficient for overwhelming host defences and permitting unimpeded colonization of bark tissue of entirely healthy, vigorous trees (Coulson, 1979). Tree death is assured by the girdling of the phloem tissue by the beetles' mining activities as well as the mass inoculation of weakly pathogenic fungi that the beetles carry into the host. Thus aggressive bark beetle species live double lives; persisting almost invisibly as scavengers in the environment when at low population densities but killing vast acreages of host trees as predators during outbreaks (Martinson et al., 2013; Raffa et al., 1993). The mass attacks required to kill a healthy tree are initiated, sustained, and potentially terminated through the beetles' pheromones (Byers, 1989b; Wood, 1982a), and semiochemicals are an essential mechanism underlying the epidemiology of these disturbance agents.

Host colonization by aggressive bark beetles has been divided into at least four stages: beetle dispersal from natal host material, selection of a host, population concentration on a host, and establishment within and commencement of reproduction by beetles in the host tissue (Raffa et al., 1993; Wood, 1982a). These categories apply to the spatial scale of a single tree; however, the most significant harm done by aggressive bark beetles occurs when mortality "spills-over" from one tree undergoing mass attack to those adjacent as these trees in turn become the new foci for attacks (Bentz et al., 1996; Powell et al., 1998; Renwick and Vité, 1970). The focus of beetle attacks moves typically from the initially attacked tree (possibly made susceptible to attack by fire/mechanical injury/disease-compromised defences) to the closest adjacent trees; these may be quite vigorous but nonetheless insufficiently defended from the overwhelming numbers of attacking beetles drawn-in by the pheromone of beetles attacking the original tree (Coulson, 1979; Coulson et al., 1985; Fargo et al., 1985; Schowalter et al., 1981). As this process is conducted from tree-to-tree, the ultimate result is a patch of contiguous killed hosts called an "infestation" or a "spot" (Ayres et al., 2011). It is through this pheromone-driven process called "switching" that aggressive pine beetles inflict most of their mortality on healthy trees and generate most of their economic and environmental impact (Bentz et al., 1996; Renwick and Vité, 1970; Schlyter et al., 1987a).

1.2 *Dendroctonus frontalis* and Idiosyncratic Aspects of its Population Dynamics

The southern pine beetle, *Dendroctonus frontalis* Zimmermann is generally considered the most economically important pest and biotic disturbance agent of pines (*Pinus* L.) within the southeastern United States (Clarke and Nowak, 2010; Pye et al., 2011; Tchakerian and Coulson, 2011), and ranges from Massachusetts to Texas, United States, and from Arizona, United States, south to Nicaragua. It will attack all species of pine within its range. It is a highly aggressive species and must kill its hosts in order to reproduce, and only occasionally is it found to attack prostrate hosts (Dixon and Osgood, 1961; Moser, 1987). As with other aggressive species of bark beetle, its status as a major pest arises from its capacity to kill healthy trees and be the primary agent of tree mortality (Wood, 1982b).

Dendroctonus frontalis can have five to seven generations per year that last from 26 to 54 days (Birt, 2011; Fronk, 1947; Hain et al., 2011) in the southern part of its range, and there is much overlap of generations particularly as summer progresses (Franklin, 1970). *Dendroctonus frontalis* is distinctive from most other aggressive species of *Dendroctonus* pine beetles in North America in that the growth of individual spots can continue for months and cover hundreds of hectares during a single season if unchecked (Clarke and Billings, 2003; Schowalter et al., 1981). Because of the short generation time of this species, it is possible for mass attacked and therefore aggregation pheromone-producing trees to still be present within a spot when the first generation of brood emerge (Franklin, 1970). It has been shown that emerging brood beetles tend to remain concentrated in the area where they emerge if a source of aggregation pheromone is present; otherwise they disperse into the surrounding forest (Cronin et al., 1999; Gara, 1967). Thus a high proportion of these emerging brood fly to the nearby mass-attacked trees [typically less than 50 m away (Ayres et al., 2011)] and continue the process of mass attack and host switching initiated by their parents (Franklin, 1970). Unlike many bark beetles (Bennett and Borden, 1971; Byers, 1989b), *D. frontalis* requires little if any flight exercise before they are responsive to their aggregation pheromone (Andryszak et al., 1982), a demand that would otherwise promote dispersal from an infestation. This behaviour starts a cycle of brood emergence and participation in the mass attack and switching at the growing infestation's "head" (ie, the zone of the infestation where mass attack is occurring and concentrations of aggregation pheromones are highest). In general, *D. frontalis* infestations can continue to grow by this process as long as suitable host pines are close enough to the infestation's head to allow switching to occur and there is sufficient replacement of parent beetles by their brood or beetles recruited from outside the infestation (Ayres et al., 2011). The infestations tend to grow in a single, predominantly downwind, direction (Coster et al., 1978). Because persisting infestation growth is a trait not shared with other major pine beetle pests in North America, procedures for direct control of beetle damage are also quite different for *D. frontalis*, with suppression of infestation growth being the primary approach taken (Clarke, 2001).

In this chapter I will review the chemical ecology of *D. frontalis*, a subject that has received intensive research interest for over 50 years. Southern pine beetle was the first bark beetle species for which a multifunction, multicomponent pheromone system was described (Renwick and Vité, 1969), and it generated the first hypotheses proposed for mediation of bark beetle host colonization by both aggregation and antiaggregation pheromones.

2. HOST SELECTION AND DISCRIMINATION

2.1 The Need for Dual Strategies

All aggressive bark beetle species must be adept at two very distinct host location and colonization strategies if they are to endure their characteristically extreme population cycles. At low numbers, they must be able to locate hosts that are sufficiently weak that colonization does not first require a conspecific mass attack to deplete host defences; under these circumstances, the chances of encountering conspecific pheromone that might guide them to a host undergoing colonization are relatively low and insufficient numbers of beetles would be "in range" to detect the pheromone signal of pioneers. During outbreaks, dispersing beetles are readily available in the environment to respond to aggregation pheromone and thereby reduce the risk posed to any courageous pioneer beetles that choose to attack a tree that they would not be able to colonize alone. Once an infestation is established, the plumes of pheromone generated from the mass-attacked trees are beacons that can be used by dispersing beetles to locate the abundant and high quality hosts being made available for colonization due to the elevated local densities of beetles. However, under both high and low populations, there must be pioneers that, without the aid of pheromones, can locate hosts that they can either colonize directly or utilize as points for initiating a mass attack that will render the tree suitable for their colonization.

2.2 Host Location by Pioneering *D. frontalis*

Some literature on aggressive bark beetles has been devoted to the mechanisms of host location/selection utilized by the first, pioneer bark beetles that arrive on a suitable host pine and initiate the process of mate attraction and/or concentration of conspecifics through release of pheromones. This cycle (ie, landing, pheromone release, and attraction of conspecifics that themselves make a quantitative contribution to the pheromone plume) is a positive-feedback loop that drives mass attacks and results in overwhelming of host defences and host colonization. Once a host is selected by a pioneer individual—by whatever means—pheromones produced by the pioneer presumably could be sufficient as long-range attraction cues for initiating this loop. In some beetle species it is evident that host-produced (or "primary") host location/discrimination cues can be sensed by the pioneer beetles at a distance (Gara et al., 1984; Heikkenen, 1977; Miller et al., 1986; Moeck and Simmons, 1991). Some bark beetles, including major tree killers, have been shown to be attracted to olfactory cues associated with naturally or artificially compromised hosts (Gara et al., 1984; Macias-Samano et al., 1998; Moeck and Simmons, 1991; Pureswaran and Borden, 2005). Alternatively, beetles may select a host by means of close-range olfactory, gustatory, or other contact cues following random landing on the bark surface. These beetles land presumably in response to the host's dark vertical silhouette and evaluate each tree before attempting to initiate a mine; they resume flight if the tree is found to be unsuitable (Byers, 1996; Hynum and Berryman, 1980; Moeck, 1978; Moeck et al., 1981; Saint-Germain et al., 2007). Evidence that beetles follow this "random landing" strategy include that (1) in some instances landing frequency by dispersing beetles on potential host trees is uncorrelated to subsequent attacks or is not stimulated by artificial treatments inducing host susceptibility (Hynum and Berryman, 1980; Moeck et al., 1981; Raffa and Berryman, 1980), and (2) models which indicate that, although long-range identification of suitable host trees by primary cues may be a more efficient host selection strategy, random landing is a sufficient means for host location by pioneers (Byers, 1996; Gries et al., 1989). Under the random landing hypothesis, any long-range selection of hosts is due to the aggregation pheromone (potentially synergized by otherwise unattractive host-produced compounds) released by the pioneers and then augmented by subsequent beetles that join them in the attack.

There is no compelling evidence that pioneering *D. frontalis* initially locate suitable hosts by primary attraction. Unlike many of their nonaggressive associates, *D. frontalis* are not attracted to whole or distilled (ie, turpentine) resin of host pines in the absence of beetle pheromones (Billings, 1985; Kinzer et al., 1969; Payne et al., 1978a; Sullivan et al., 2007b; Vité and Renwick, 1968) and apparently not to pine logs screened from beetle attacks (Svihra, 1982). However, α-pinene alone attracted both sexes (McCarty et al., 1980) or male (Niño-Domínguez et al., 2015b) *D. frontalis* in ambulatory laboratory olfactometers, suggesting that the compound alone may affect activity of beetles post-landing if not in flight. Pines located at points distant from active infestations (and thus free from the influence of semiochemicals from attacked trees) have been weakened artificially and this resulted in their nearly simultaneous colonization by *D. frontalis* together with associated bark beetles (Heikkenen, 1977; Rykiel et al., 1988). Tree disturbance methods included severing a pine while maintaining its vertical orientation (Heikkenen, 1977), detonating a blast cord wrapped around the bole to imitate the physical damage produced by a lightning strike (Coulson et al., 1986), or mechanical removal of a 15 cm-wide strip of bark from crown to base (Payne, 1986). However, these tree boles were not enclosed by screen to prevent bark beetle attacks. Hence it cannot be ruled out that randomly landing *D. frontalis* attacking the artificially susceptible trees could have subsequently released pheromone thereby initiating secondary attraction, or that *D. frontalis* were responding to secondary attractants of other insects.

Lightning strikes appear to be the most common disturbance that weakens individual trees sufficiently to allow colonization by small numbers of *D. frontalis*, and 10% to 75% of *D. frontalis* infestations have their origins in a lightning-struck tree (Coulson et al., 1983; Lovelady et al., 1991). Susceptibility is due at least in part to a temporary reduction of the struck tree's constitutive defences (Blanche et al., 1985; Hodges and Pickard, 1971). It has been noted that the distribution of lightning strikes in both space and time puts them within the dispersal range of southern pine beetles (Kinn, 1986; Turchin and Thoeny, 1993). The resulting reliability of lightning-struck pines assures the continuous availability of host material for *D. frontalis* within which low-level populations can be sustained or build to levels capable of initiating an infestation (Coulson et al., 1999; Lovelady et al., 1991; Rykiel et al., 1988). The speed and dependability with which lightning-struck trees can be attacked by *D. frontalis* and associated bark beetles have led to speculation about the existence of a primary attractant generated by the strike (Hodges and Pickard, 1971). However, preliminary coupled gas chromatography-electroantennographic detection (GC-EAD) analyses of *D. frontalis* responses to aeration samples of recently struck pines (Fig. 1) have not indicated the presence of olfactory stimulants other than major volatile constituents of *P. taeda* resin, which have not been shown to be attractive to *D. frontalis* in the absence of beetle-produced

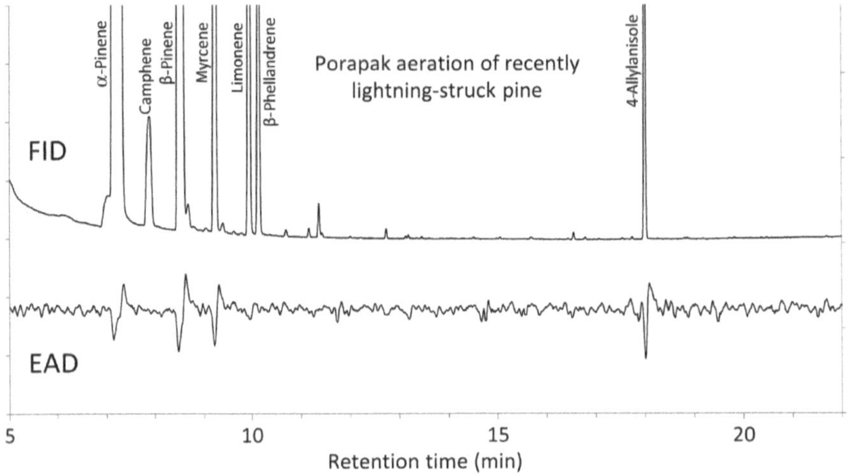

Fig. 1 Coupled gas chromatography-electroantennographic detection (GC-EAD) analysis of two female *D. frontalis* antennal clubs (GC-EAD analyses summed digitally) in response to a concentrated pentane extract of a 3 h Porapak Q (Waters; Milford, Massachusetts, United States) sampling of air from the bole of a *P. taeda* that had been recently (within days) struck by lightning. A 1 m length of the bole including areas with sapwood exposed by the strike had been enclosed in PTFE sheeting to concentrate odours at the air intake of the adsorbent cartridge. The tree had not been attacked by subcortical insects except for a single *D. terebrans* entrance at the soil-line, and I sampled >1 m above this attack. The GC-EAD apparatus and antennal preparation methods were as in Sullivan (2005). *Antennograms courtesy of Dr. William Shepherd, USDA Forest Service, Southern Research Station.*

compounds (Billings, 1985; Kinzer et al., 1969; Payne et al., 1978a; Sullivan et al., 2007b). The aforementioned experiments with blast cord were intended to induce host suitability and *D. frontalis* attacks by duplicating, at least in part, the effects of a lightning strike on a host tree (Miller, 1983). Payne (1986) hypothesized that the very large release of resin odours from a lightning-caused wound at the approximate height of preferred flight for *D. frontalis* (ie, mid-bole) might generate close-range arrestment and thereby provide a short-range primary semiochemical cue for *D. frontalis* otherwise searching randomly.

However, lightning-struck or otherwise disturbed trees may be rapidly attacked by a variety of bark beetle species within the southern pine bark beetle guild and not merely *D. frontalis* (Anderson and Anderson, 1968; Coulson et al., 1985; Flamm et al., 1993). At least two of these species [*Ips grandicollis* (Eichhoff) and *Dendroctonus terebrans* (Olivier)], unlike

D. frontalis, unquestionably respond to primary attractants including compounds (eg, host resin monoterpenes) (Erbilgin and Raffa, 2000; Fatzinger, 1985; Fatzinger et al., 1987; Miller and Rabaglia, 2009; Phillips et al., 1988; Siegfried et al., 1986; Werner, 1972) that are released from the exposed sapwood and phloem tissue of lightning-struck pines. Both species release pheromones post-landing, and it is a conceivable strategy for *D. frontalis* to rely on these other species to locate lightning-struck trees and to then exploit the pheromones of these species as kairomones in locating breeding material. Such "third-party" secondary attraction seems unlikely in the case of *I. grandicollis* since *D. frontalis* are not attracted to logs infested with *Ips grandicollis* or *Ips* pheromone components (Birch et al., 1980; Svihra et al., 1980). However, as suggested by Hodges and Pickard (1971), *D. terebrans* could play a role in attracting *D. frontalis* to trees struck by lightning or receiving other types of injury. *Dendroctonus terebrans* is typically the first bark beetle species to arrive on a lightning-struck tree (Hodges and Pickard, 1971), whereas attacking *D. terebrans* pairs generate all components of the *D. frontalis* aggregation attractant (Payne et al., 1987; Phillips et al., 1989): frontalin and *trans*-verbenol (by females), both *endo*- and *exo*-isomers of brevicomin (by males), and resin odors from the damaged host tissue. Thus, as suggested by Payne et al. (1987), a kairomonal response by *D. frontalis* to trees attacked by *D. terebrans* could assist the former in locating lightning-struck host trees. *Dendroctonus terebrans* restrict their attacks to the lower 1–2 m of the host bole hence they compete minimally for phloem resources with *D. frontalis* (Thatcher, 1960). As an additional conceivable benefit to *D. frontalis*, *D. terebrans* attacks might also weaken the tree or in some other way increase its susceptibility to *D. frontalis* colonization.

3. CONCENTRATION OF CONSPECIFICS FOR MASS ATTACK

A female southern pine beetle that has selected a host (which she has identified through the presence of conspecific pheromones or its own evaluation of host quality upon contact and inspection) will release pheromones with at least two potential functions: attracting a mate and, if the host is producing resistance, attracting conspecifics of both sexes to overcome host defences. As with all bark beetles, pheromone components are accumulated in the hindgut and released from the anus.

3.1 The Aggregation Attractant

The aggregation attractant for *D. frontalis* appears to be composed of three major components (Fig. 2): frontalin (1,5-dimethyl-6,8-dioxabicyclo [3.2.1] octane) produced by females (Kinzer et al., 1969), male-produced *endo*-brevicomin [*endo*-7-ethyl-5-methyl-6,8-dioxabicyclo[3.2.1]octane (Vité and Renwick, 1971)], and host resin odours, particularly α-pinene [2,6,6-trimethylbicyclo[3.1.1]hept-2-ene (Renwick and Vité, 1969)], and perhaps other monoterpenes. Female-produced *trans*-verbenol (*trans*-4,6,6-trimethylbicyclo[3.1.1]hept-3-en-2-ol) may also participate but the insufficient studies to date suggest that its presence may not be essential (see below). The ternary blend is the most attractive lure yet identified for this species [at least when released in the absence of natural or artificial competing sources of attractant (Moreno et al., 2008; Sullivan and Mori, 2009; Sullivan et al., 2007b, 2011; Vité et al., 1985)], and the species possesses exceptional olfactory sensitivity to the two insect-produced components (Payne, 1975; Sullivan, 2005; Sullivan et al., 2007b). The pheromone component combination parallels that of closely related, aggressive species, *D. brevicomis* LeConte and *D. adjunctus* Blandford in which each sex contributes a single, different bicyclic ketal to the aggregation pheromone (ie, brevicomin or frontalin) that—at least in the case of *D. brevicomis*—acts synergistically (Bedard et al., 1980b; Browne et al., 1979; Byers et al., 1984; Hughes et al., 1976). This is not true for aggressive pine beetles *D. ponderosae* Hopkins in which females produce no bicyclic ketals (Pitman and Vité, 1969), and the close sibling of *D. frontalis*, *D. mesoamericanus* Armendáriz-Toledano and Sullivan, in which females produce both brevicomin and frontalin (Armendáriz-Toledano et al., 2015; Sullivan et al., 2012). There have as yet been no direct, experimental

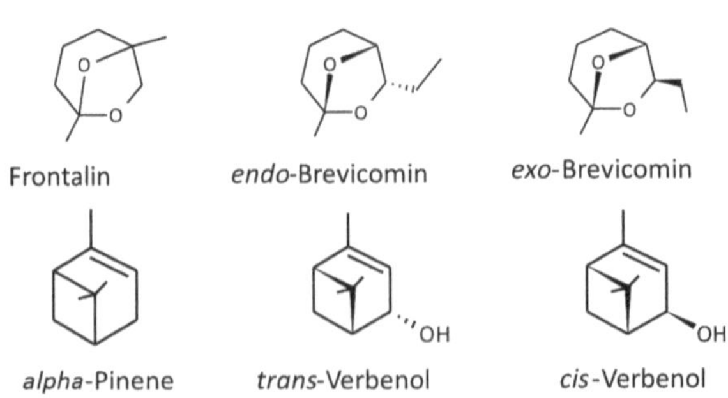

Fig. 2 Semiochemicals capable of attracting or synergizing the attraction of flying *D. frontalis* and therefore playing a role in mediating mass-aggregation.

contrasts of attraction between trees undergoing *D. frontalis* mass attack and the aforementioned three-component "best lure", hence it is not yet known whether the blend of three semiochemicals (with host odours considered here as a single semiochemical) truly duplicates the activity of the natural attractant.

3.2 Frontalin

The bicyclic ketal frontalin can arguably be named the "major component" of the aggregation pheromone/attractant for *D. frontalis* as no mixture of semiochemicals has been identified to be attractive to flying *D. frontalis* in its absence (Smith et al., 1993). It is present in newly emerged, host-arriving, and gallery-establishing females (Kinzer et al., 1969; Pitman et al., 1969; Pureswaran et al., 2007; Renwick and Vité, 1968; Sullivan et al., 2007b, 2012) and generally is undetectable in males (although see Grosman, 1996; Grosman et al., 1997; Pureswaran et al., 2006; Rudinsky et al., 1974). Evidence of its uniqueness to females is consistent with the observation that logs infested with solitary *D. frontalis* females are attractive whereas those with forced male attacks are not (Coster et al., 1977). Alone, frontalin can attract both sexes to traps; hence, it is a true aggregation pheromone (Kinzer et al., 1969; Payne et al., 1978a; Sullivan et al., 2007b). Since it is produced by gallery-establishing females and alone can attract males to females—both flying and walking—it can be described additionally as a sex pheromone component for this species (McCarty et al., 1980; Niño-Domínguez et al., 2015b; Payne et al., 1978a). It is more attractive to males than females in ambulatory bioassays (McCarty et al., 1980), and trap responses by beetles to frontalin/host-odour lures tend to be strongly male-skewed (Moreno et al., 2008; Renwick and Vité, 1969). Its attractive capacity for flying beetles is very low in the absence of host-associated synergists or *trans*-verbenol (discussed later), and it is possible that monoterpenes released by pines in the environment in which field bioassays were completed may have acted synergistically with frontalin in trapping trials of lures with frontalin alone (Renwick and Vité, 1970).

Dendroctonus frontalis has been reported to produce predominantly the minus enantiomer of frontalin: 85% for an unstated collection site (Stewart et al., 1977); 95% in Mississippi, United States (Sullivan et al., 2007b), and 94% in Chiapas, Mexico (Niño-Domínguez et al., 2015b). Compositions of 25–35% of the (−)-enantiomer were reported for beetles from South Carolina, North Carolina, and Texas, United States (Grosman et al., 1997), although this study had methodological problems (discussed below). EAG (electroantennogram) and GC-EAD studies indicated that antennae of both sexes of *D. frontalis* are substantially more sensitive to the (−)-enantiomer (Payne et al., 1982; Sullivan et al., 2007b). Adaptation studies indicated that beetles possess different receptors for each enantiomer, although single cell studies showed that the enantiomers stimulated the same olfactory neurons but produced differing spike frequencies (Payne et al., 1982). Adaptation to frontalin eliminates EAG responses to other semiochemicals including *endo*- and *exo*-brevicomin, *trans*-verbenol, verbenone, and α-pinene, suggesting that frontalin can interact with olfactory receptors for all of these semiochemicals (Dickens, 1979; Dickens and Payne, 1977). Both enantiomers (≥98% enantiomeric purity) were attractive to *D. frontalis* both in ambulatory laboratory olfactometers and field trapping bioassays, with beetles being generally more responsive to the (−)-enantiomer (Payne et al., 1982). No difference in attraction was detected between the pure (−)-enantiomer, the approximate beetle-produced proportion of enantiomers [ie, 85% (−)], or the racemate, indicating the lack of preference for a specific ratio of the enantiomers (Payne et al., 1982). Analogues of frontalin which involved either repositioning or elimination of one or both methyl groups decreased EAG responses with the degree of response reduction being roughly correlated to the degree of modification of the analogue; however, beetles in walking olfactometers responded to all analogues (Payne et al., 1988). A weak attraction to traps was registered to only a single frontalin analogue on which the methyl group at the 1 position was moved to the 7 position (the *endo*- but not *exo*-isomer) (Payne et al., 1988; Renwick, 1970). This analogue somewhat resembles the important pheromone component *endo*-brevicomin except that in the latter compound the 7-position methyl is replaced with an ethyl group (Renwick, 1970). Opening of the ring structure also apparently eliminates the behavioural activity of frontalin as a flight lure (Renwick, 1970).

Early studies reported a rapid reduction of the quantities of frontalin in hindguts of females forced to attack logs, and this apparently coincided with dissapearance of droplets of *trans*-verbenol-dominated liquid from the hindgut (Coster and Vité, 1972; Vité and Pitman, 1968). These authors inferred that *D. frontalis* release most of their aggregation pheromone prior to entry into the host. Contrary to this, Hughes (1973) found no change in frontalin content of hindguts due to feeding. Furthermore, pine posts infested with virgin female *D. frontalis* reached maximum attractiveness after 48 h (Coster and Vité, 1972), and pine logs either artificially infested with beetles or cut from mass-attacked trees could maintain some attractiveness for multiple days (Coster et al., 1977; Franklin, 1970). Additionally, static headspace aerations (ie, involving beetles confined in still air with chemical adsorbent) indicated that emergent females released frontalin at a lower rate than females that were excised from logs after feeding for ~1 d (Sullivan et al., 2007b; Pureswaran et al., 2008b). The opposite was observed for *trans*-verbenol, which is by far the dominant volatile compound present in the hindguts of emergent females. These data suggest that females continue to produce and release frontalin after commencement of feeding although the bulk of other volatile compounds in the hindgut at arrival (eg, *trans*-verbenol, myrtenol, *cis*-verbenol) apparently are released quickly and are not or only partially replaced (Sullivan, 2005; Sullivan et al., 2012). Mating may reduce production of frontalin (Sullivan et al., 2007b, 2012), although these studies confounded pairing with the female's residence time in the host. However, in an experiment in which virgin female beetles were infested onto freshly cut pine posts, attraction rapidly declined after a peak at 2 days whether or not males were added to the females' galleries (Coster and Vité, 1972). This loss in attraction may well have been due to a reduction in frontalin emission as well as to a decline in release of host odours by the ageing posts (Vité and Crozier, 1968). Frontalin emissions from gallery entrances of mass-attacked trees are significantly higher for entrances of solitary females than for entrances of pairs either before or after beginning oviposition (Pureswaran and Sullivan, 2012).

3.3 *endo*-Brevicomin

endo-Brevicomin is produced by newly emerged *D. frontalis* males, those alighting on mass-attacked trees (Pureswaran et al., 2006; Vité and Renwick, 1971), and males paired with a female in a gallery (Pureswaran and Sullivan, 2012; Sullivan et al., 2007b). For populations examined in Mississippi, United States, solitary males feeding in phloem (ie, forced to initiate a gallery on a host log, a behaviour that does not occur in nature), did not produce significantly greater amounts than newly emerged beetles whereas pairing caused production to increase nearly 10-fold (Sullivan et al., 2007b). Since production of *endo*-brevicomin has been shown to be stimulated in callow adult *D. frontalis* males by exposure to juvenile hormone II or its analogue methoprene (Bridges, 1982), it appears that pairing or mating may induce production of juvenile hormone in males. Male *D. frontalis* have been reported to produce 3% (Redlich et al., 1987) and

9–21% [newly emerged beetles from Texas and the Carolinas, United States (Grosman et al., 1997)], of the (−)-enantiomer. However, studies by the author and collaborators using enantioselective GC-MS have failed to detect the (−)-enantiomer in any hindgut or aeration sample of solitary or paired *D. frontalis* males, including paired males from Alabama, United States, and Chiapas, Mexico, as well as newly emerged, solitary feeding, or paired individuals from Mississippi, United States (Sullivan et al., 2007b, author's unpublished data). These analyses included at least 18 samples in which the mass selective detector's threshold of detection for the (−)-enantiomer was less than 1% of the detected abundance of the (+)-enantiomer (Sullivan et al., 2007b). Based on these data, it can be concluded that *D. frontalis* produces little or no (−)-*endo*-brevicomin, although I caution that this conclusion requires broader sampling of the geographic range of *D. frontalis*.

Dendroctonus frontalis appears to have greater olfactory responses to (+)-*endo*-brevicomin than any semiochemical tested to date in GC-EAD and EAG studies. Dose-response studies indicated that olfactory sensitivity (=concentration threshold of response) of both sexes of *D. frontalis* was four orders of magnitude lower for (+) than (−)-*endo*-brevicomin, and sensitivity was greater to (+)-*endo*-brevicomin than either enantiomer of frontalin (Sullivan et al., 2007b). When antennae of both sexes of *D. frontalis* were exposed in a GC-EAD analysis to 0.1 insect equivalents of a pooled hindgut extract of emerged males, *endo*-brevicomin generated a higher amplitude EAD response than any other compound present (Sullivan 2005). In sensory adaptation studies, *endo*-brevicomin occupied 75–87% of receptors on *D. frontalis* antennae; of tested semiochemicals only frontalin occupied more than this (Dickens, 1979; Dickens and Payne, 1977). The exceptional sensitivity of *D. frontalis* to *endo*-brevicomin is consistent with its having important and likely long-distance influence on the biology of this species.

Earlier behavioural studies with racemic *endo*-brevicomin (often released in combination with its *exo*-isomer) deployed in traps generally indicated that it was a potent attraction or landing inhibitor for both sexes of *D. frontalis* (Payne et al., 1977, 1978a; Richerson and Payne, 1979; Salom et al., 1992b; Vité and Renwick, 1971). Furthermore, across a wide range of concentrations it prevented arrestment by walking males over artificial gallery entrances releasing female-associated attractant (Rudinsky et al., 1974). In the same study it induced males to stridulate with their "rivalry" chirp, a sound which they produce during aggressive same-sex encounters (Rudinsky and Michael, 1974; Rudinsky et al., 1974; Ryker,

1988). Thus, *endo*-brevicomin apparently serves as a signal of the presence of a male in a female's gallery.

Due to its consistent inhibition of attraction and production solely by the second-arriving sex, *endo*-brevicomin was originally classified as an antiaggregation pheromone for this species (Rudinsky et al., 1974). However Vité et al. (1985) discovered that capillaries releasing the purified (+)- or (−)-enantiomers of *endo*-brevicomin either enhanced or inhibited, respectively, beetle response to traps baited with attractant. They also observed some enhancement by the racemic mix of *endo*-brevicomin [but less than produced by pure (+)]. Additionally, they hypothesized that the failure of previous studies to detect the attractant-synergistic properties of *endo*-brevicomin was due to the inhibitory/repellant properties of the (−)-enantiomer of *endo*-brevicomin surpassing the activity of the (+)-enantiomer when the racemic mixture was released at a high rate. Sullivan et al. (2007b) confirmed the strongly synergistic effect of pure (+)-*endo*-brevicomin on attraction of *D. frontalis* (ie, a ~40-fold catch increase) tested within uninfested stands, however, these authors were unable to replicate the reported inhibitory effect of pure (−) in field studies (unpublished data). In sum, these data seemed to be compelling evidence that (+)-*endo*-brevicomin was a key component of the aggregation pheromone for *D. frontalis*, and, given the apparent absence of production of the antipode, that *endo*-brevicomin might act purely as an attractant synergist in intraspecific communication by *D. frontalis*.

However, a dose–response trapping test inside a *D. frontalis* infestation with either (+)-*endo*-brevicomin or a doubled dose of the racemate [so lure release of the (+) enantiomer was equal for both treatments] found no difference in responses to either enantiomeric composition (Sullivan et al., 2011). For both (+) and racemic devices, *endo*-brevicomin added to frontalin and host odours did not increase attraction at any dose spanning 0.005–3 mg/d, whereas at high doses (>~1 mg/d of each enantiomer) it significantly inhibited beetle catches (Sullivan et al., 2011). These data suggested that the attractive effects of *endo*-brevicomin might not be detectable or not occur inside active infestations. This hypothesis was confirmed directly in an experiment in which frontalin/host odour-baited traps were erected simultaneously inside as well as 100 and 200 m outside active heads of growing *D. frontalis* infestations. Addition of an identical *endo*-brevicomin device to these traps caused catches of *D. frontalis* to be significantly reduced within the infestations but increased outside (Sullivan et al., 2011).

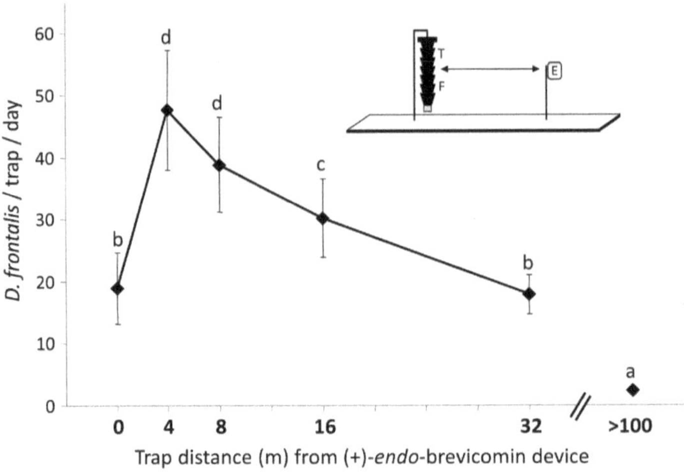

Fig. 3 Trap catches (mean ± SEM) of *D. frontalis* in a multiple-funnel trap baited with frontalin (F) and turpentine (T) and with a single device releasing 0.23 mg/d (+)-*endo*-brevicomin (E) positioned varying distances from the trap. Traps were spaced >100 m apart and the direction of the releaser relative to the trap was randomized. *Figure from Sullivan and Mori (2009).*

I believe that this remarkable phenomenon is at least partially explained by the fact that (+)-*endo*-brevicomin released from a point source at a rate approximately equal to a single mass-attacked tree can have synergistic effects on frontalin/host odour-baited traps located within a radius of at least 32 m (Fig. 3), a zone that would encompass many currently attacked and adjacent unattacked trees near the head of a beetle infestation (Sullivan and Mori, 2009). Furthermore, the synergistic effect in this study was significantly greater (ie, catches were higher) when the *endo*-brevicomin device was 4–16 m distant rather than on the trap. This could be due at least in part to *endo*-brevicomin having a "multifunctional"-type (Rudinsky, 1973a) dose–response for *D. frontalis* as illustrated in Fig. 4. In the experiment of Fig. 4, synergism by *endo*-brevicomin increased up to approximately 0.2 mg/d release from the trap, and then declined and finally switched to inhibition at a rate above ~5 mg/d. Airborne concentrations of a semiochemical decline with increasing distance from a point source. Hence one possible outcome of the multifunctional dose–response curve for *endo*-brevicomin would be relatively greater beetle response to release points of frontalin/host odours located at greater distances from the *endo*-brevicomin release point as the release rate of the latter is increased above the level for optimal synergism (eg, in the curve of Fig. 4, a rate

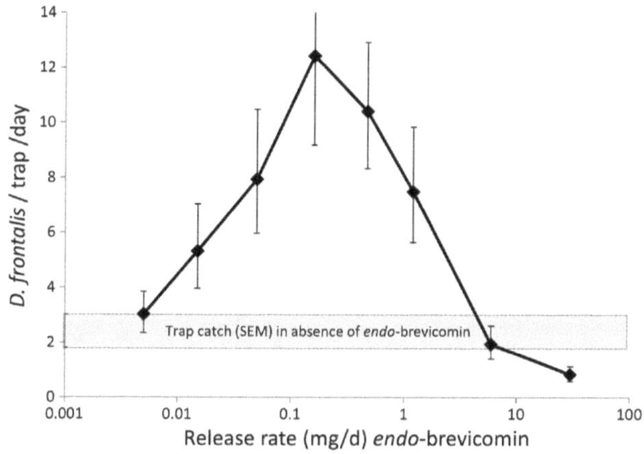

Fig. 4 Trap catches (±SEM) of *D. frontalis* in multiple-funnel traps baited with α-pinene [1–2 g/d; 23% (+)], (±)-frontalin (6–8 mg/d), and a device releasing (±)-*endo*-brevicomin at each of eight release rates (open-vial or capillary-type devices with different diametre openings). *endo*-Brevicomin devices were placed directly on traps which were located >200 m apart. The grey band in the figure encloses the upper and lower bound of the SEM for the *endo*-brevicomin-lacking control traps. Tests were performed in mixed pine/hardwood stands during spring 2014 in Homochitto National Forest, western Mississippi, United States. Means and SEMs calculated with log-transformed catch data then back-transformed for the figure ($n=18$).

>~0.2 mg/d). Furthermore, within the active radius of "preexisting" local sources of *endo*-brevicomin (such as within a *D. frontalis* infestation or in the vicinity of artificial release devices) the effect of addition of *endo*-brevicomin releasers directly to a trap might be redundant and ineffectual, or, by augmenting the summed area-wide concentration, cause a catch reduction even though increased attraction might have been the effect in the absence of preexisting *endo*-brevicomin (Sullivan and Mori, 2009, author's unpublished data).

It is possible that early trapping studies did not detect the synergistic properties of *endo*-brevicomin for *D. frontalis* because of insufficient distance among traps and the practice by researchers of performing lure comparison trials inside or near active infestations because these locations could more readily provide statistically meaningful numbers of beetle trap catches (Payne et al., 1978a,b; Salom et al., 1992b). One implication of these findings more generally for bark beetle semiochemical research is that field experiments with unrecognized multifunctional semiochemicals could be interpreted as indicating either inhibitory/repellent or synergistic/attractive properties for the test compound depending on trap spacing (Sullivan and Mori, 2009). A similar problem might arise as a result of natural sources of pheromone in the environment where tests are performed. It is worthwhile noting that it is possible in bark beetles for certain pheromone components determined to be inhibitors or inactive in trapping tests to induce or increase the risk of mass attacks when deployed alone on host trees (Borden et al., 1990; Chatelain and Schenk, 1984). This implies that by their very nature trap tests can conceal potentially undesirable attractive or attack-inducing effects.

Relatively few studies have been performed on spatial effects of multifunctional or other bark beetle pheromone components, and no close parallels to *endo*-brevicomin's activity with *D. frontalis* have been identified in the semiochemistries of other bark beetles. Spatial separation of the releasers of the female (*exo*-brevicomin) and male (frontalin) contributions to the aggregation pheromone of *D. brevicomis* caused an immediate and strong decline in attraction, with a fivefold reduction in responses occurring with a mere 4 m separation (Byers, 1987). However, dose-dependent broadening of the zone of attraction and landing is reported to occur with the aggregation pheromones of several bark beetles. In both *Ips paraconfusus* Lanier and *I. typographus* L. the gallery-initiating, aggregation pheromone-producing sex (males) tend to land further from a source of aggregation pheromone as the concentration is increased (Byers, 1983; Schlyter et al., 1987a,b); this presumably serves to reduce intraspecific competition by allowing males to avoid areas where conspecific male densities are already high (Byers, 1989a). A similar behaviour occurs for *D. frontalis* with frontalin, where females (the gallery-initiating sex in this case) tended to land above or below the release point of frontalin whereas males landed close to the source (Hughes, 1976). This author believed that this behaviour might explain the strongly male-skewed sex ratios caught in frontalin-baited traps possessing short vertical profiles. In *D. brevicomis*, progressively higher rates of release of the complete aggregation attractant caused beetles to land on traps in greater relative numbers at increasing distances from the release point, and, at yet higher rates, beetle catches were reduced at the release point itself (Wood and Bedard, 1977). Similarly, frontalin will cause higher proportions of landings of *D. frontalis* at progressively greater horizontal distances from the pheromone release point as the release rate is increased (Vité, 1970).

However, a property that distinguishes *endo*-brevicomin and its behaviour from the above listed aggregation attractants which likewise cause displaced landing effects at high concentrations (besides it being a synergist and not an attractant) is that at no release rate does *endo*-brevicomin appear to direct landing to its point of release more than to the immediately surrounding area (Sullivan and Mori, 2009); rather, the compound apparently produces a zone of synergistic activity, with the release rate determining the size of the zone and the radius of maximum synergism (Sullivan and Mori, 2009, author's unpublished data). This behavioural effect also appears to occur about equally for both sexes of *D. frontalis* rather than predominantly the first arriving sex. Frontalin by contrast appears to cause landings to focus at or near its point of release, at least when it is released at low to moderate rates (Sullivan and Mori, 2009, author's unpublished data). Thus with regard to the natural colonization behaviour of *D. frontalis*, *endo*-brevicomin at low concentrations (as should be produced by a few arriving or paired males on the host) should enhance the attractiveness of all solitary female attacks within a certain, presumably small radius, thereby enhancing landings near these females. With increasing landings and pairings of arriving males, levels of *endo*-brevicomin should increase to produce synergistic effects for attacks by solitary females on trees at greater distances determined by the numbers of attacking males and thus the total release rate. Simultaneously, the increased levels of *endo*-brevicomin at the source trees should inhibit attraction and landing on them. Such a pheromone effect should promote and enhance switching of the attack focus by increasing the attractiveness of adjacent trees receiving pioneer female attacks while simultaneously inhibiting attraction to the previous attack focus trees.

The *exo*-isomer of brevicomin also occurs in small quantities in *D. frontalis*, and has been detected in a 1:99 to 6:94 ratio with *endo*-brevicomin (populations in Mississippi, Arizona, and Chiapas) (Pureswaran et al. 2008a; Sullivan et al., 2012). The quantities of *endo*- and *exo*-brevicomin produced by individual *D. frontalis* males are strongly correlated, thus their synthesis and regulation appear to be similar. As with *endo*-brevicomin, reported behavioural responses by *D. frontalis* to *exo*-brevicomin have varied. It has alternately been shown to reduce (Vité and Renwick, 1971), enhance (Hofstetter et al., 2008, 2012; Pureswaran et al., 2008a), or not significantly alter (Payne et al., 1978a) responses of *D. frontalis* to traps baited with frontalin and host odours. As with the *endo*-isomer, it has not demonstrated attractive properties when presented alone or with host odours. The causes of the variability in response to *exo*-brevicomin are unknown but could be similar as those for *endo*-brevicomin (ie, multifunctionality). Antennal sensilla of *D. frontalis* are quite sensitive to *exo*-brevicomin (Payne, 1975; Pureswaran et al., 2008a). However *endo*-brevicomin can adapt the antennae completely to the *exo*-isomer whereas the reverse does not occur, suggesting that *D. frontalis* have the olfactory capacity to distinguish (and therefore behave differently) to the two forms and are more sensitive to *endo*- than *exo*-brevicomin (Dickens and Payne, 1977). Given the very small amounts of *exo*-brevicomin present in

D. frontalis, the compound may not function as a pheromone per se, but the beetles' responses may reflect interspecific interactions with species that produce larger quantities. As previously mentioned, the black turpentine beetle *D. terebrans* releases *exo*-brevicomin as part of its pheromone (Payne et al., 1987; Phillips et al., 1989), and *D. frontalis* cross-attraction to *exo*-brevicomin from this species could aid it in locating and exploiting lightning struck or other compromised hosts discovered by *D. terebrans* (Smith et al., 1990). Similar cross-attraction may occur in the sympatric zone of *D. frontalis* and *D. brevicomis* located in the southwestern United States, as the latter produces *exo*-brevicomin as an important component of its aggregation pheromone and the two species frequently cohabit the same host trees (Davis and Hofstetter, 2009; Hofstetter et al., 2008; Pureswaran et al., 2008a).

3.4 Host Monoterpenes

Aggregation pheromone components of *D. frontalis* (frontalin alone or in combination with *endo*-brevicomin) are strongly synergized by odours of host resin. Raw pine oleoresin strongly synergized responses of flying *D. frontalis* to frontalin (Kinzer et al., 1969), and distilled pine oleoresin (turpentine) from host pines of *D. frontalis* also had a strong synergistic attractive effect on frontalin alone as well as the combination of frontalin and *endo*-brevicomin (Payne et al., 1978a; Sullivan et al., 2007b). Addition of α-pinene to lures of frontalin and *trans*-verbenol caused a greater increase in responses of flying *D. frontalis* than did addition of other major monoterpene components of the oleoresin of *D. frontalis* host species (ie, β-pinene, camphene, myrcene, limonene, 3-carene, 4-allylanisole, and terpinolene), although this result requires confirmation as no statistical data were presented (Renwick and Vité, 1969). α-Pinene was likewise found to be a superior synergist to myrcene in frontalin lures for *D. frontalis* in Arizona (Hofstetter et al., 2008). This preference may reflect α-pinene being the predominant component of the resins of the common host species for *D. frontalis* including (in the eastern United States) *P. taeda* L., *P. echinata* Mill., *P. pallustris*, *P. ellioti* Engelm., *P. strobus* L., *P. rigida* Mill., and *P. virginiana* Mill. (Mirov, 1961). The (+)-enantiomer of α-pinene is a better synergist than the (−)-enantiomer, and EAG cross-adaptation studies indicated that both sexes possess at least some receptors with differing affinities for the two enantiomers (Staeben et al., 2015). The antennae also have a lower response threshold to (+) than (−)-α-pinene. The capacity to distinguish α-pinene enantiomers may mediate selection of species of host or individual trees with a higher (+)-content to their resin, but this possibility and its possible biological significance have not been investigated. The synergistic effects of high rates of host odours are quite dramatic. Billings (1985) observed that a high release rate of turpentine (3.6 mg/d) could enhance mean *D. frontalis* trap catches with a frontalure bait (ie, a 1:2 blend of frontalin and α-pinene released at 50 mg/d) almost 30-fold. In Mississippi, an approximate 100-fold increase in the release rate of turpentine from traps caused a nearly 10-fold increase in catches both when the pheromone lure was either frontalin and *endo*-brevicomin or frontalin alone (Fig. 5). Addition of a device releasing *P. taeda*-derived turpentine at 7 g/d to traps with a frontalin/(+)-*endo*-brevicomin lure enhanced catches 45-fold (Sullivan et al., 2007b). This response may reflect the fact that pines undergoing mass attack by *D. frontalis* release very large amounts of host odours, with individual attacks releasing an average of 14.4 mg/d α-pinene and with beetle attacks typically occurring in the thousands on mass-attacked trees (Pureswaran and Sullivan, 2012). These quantities decline substantially once

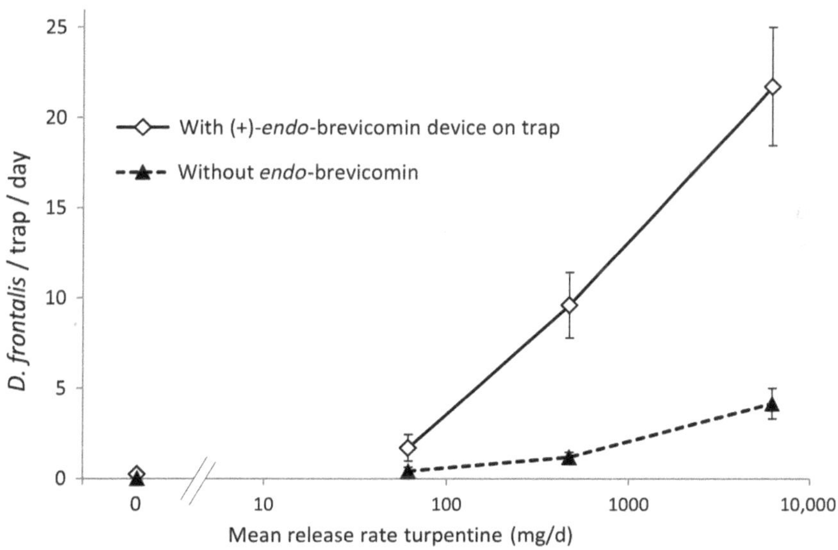

Fig. 5 Effect of varying the release rate of a blend of host odours (ie, distilled *P. taeda* turpentine) on catches of *D. frontalis* in multiple-funnel traps baited with (±)-frontalin (~5 mg/d) released either alone or with a (+)-*endo*-brevicomin device (releasing 0.23 mg/d) on the trap. The test was performed in mixed pine/hardwood stands during late winter/spring of 2007 in the Homochitto National Forest, western Mississippi, United States ($n = 24$), and traps were located >150 m apart. Release rates for lures were determined gravimetrically in a fume hood.

mass attack is complete. Individual entrances of attacking *D. frontalis* release the attractive synergist α-pinene at three to four orders of magnitude greater amounts than the beetles release their pheromone components (Pureswaran and Sullivan, 2012).

α-Pinene is abundant in the atmosphere of pine forests (Seybold et al., 2006) and is produced by a great diversity of tree species. Therefore, the mere presence of α-pinene should provide little if any specific information to a dispersing, host/mate-seeking bark beetle. Relatively low olfactory sensitivity to α-pinene by *D. frontalis* (Dickens and Payne, 1977; Payne, 1975) is likely a reflection of this. Higher concentrations may provide information regarding host species (ie, high levels of α-pinene are associated with conifers) and its release could signal the presence of an insect attack or an injury that might render the tree more susceptible. It may also communicate the defensive capabilities of the potential host, as has been suggested to occur for less aggressive bark beetle species that are attracted to modest concentrations of host monoterpenes but repelled by higher ones (Erbilgin et al., 2003). Host monoterpenes are toxic to bark beetles (Cook and Hain, 1988; Everaerts et al., 2012), and presumably airborne resin terpene concentrations that exceed a certain threshold should signal a host whose defensive response may be greater than can be withstood by the host-seeking species (Erbilgin et al., 2007b). The very high rates of turpentine odours preferred by *D. frontalis* are simultaneously inhibitory to their significant competitor *I. avulsus* (Billings, 1985). Attractive synergism by very high amounts of α-pinene may be a means for outbreak-level *D. frontalis* to select trees that are too vigorous to be colonized by less aggressive competitors, and thus this behaviour may reduce interspecific competition.

3.5 *trans*-Verbenol

In addition to frontalin, *D. frontalis* females produce *trans*-verbenol, which is found in very large quantities in emergent brood females (ie, microgram amounts) as well as in those initially arriving on the host (Pitman et al., 1968; Pureswaran et al., 2006, 2008a; Renwick, 1967; Sullivan et al., 2007b). It apparently is released quickly upon landing or gallery initiation since feeding females—both solitary and paired with males inside galleries—contain and release much smaller amounts, even within a few hours after entering a host (Coster and Vité, 1972; Hughes, 1973; Pureswaran et al., 2006, 2008b; Sullivan et al., 2012). Its presence is strongly sexually dimorphic in newly emerged beetles (Renwick, 1967) with females containing 10 to >1000 times more than emerged males (Grosman et al., 1997; Pureswaran et al., 2008a; Sullivan et al., 2012). Its production is stimulated in newly enclosed adults of both sexes by exposure to juvenile hormone II or its analogue methoprene (Bridges, 1982). In studies which lacked statistical analysis, *trans*-verbenol sprayed onto the boles of trees under attack did not alter beetle landings (Vité and Crozier, 1968), but it increased responses to field "sleeve" olfactometers baited with frontalin and a low release rate of verbenone (Renwick and Vité, 1969). Payne et al. (1978a) found that *trans*-verbenol significantly increased (~2.5-fold) *D. frontalis* response to traps baited with frontalin, although its enhancing effect was duplicated by releasing host turpentine at a 12-fold greater rate than either compound, and their data likewise indicated a similar redundancy of effects between *trans*-verbenol and α-pinene. This finding is consistent with the observation that addition of oleoresin, which is a potent synergist of frontalin (Kinzer et al., 1969), had no effect in the field on increasing the attractiveness of crushed beetles which presumably would have released large quantities of *trans*-verbenol (Vité and Renwick, 1968). The apparently redundant effects of *trans*-verbenol and host odours on attraction of flying *D. frontalis* may function in allowing first-arriving *D. frontalis* to initiate concentration on a host before the phloem is penetrated and host resin is released (Renwick and Vité, 1969, 1970). This hypothesis is consistent with the observation that in the first stage of mass attack *D. frontalis* will congregate under bark flakes of a target tree prior to mining into the bark in apparent synchrony (Hopkins, 1909; Vité and Crozier, 1968; Vité and Renwick, 1968). This hypothesis is provocative since if beetle concentration were completed prior to entry of the phloem, much of the selective pressure against pioneering behaviour (Latty and Reid, 2010; Pureswaran et al., 2006) would be alleviated. Once resin exudation by the damaged host has been triggered, it is presumed that *trans*-verbenol is no longer relevant to mass attack (Renwick and Vité, 1969). It is likely for this reason that *trans*-verbenol has received very little attention or additional research in efforts to develop semiochemical-based management for *D. frontalis*. Given that most of the female's large "store" of *trans*-verbenol appears to be released before entry into the bark, it is possible that *trans*-verbenol may play a more important role in close-range interactions by beetles on the bark surface (Bunt et al., 1980) rather than or in addition to functioning as a long-range component of the aggregation pheromone. The enantiomeric ratio produced by females has been measured as 60:40 and 25:75 (+)/(−) (Grosman et al., 1997; Plummer et al., 1976), although this ratio is likely influenced by the enantiomeric composition of α-pinene of the host tree. In the mountain pine beetle, *D. ponderosae, trans*-verbenol is the major component of the aggregation pheromone released by solitary females attacking a host and acts synergistically rather than redundantly with host odours (Pitman and Vité, 1969; Skillen et al., 1997). In the western pine beetle, *D. brevicomis*, it is released by arriving females but appears to act as a multifunctional pheromone component, enhancing attraction at a low release rate but decreasing it at a high rate (Bedard et al., 1980a; Byers et al., 1984).

cis-Verbenol is also produced by *D. frontalis* females in modest quantities (Grosman et al., 1997; Hughes, 1973; Pitman et al., 1969; Renwick et al., 1973), and one trapping study indicated that a high release rate (~50 mg/d) of the compound could enhance *D. frontalis* responses to traps baited with frontalin and a low release of α-pinene (Sullivan, 2005). However, significant levels of *trans*-verbenol contamination (1–2%) in the *cis*-verbenol lures in this study could have produced some or all of the attractive effect. *cis*-Verbenol is produced by and has been shown to enhance aggregation in *D. ponderosae* (Miller and LaFontaine, 1991) although it appears to be a far more important component in the pheromone composition of *Ips* spp. (Lanier and Wood, 1975; Vité et al., 1972).

3.6 Role of the Two Sexes in Mediating Mass Aggregation

Early papers on the chemical ecology of *D. frontalis* implied that females were entirely responsible for producing the aggregation pheromone whereas males produced compounds that mediated termination of aggregation and switching of the focus of mass attack (Renwick and Vité, 1970; Vité and Francke, 1976). This conclusion arose because the two major active compounds which appeared to distinguish males (ie, verbenone and *endo*-brevicomin) inhibited responses to attractant-baited traps (Payne et al., 1978a; Renwick and Vité, 1969; Vité and Renwick, 1971). However, attractiveness of artificially infested logs was found to not differ significantly depending upon whether females alone or pairs were present (Coster and Vité, 1972; Coster et al., 1977; Svihra, 1982). These results were inconsistent both with the original hypothesis that the males produce attraction inhibitors and the subsequent discovery that *endo*-brevicomin can be a potent synergist of beetle attraction (Sullivan et al., 2007b; Vité et al., 1985). However, when healthy pines spaced >100 m apart and screened from wild attacks were infested with groups of 100 females, addition of

100 male beetles increased conspecific attraction approximately sixfold over trees to which no males were added (Sullivan et al., 2007b). The inconsistency between the standing tree and log experiments could be related to the relatively higher resin exudation from attacks on the live, vigorous trees (Vité and Crozier, 1968; Vité and Renwick, 1968) and the fact that logs are an unnatural host substrate for *D. frontalis*. Additionally, two of the aforementioned bolt studies were performed in close proximity to *D. frontalis* infestations (Coster and Vité, 1972; Coster et al., 1977); the third did not indicate the relative proximity of the study site to infestations (Svihra, 1982). As discussed earlier, close proximity to active infestations may conceal the attractive effects of *endo*-brevicomin (Sullivan et al., 2011), apparently due to the pheromone component's area-wide synergistic activity.

4. SEMIOCHEMICAL INHIBITION OF CONCENTRATION (ANTIAGGREGANTS)

This list (Figs 6 and 7) includes all compounds (except *endo*-brevicomin, the multifunctional pheromone component which has already been thoroughly discussed) produced by a range of sources, including the beetles themselves, that individually have exhibited the capacity to reduce responses of *D. frontalis* to attractive lures. For compounds produced by the beetles themselves that reduce response to aggregation attractant, the presumed function is as a pheromone which communicates that the host resource has been fully exploited (ie, that there is not sufficient quantity

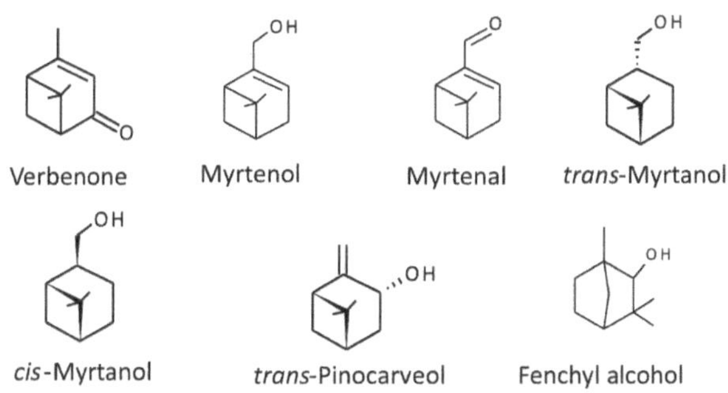

Fig. 6 Oxygenated monoterpenes released by one or both sexes of *D. frontalis* which are capable at some concentration of reducing flying *D. frontalis* response to attractants, and thus these may play a role in terminating mass attack, switching of attack focus to adjacent trees, or indicating exhausted and unsuitable hosts. Both verbenone and myrtenol have been classified as "multifunctional" on the basis of their attractive synergism when presented in low doses in ambulatory olfactometer assays.

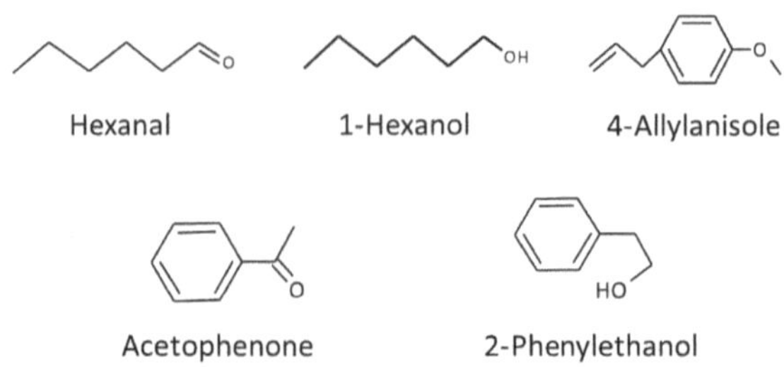

Fig. 7 Non-terpene semiochemicals shown capable individually of inhibiting response of flying *D. frontalis* to attractants. Hexanol and hexanal are "green leaf volatiles" common to deciduous trees; 4-allylanisole is a component of host pine resin. Acetophenone and 2-phenylethanol are produced in small quantities by both sexes of *D. frontalis* and occur in other species of bark beetles and beetle-associated microbes.

or quality of host tissue left available, or—for males—that all available females have paired). Presumably, host/mate-availability-communicating or attack spacing (epideictic) pheromones might be produced by either or both sexes; this is because the outcome should benefit both sexes (and both the signal sender and receiver) by reducing competition for resources by their offspring (Borden, 1997; Byers, 1989a). Semiochemicals that indicate saturation or onset of host tissue unsuitability on a tree undergoing attack could arise from the host tree itself or from other host tissue-colonizing associates including arthropods and microbes.

4.1 Verbenone

The occurrence of verbenone (4,6,6-trimethylbicyclo[3.1.1]hept-3-en-2-one) in male *D. frontalis* has much similarity to *trans*-verbenol in females: emergent males contain quantities which are so great (ie, microgram amounts) as to be apparent as verbenone-dominated droplets of oil in excised hindguts whereas emergent females contain merely nanogram quantities of verbenone (Vité and Crozier, 1968). Likewise these droplets disappear and the quantities measured in hindgut extracts decline rapidly following the male's entry into the bark (Pitman et al., 1968; Pureswaran et al., 2007; Vité and Crozier, 1968). Forcing a solitary male to chew into the bark by confinement appears to cause a similar drop in presence of verbenone as does pairing with a female, suggesting the reduction is due to defecation of the hindgut-stored quantities of the compound and not mating (Sullivan et al., 2007b). Nonetheless, despite this apparent depletion of verbenone amounts in the gut and reduction of amounts in headspace aerations of isolated male beetles before vs after feeding (Pureswaran et al., 2007), aerations of trees undergoing mass attack indicated that entrances of beetle pairs released about twice as much verbenone as entrances of solitary females (Pureswaran and Sullivan, 2012).

Verbenone is frequently called an "antiaggregation pheromone" for *D. frontalis* [and several other species of *Dendroctonus* (Skillen et al., 1997)] based largely on its capacity to reduce beetle responses to attractant-baited traps, and the endurance of this label may be due in part to verbenone's being the first bark beetle-produced compound discovered with anti-attractive effects and therefore to be hypothesized to play a role in terminating mass attack in a bark beetle (Renwick and Vité, 1969). The activity originally proposed for this compound in nature (Renwick and Vité, 1969, 1970) was that at relatively low concentrations (ie, as a mass attack was building) it reduced responses of males thereby bringing the strongly male-biased sex ratio attracted by frontalin closer to 1:1. Higher concentrations inhibited catches of both sexes and were presumed to coincide with host resource depletion, the termination of frontalin production by females, as well as the cessation of resin exudation by the host (Renwick and Vité, 1969). The dose–effect of verbenone on responses to *D. frontalis* to attractant was confirmed in field trials which produced a progressively lower representation of males and then an overall reduction in responding beetles as release rates were increased from 12 up to 240 mg per day (Payne et al., 1978a).

However, the lowest release rate devices that have been shown to reduce *D. frontalis* responses to attractant-baited traps (~25 mg/d) have a greater rate than that of verbenone expected from a single tree undergoing mass attack (~10 mg/d) (Pureswaran and Sullivan, 2012; Salom et al., 1992b). Furthermore, sixteen 5 mg/d devices distributed evenly about the bole and releasing 80 mg/d in total failed to stop mass attack, reduce landings, or hinder brood production by *D. frontalis* in the treated pines (Richerson and Payne, 1979). These data imply that the levels of verbenone released by males on a mass-attacked tree should not be sufficient to alter the progression of a mass attack. Furthermore, aforementioned evidence suggesting that males release much of their verbenone between landing and pairing (ie, when mass attack is ongoing) indicates that the timing of release is not consistent with a major role in shutting down mass attack, an observation that was also made for verbenone as an attraction inhibitor in *D. brevicomis* (Byers et al., 1984).

Verbenone was also classified as a "multifunctional" pheromone because in laboratory tests low concentrations of verbenone increased the frequency that males were arrested over an artificial female entrance releasing female pheromone components; higher concentrations of verbenone caused males to pass the entrance (Rudinsky, 1973b). However, field trapping trials have never demonstrated attraction enhancement of males at any dose tested [although a sufficiently broad "dose–response" test has never been performed (Payne et al., 1978a; Salom et al., 1992a; Sullivan et al., 2007a)]. However, in a single trial of one study (Salom et al., 1992b), a verbenone lure with a high proportion of the (+)-enantiomer significantly enhanced responses of females to attractant-baited traps.

There is also some imprecision in applying the name "pheromone" to verbenone (and, actually, to many oxygenated monoterpenes produced by bark beetles) because it has many sources in nature. Verbenone (and verbenol) can be generated through autoxidation of host-released α-pinene coming into contact with atmospheric oxygen (Hunt et al., 1989; Moore et al., 1956). Microbes including bark beetle symbionts and incidental associates are also capable of oxidizing α-pinene to verbenol and/or verbenol to verbenone (Borden et al., 1986; Brand et al., 1975; Hunt and Borden, 1990; Xu et al., 2015). A fungus carried in the mycangium of female *D. frontalis* (*Ceratocystiopsis ranaculosus* J.R. Bridges and T.J. Perry) was shown to convert verbenol to verbenone (Brand et al., 1976). However, no tests have been reported which demonstrate that such microbes make a significant contribution to semiochemical production by *D. frontalis*-attacked trees in nature, or whether microbial activities have an impact on the species' chemical ecology more generally.

4.2 Other Oxygenated Monoterpenes

In addition to verbenol and verbenone, *D. frontalis* produce a number of other oxygenated monoterpenes some which have demonstrated behavioural activity with conspecifics (Smith et al., 1993) and appear to be derived from precursors in the host oleoresin (Hughes, 1973; Renwick et al., 1973). It is likely that many if not most of these oxygenated monoterpenes arise through a non-specific enzymatic oxidation of unsaturated hydrocarbon molecules at an allylic position, and thus a process that may not necessarily be unique to monoterpene substrates (Renwick and Hughes, 1975). These non-specific oxidations of host monoterpenes were demonstrated through exposure of *D. frontalis* (and other bark beetles) to vapours or cuticular contact with different individual monoterpenes; this resulted in appearance or increase in concentrations of the allylic oxidation products of those specific terpenes (Renwick et al., 1973, 1976a). Such direct oxidations of host monoterpenes have been confirmed in *D. ponderosae* using labelled substrates (Gries et al., 1990a). Quantities produced were dependent on the length of time of exposure, and occurred despite removal of chemical sensory organs, suggesting that the production was not under regulation by feedback through the central nervous system. The existence of non-specific terpenoid oxidation mechanisms likely has its origins in the beetle's need to detoxify the monoterpenes to which they are exposed during invasion of the host (Blomquist et al., 2010; Seybold et al., 2006), as oxidation increases solubility and facilitates elimination (Harborne, 1988). Because of the shared need to detoxify resin of similar monoterpene compositions, pine beetle species tend to produce many of the same oxygenated monoterpenes "randomly" (Francke and Vité, 1983), and thus these products can provide little signal specificity or consistency. However, the spectrum and prominence of specific oxidation products may differ among species and between sexes [for example, *D. frontalis* males produce almost entirely monoterpene aldehydes and ketones whereas females appear to be limited largely to producing alcohols (Renwick et al., 1973)]. This implies that oxidases with differing degrees of substrate affinity vary in their abundance among bark beetle species and sexes.

A diversity of such oxygenated terpenes has been identified from *D. frontalis*; most occur in relatively small quantities, and several have exhibited behavioural activity (Sullivan, 2005). Myrtenol occurs in both sexes of *D. frontalis* (Grosman et al., 1997; Renwick et al., 1973; Sullivan et al., 2012), and laboratory assays indicated a "multifunctional" behavioural response to this compound, with low concentrations increasing beetle arrestment at artificial female entrances and with higher concentrations inhibiting arrestment (Rudinsky et al., 1974). Devices releasing 1.5–3 mg/d significantly reduced *D. frontalis* catches in traps baited with frontalin and α-pinene (Sullivan et al., 2007a). Myrtenal, *trans*-pinocarveol, *cis*-myrtanol, *trans*-myrtanol, and fenchyl alcohol are produced by both emergent and feeding *D. frontalis* and elicit EAD responses; at a release of $\leq$66 mg/d all significantly reduced *D. frontalis* responses to traps baited with frontalin and α-pinene (Sullivan, 2005). Several of these compounds are produced by the beetles in very small quantities and it is unclear whether the inhibition of attraction reflects their having a function in mediating intraspecific interactions in nature, a requirement for classifying these compounds as "genuine" pheromone components. Rather, these and additional common oxygenated monoterpenes may be more general infochemicals which indicate the early stages of a decaying and thus unsuitable host

(Flechtmann et al., 1999; Lindgren and Miller, 2002). As with verbenone, many may be generated through the enzymatic oxidation activities of microbes in the host tissues or autoxidation of host resin (Flechtmann et al., 1999; Hunt et al., 1989; Sullivan, 1997; Sullivan et al., 2000).

4.3 Non-Monoterpene Aromatics

Both sexes of *D. frontalis* produce both acetophenone and 2-phenylethanol in small amounts whether newly emerged, feeding solitary, or paired (Sullivan, 2005; Sullivan et al., 2012). Both compounds inhibit responses of flying beetles to traps baited with frontalin and α-pinene (Sullivan, 2005; Sullivan et al., 2007a), and 2-phenylethanol reduced responses of beetles to attractant in an ambulatory olfactometer (Brand et al., 1977). These compounds have been detected in other species of *Dendroctonus* and *Ips* (Kohnle et al., 1987; Pureswaran and Borden, 2004; Pureswaran et al., 2000; Renwick et al., 1976b; Zhang et al., 2007), and have demonstrated behavioural activity with some (Erbilgin et al., 2007a, 2008; Pureswaran et al., 2000). It was shown in *Ips pini* that 2-phenylethanol is generated from the amino acid phenylalanine, and both this compound and acetophenone appear to be amino acid derivatives (Gries et al., 1990b; Seybold and Vanderwel, 2003). 2-Phenylethanol is also generated by microbial associates of *D. frontalis* growing in culture (Brand et al., 1977).

5. ATTACK SEQUENCE

Fig. 8 illustrates the process by which a *D. frontalis* infestation becomes initiated and begins to grow through the mediation of semiochemicals discussed above. The functional roles of most individual compounds are not well studied and what we know of them is based on inferences from field trapping experiments and walking olfactometer trials in the laboratory. Therefore this "model" like those that have been published in the past, eg (Payne, 1980; Renwick and Vité, 1969, 1970) is largely conjecture beyond the information detailed earlier in this review, and the model should be treated merely as a likely scenario given the limited available data.

5.1 Establishment and Initiation of Spot Growth

Infestations most frequently originate in the spring which is the major period of *D. frontalis* dispersal (Thatcher and Pickard, 1964, author's unpublished

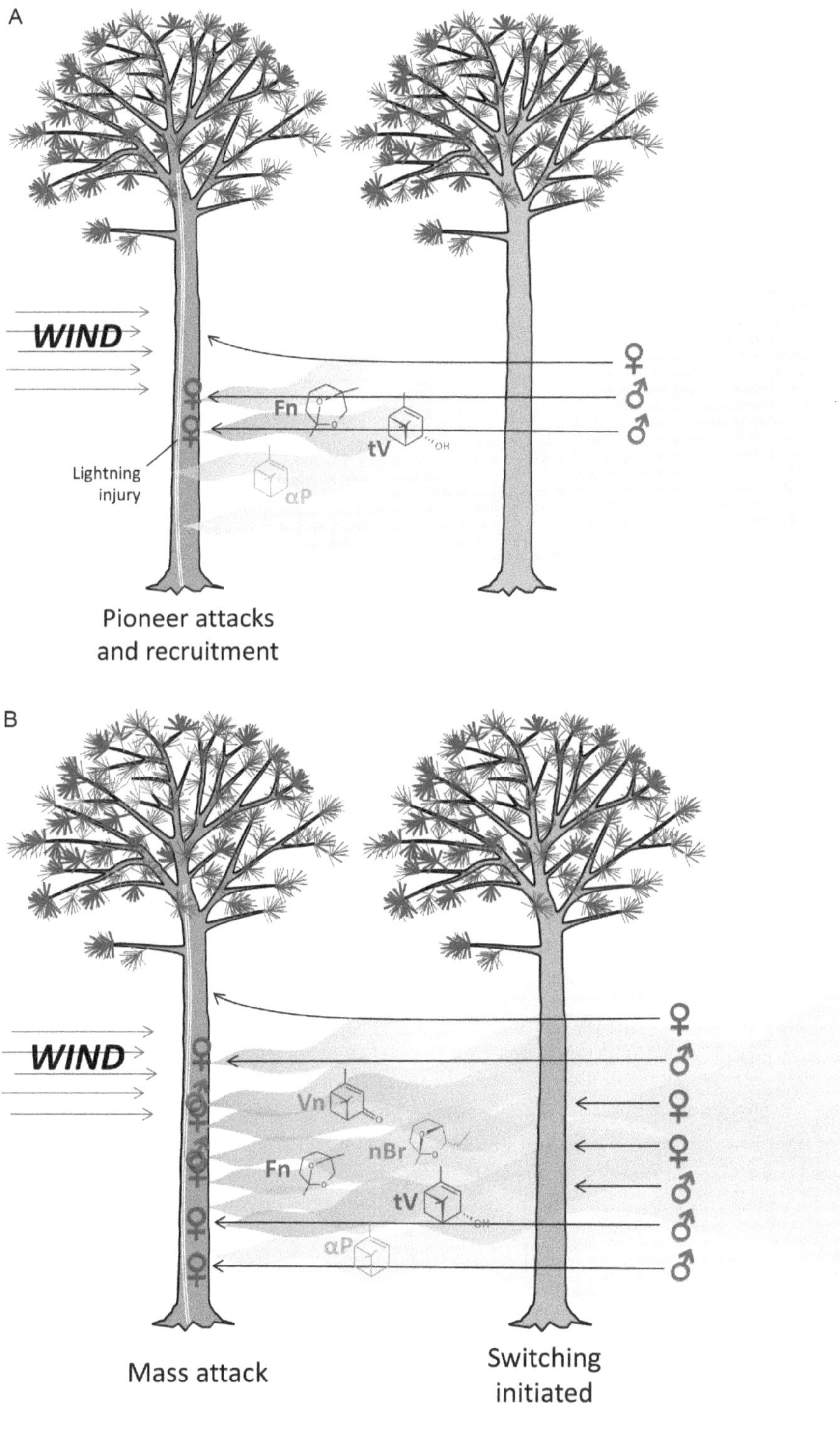
A
WIND
Fn
tV
OH
Lightning
injury
αP
Pioneer attacks
and recruitment
B
WIND
Vn
nBr
Fn
tV
OH
αP
Mass attack
Switching
initiated

Fig. 8—Cont'd

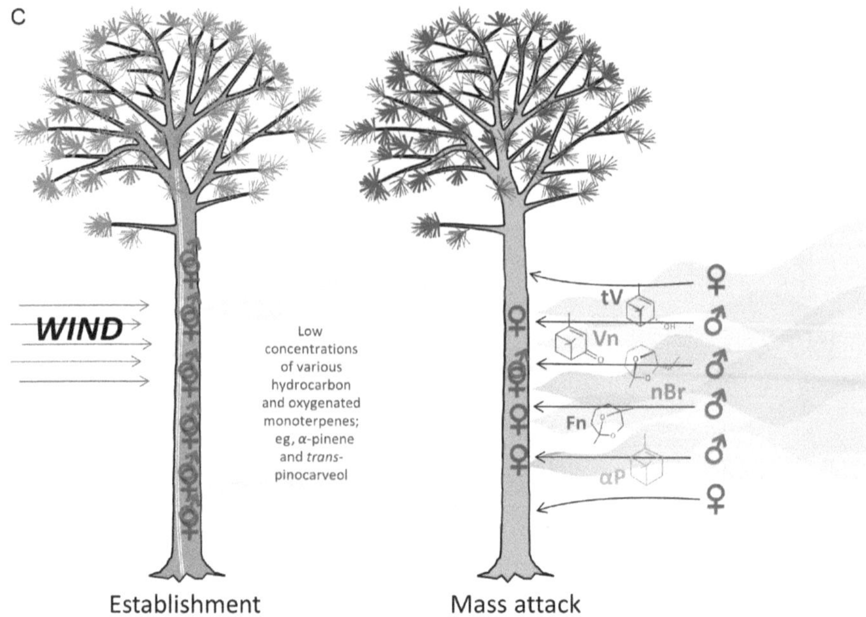

Fig. 8 (A) Stage I: Initiation of a *D. frontalis* infestation. Dispersing pioneer female beetles locate a lightning-weakened (or otherwise susceptible) pine apparently by random searching. They then release frontalin (Fn) and *trans*-verbenol (tV) which function synergistically to attract conspecifics of both sexes. Host-produced α-pinene (αP) from the lightning wound (and later from the females' entrances once they penetrate the phloem) may act as a synergist that duplicates the effects of the *trans*-verbenol. Attacking females establish nuptial chambers if resin defences are sufficiently low; otherwise they may expel resin from the gallery entrances to form pitch-tubes while continuing to release aggregation pheromone. *endo*-Brevicomin (not shown) from the first arriving males synergizes attraction to the pheromone-releasing females. Further arriving females, deterred by high concentrations of frontalin odours, tend to land above and below the area of initial female attack at mid-bole. (B) Stage II: Mass attack and initiation of switching of attack focus to an adjacent tree. Large numbers of beetles of both sexes arrive on and attack the original pine in response to the aggregation attractant of the first and subsequent arrivers. Males locate attacks of females and pair with them; they assist the female with clearing the gallery entrance while they simultaneously release both verbenone (Vn) and *endo*-brevicomin (nBr). These two compounds indicate to mate-seeking males on the bark surface that a female in an entrance has already paired, in which case a searching male must either remove the occupying male forcefully or continue looking for an unpaired female. At high concentrations (ie, with a high density of males), *endo*-brevicomin may also deter beetle landing, as may verbenone. However, male-produced *endo*-brevicomin strongly synergizes responses of flying beetles to solitary female attacks within a distance of at least several metres of the releasing males, and this aids in synergizing responses of beetles to "pioneer" female attacks on surrounding trees. Paired females on the original tree continue to release frontalin (Fn) and *trans*-verbenol (tV) at reduced levels, and synergistic α-pinene continues to be

data). Dispersing females may locate a standing tree weakened by a disturbance (such as a lightning strike) or other cause, land near mid-bole (Coster et al., 1977), and begin releasing pheromone and perhaps initiate attack on the bark (Fig. 8). High levels of α-pinene released from the tree's injuries may arrest dispersing beetles post-landing, but there is no evidence that a primary attractant draws *D. frontalis* pioneer females to a susceptible tree. However, the α-pinene released from the injury should nonetheless act as a synergist for the frontalin being released by the pioneer females and thus increase attraction. The pheromone release by the pioneer females (frontalin and to a lesser extent *trans*-verbenol) plus the host odours from the tree attract dispersing beetles of both sexes to the tree which, through the positive feedback of increasing numbers of attacking beetles and associated increasing pheromone, trigger mass attack. Males that land and pair with females release large quantities of verbenone and lesser quantities of *endo*-brevicomin. The *endo*-brevicomin released by the initially arriving males synergizes the attraction of both sexes to female attacks on the tree. Arriving females avoid landing close to areas of high release of frontalin produced by earlier female arrivals and instead land above or below the portions of the bark already colonized by females. This may in part cause the expansion of the infested portions of the bole both upward and downward (Hughes, 1976). For reasons already discussed, the precise role of verbenone in regulating mass attack—if any—is uncertain.

Fig. 8—Cont'd released from damaged host tissue. A plume of attractant and synergists encompasses downwind, unattacked trees; upwind-orienting (anemotactic) females land on them and may begin to bore if the semiochemical concentration of the engulfing plume is sufficiently elevated. These attacks initiate the shift of the focus of attack to the downwind tree. (C) Stage III: Termination of attack on the original tree and re-focus of attack onto the "switch tree". After the original tree ceases to produce a constitutive resin response and has reached its carrying capacity of parent beetle pairs, each pair plugs its gallery entrance with frass and begins to extend the gallery in the phloem while laying eggs. This event coincides with a substantial reduction in semiochemical release (particularly of beetle pheromone components) from the original tree. Nonetheless, the tree continues to release odours for many weeks, including a variety of possibly inhibitory/repellant oxygenated monoterpenes generated from spontaneous and microbially mediated oxidation of the dead host's residual resin monoterpenes. The foliage of the tree crown begins to fade at this time. Meanwhile, a mass attack proceeds on the new focus tree as it had on the original tree, with its resulting semiochemical plume likewise being capable of starting attacks on nearby, particularly downwind trees. Parent adults reemerging from the original tree (and later, the emerging, developed brood) may participate in mass attack of new focal trees, and thus these beetles may initiate and then maintain a continuous cycle of new attacks by both re-emergent parents and newly emerged brood beetles that can propel a steadily growing infestation and continue to consume trees over a period of months. (See the colour plate.)

The plume of pheromone components is carried downwind where, if it engulfs a nearby host pine and the frontalin in the plume is of sufficiently high concentration, upwind-flying females responding to the attractant will be stimulated to land, release pheromone, and possibly begin mining into the tree (Gara, 1967; Gara et al., 1965) (Fig. 8B). *endo*-Brevicomin released by the growing number of males on the original tree should tend to synergize frontalin released from increasingly more distant attacks by solitary females (ie, including downwind trees) while reducing synergism and attraction at and close to the point of release. The combined effects of the high concentrations of frontalin and resin odours (and possibly also *trans*-verbenol) engulfing the adjacent trees, and the synergistic effects of *endo*-brevicomin, result in increasing attacks on one or more adjacent and typically downwind trees (Gara and Coster, 1968). Landings and attacks on the original tree decline as all of the available host tissue becomes colonized (ie, the host reaches saturation), and established beetle pairs cease producing pheromone components as they proceed to mine into the phloem and close-off the gallery entrance with their frass. This is accompanied by a decrease in resin exudation by the host tree (Vité and Crozier, 1968), an event that reduces release of synergistic α-pinene (Pureswaran and Sullivan, 2012). The newly attacked trees then exceed the attractiveness of original tree and therefore become the new foci of attack. The process of "switching" is then complete (Fig 8C). The pheromone plume from the newly mass-attacked trees then in turn induces attacks on adjacent healthy trees and propels the cycle of spot growth.

5.2 Persistence and Termination of Spot Growth

As implied by the above model, spot growth can persist only as long as there are (1) attacked trees releasing pheromone and host odours at sufficient rates to induce attacks on adjacent trees (Gara and Coster, 1968) (2) suitable uninfested trees adequately close to the pheromone source tree(s) that landing and mass attack can be induced on them by the semiochemical plume (Johnson and Coster, 1978; Schowalter et al., 1981) and (3) local beetle population levels that are sufficient to successfully mass attack the "next" tree(s) (Clarke, 2012; Gara and Coster, 1968).The obstacle posed to the beetles by requirement (3) is at least partly addressed once brood begin emerging from earlier-attacked trees since these brood beetles fly to the spot head and contribute to sustaining fresh mass attacks and uninterrupted presence of the aggregation attractant (Gara, 1967). Additionally, parent adult beetles typically reemerge and contribute to new mass attacks.

Switching will be prevented and infestations will collapse (or never become initiated) if the above three conditions are not met, and elimination of one or more of these conditions is the biological basis for methods currently being applied for direct control of *D. frontalis*, which are aimed primarily at stopping spot growth. It has been demonstrated that removal of the currently mass-attacked trees (ie, the source of aggregation pheromone) within an infestation can often be a sufficient measure to stop spot growth (Clarke et al., 1999; Cronin et al., 1999; Gara, 1967). In addition, the recommended cutting of a "buffer" of unattacked trees in advance of the mass-attacked trees increases the distance of the nearest suitable host tree to the natural source of *D. frontalis* attractant. If infested trees with beetle brood are removed from the site after cutting, this reduces the local densities of beetles available to sustain mass attacks.

6. GEOGRAPHIC VARIATION

Intraspecific geographic variation in the pheromone composition of bark beetles is common and may occur across very small distances (Lanier and Burkholder, 1974; Lanier et al., 1972; Miller et al., 1989; Seybold, 1993). Cross-mating experiments among *D. frontalis* populations from the southeastern United States, Arizona, and Nuevo Leon, Mexico as well as from Texas and Honduras (Lanier et al., 1988; Vité et al., 1974) suggest the lack of post-mating reproductive isolation mechanisms among these populations. Furthermore, consistency in seminal rod morphology and karyotype further supports the validity of *D. frontalis* as a single species throughout its extensive range (Armendáriz-Toledano et al., 2014; Lanier et al., 1988). However, there is evidence of some variation in pheromone production within the species. Newly emergent males in Honduras and Chiapas, Mexico, and Arizona contain substantially larger quantities of *endo*-brevicomin than beetles sampled in Texas or Mississippi (Pureswaran et al., 2008a; Sullivan et al., 2007b, 2012; Vité et al., 1974). The biological implications of this difference are unknown. A study which examined volatiles from emergent male and female *D. frontalis* discovered some quantitative differences in pheromone component production by beetles in North Carolina, South Carolina, and Texas (Grosman et al., 1997). Nevertheless, in general the pheromone component blends produced by *D. frontalis* are remarkably consistent across its vast range (Fig. 9) and, at least for females, are distinguished from those of sibling species that have been studied (Niño-Domínguez et al., 2015a; Pureswaran et al., 2008a; Renwick et al., 1975; Sullivan et al., 2012; Vité et al., 1974).

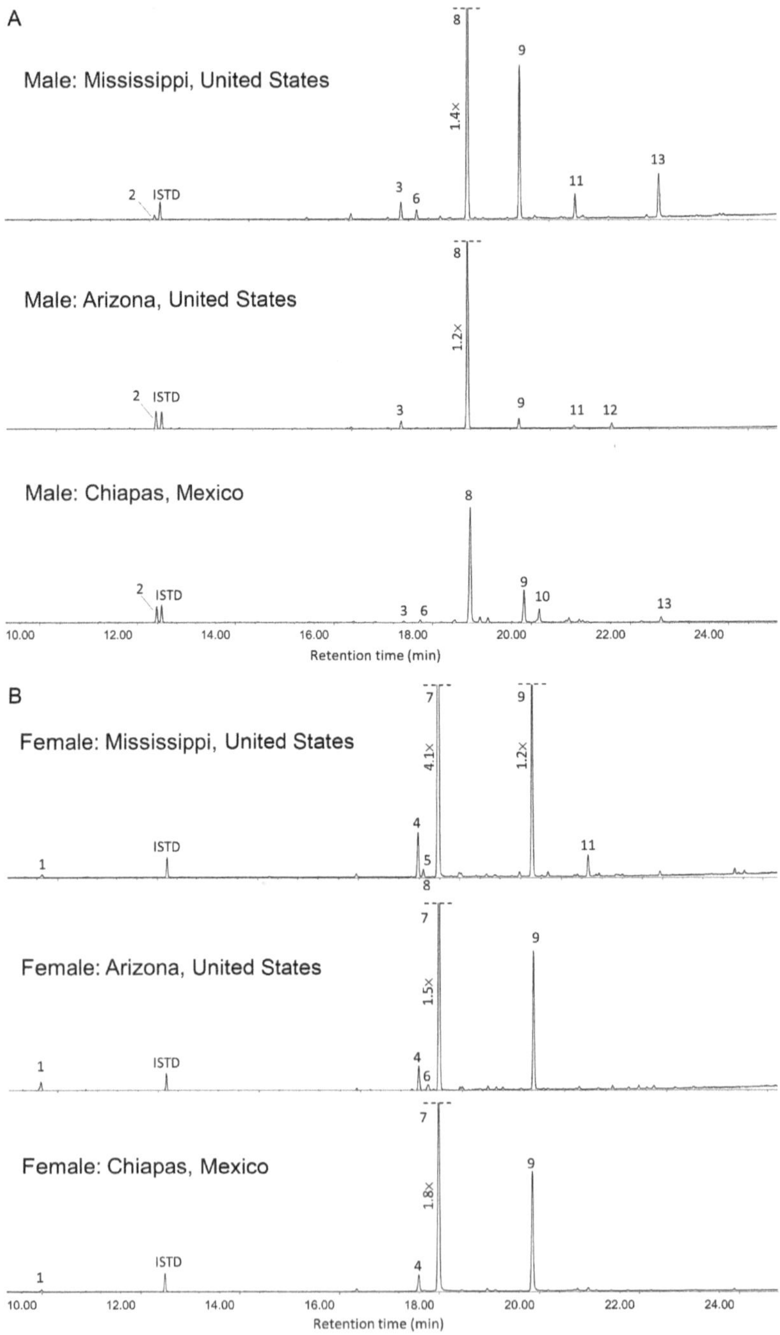

Fig. 9 See legend on opposite page.

In the only experiment in which geographic variation in *D. frontalis* pheromone responses was directly investigated, *D. frontalis* in a walking olfactometer responded more frequently to odours from logs artificially infested by beetles from the same population than distant populations, with these populations being derived from Georgia, Texas, and Virginia, United States (Berisford et al., 1990). However, the chemical composition of the volatiles released by these infested logs was not examined for regional differences. In general, responses by *D. frontalis* to semiochemical lures appear to be similar wherever in the range of the insect that trapping experiments have been performed. For example, in both Chiapas, Mexico, and in Mississippi, United States (in trapping experiments performed in portions of the forest located away from infestations), *D. frontalis* were strongly attracted to the combination of frontalin and turpentine, and this combination was synergized by the addition of *endo*-brevicomin but unattractive without frontalin (Moreno et al., 2008; Niño-Domínguez et al., 2015a; Sullivan et al., 2007b). Tests performed near *Dendroctonus* infestations in Honduras indicated *D. frontalis* attraction to frontalin and host odours but inhibition by *endo*-brevicomin (Vité et al., 1974); the same outcome was observed in studies performed in *D. frontalis* infestations in the southern United States (Payne et al., 1978a; Salom et al., 1992b; Sullivan et al., 2011). It is possible that the "reversing" of the attraction-enhancing effects of *endo*-brevicomin lures associated with close proximity of infested trees or other sources of *endo*-brevicomin is consistent between regions.

7. MATE FINDING AND SELECTION

7.1 Aggregation vs Sex Pheromones

Frontalin appears to be the critical pheromone component which attracts alighted male *D. frontalis* to a gallery entrance occupied by a solitary female (Niño-Domínguez et al., 2015b). When an odour blend associated with solitary female entrances (ie, frontalin, *trans*-verbenol, and α-pinene) was

Fig. 9 Representative total ion chromatogram traces of static headspace aeration (18 h) plus hindgut contents for newly emerged male (A) and female (B) *D. frontalis* collected from three distant populations. Each trace is the analysis from a single insect, and the sample chosen for presentation here was selected randomly from a pool of sample analyses. Compounds are (1) frontalin, (2) *endo*-brevicomin, (3) myrtenal, (4) *cis*-verbenol, (5) *trans*-pinocarveol, (6) acetophenone, (7) *trans*-verbenol, (8) verbenone, (9) myrtenol, (10) 1-phenylethanol, (11) *trans*-myrtanol, (12) 2-phenylethanol, (13) unknown—likely a dioxygenated monoterpene. "ISTD" is the internal standard (heptyl acetate at 175 ng/sample). *Procedures as Sullivan et al. (2012).*

released from an artificial gallery entrance covered by a screen (Ryker, 1988), males were arrested above the screen and stridulated producing the "courtship chirp" which they also produce when digging through frass at the gallery entrance of a solitary female (Rudinsky and Michael, 1974). Hence compounds in the female portion of the aggregation-stimulating blend also release male behaviours associated specifically with courtship and pairing. Furthermore, when the female-associated blend was combined with the male pheromone components *endo*-brevicomin (released over a wide range of doses) or verbenone (at high concentrations), males were less frequently arrested at the entrance and produced a "'rivalry chirp" associated with male–male encounters (Rudinsky, 1973b; Rudinsky et al., 1974). Presumably these pheromone components signal the presence of a male in the gallery (Rudinsky and Michael, 1974). The contrast between the consistent inhibitory effects of *endo*-brevicomin in laboratory trials with walking male *D. frontalis* (Niño-Domínguez et al., 2015a; Rudinsky et al., 1974) and the attractive effects of *endo*-brevicomin on flying males (ie, at low doses and away from infestations) indicates that the signal may have a different, context-dependent meaning during search for either an active infestation or a host tree (where *endo*-brevicomin could signal that successful colonization by conspecifics is occurring) or a mate (in which case it indicates a prospective mate is already paired).

7.2 Pheromones in Reproductive Isolation

Little research has been done on the role of *D. frontalis* pheromones in mediating reproductive isolation from closely related species, although several such species may occupy the same portions of hosts (ie, exist in syntopy), utilize the same aggregation pheromone components, and be capable of pairing in the laboratory (Armendáriz-Toledano et al., 2014, 2015; Davis and Hofstetter, 2009; Hofstetter et al., 2008, 2012; Lanier et al., 1988; Moser et al., 2005; Niño-Domínguez et al., 2015a; Sullivan et al., 2012). In olfactometer studies of *D. frontalis* and its sibling *D. mesoamericanus* which jointly colonize the same hosts in the Central American region, it was found that males could readily distinguish odours of conspecific and heterospecific female gallery entrances, and ipsdienol and *endo*-brevicomin (compounds produced by *D. mesoamericanus* but not *D. frontalis* females) were identified as the species-specific cues that mediated this discrimination (Niño-Domínguez et al., 2015a).

The seemingly small amount of variation in pheromone composition and production across widely separated populations of *D. frontalis* contrasts with a high degree of individual variation in pheromone production within populations (Renwick and Vité, 1970), such that the coefficient of variation for female-produced pheromone components including frontalin and *trans*-verbenol typically exceeds 100% (Pureswaran et al., 2007). It is possible that aggregation behaviour and communal production of pheromones decreases the importance of any single insect's contribution to the aggregation pheromone plume and thus to successful mass attack. Thus natural selection may be limited in its influence on the composition of the aggregation pheromone (Pureswaran et al., 2007). However, male (and possibly female) discrimination and selection of mates at gallery entrances on the basis of pheromone composition should not be subject to the diluting effects of communal production of aggregation pheromones, and thus there may be stronger selection on composition of the pheromone blend in its function as a sex pheromone than as an aggregation pheromone.

A significant degree of cross-attraction between *D. frontalis* and syntopic species of *Dendroctonus* to their respective aggregation pheromones (Armendáriz-Toledano et al., 2014; Hofstetter et al., 2008, 2012; Niño-Domínguez et al., 2015a) would seem to be at odds with pressures to maintain reproductive isolation between such species. However, it is likely that mass attacks by multiple-species may be more successful than single-species attacks if a single species is at insufficient numbers to mass attack a tree (Økland et al., 2009, 2011). Thus sharing of aggregation pheromone components may mediate such joint mass attacks and be a selective force that sustains overlap in pheromone composition among species (Symonds and Elgar, 2004). Differences in pheromone production (and other traits) in solitary females of differing species would confer reproductive isolation once beetles had landed.

8. SEMIOCHEMICAL INTERACTIONS WITH ARTHROPOD ASSOCIATES

8.1 Interspecific Pheromone Interactions with Other Bark Beetles

Within the southeastern United States, *D. frontalis* belongs to a guild of five species of bark beetles that may directly compete by infesting the same or adjacent tissues of the same pines (Coulson et al., 1986; Flamm et al., 1993; Nebeker, 2011). The "southern pine bark beetle guild" includes *Ips avulsus* (Eichhoff), *I. grandicollis* (Eichhoff), *I. calligraphus* (Germar), and *D. terebrans* in addition to *D. frontalis*. Each produces a unique aggregation/sex pheromone blend that functions in attracting mates and possibly in aggregating conspecifics for mass attack (Smith et al., 1993). The species tend to partition the host tree by colonizing different portions of the bole but there is nonetheless substantial niche overlap and competition among them (Paine et al., 1981). Within established, growing *D. frontalis* infestations, *D. frontalis* represents the pioneer species that selects trees and initiates mass attack on new trees; other members of the guild typically follow (Stephen, 2011). However, during non-epidemic periods or initiation of infestations (ie, during first attacks on an isolated, susceptible tree) the order of arrival of species within the guild and their relative abundance may vary (Birch et al., 1980; Coulson et al., 1986; Flamm et al., 1993; Paine et al., 1981; Svihra et al., 1980). One might expect pheromones produced by each member of such a guild to mediate at least one of three potential interspecific responses: (1) synchronized, multispecies mass attack in which species are simultaneously engaged in overcoming host defence, (2) sequential species attack due to species that "easvesdrop" on pheromones of the earlier-arriving, mass-attacking species and then colonize only after host defences are rendered sufficiently neutralized, (3) cross-inhibition to reduce competitive interactions through host partitioning. Given the seeming fitness benefits of any or all of the three responses above, it is surprising both that (1) *Dendroctonus frontalis* shows no attraction to the pheromone blends associated with any of the *Ips* species in the southern guild or to *Ips* more generally (Birch et al., 1980; Hofstetter et al., 2012; Svihra et al., 1980) and (2) the other *Ips* species display no attraction to *D. frontalis* with the possible exception of a weak positive response by *I. grandicollis* (Svihra et al., 1980). Cross-inhibition of the *Ips* spp. with *D. frontalis* is also limited. Flying *D. frontalis* appeared to be less attracted to logs infested simultaneously by both conspecific females and *I. grandicollis* males, and the paper's authors argued that this response could relate to interspecific bole partitioning, with *I. grandicollis* infesting lower branches and *D. frontalis* infesting the bole (Svihra et al., 1980). In both laboratory and field experiments in Chiapas, Mexico, the *Ips* pheromone component ipsdienol strongly inhibited responses by *D. frontalis* to attractant (Niño-Domínguez et al., 2015a,b), although in Chiapas ipsdienol is produced by potentially sympatric *D. mesoamericanus* and possibly male *D. frontalis* (Sullivan et al., 2012). As discussed earlier, responses to mutually produced, attractive pheromone components by *D. terebrans* and *D. frontalis* could potentially mediate cross-attraction and joint host colonization (Smith et al., 1990), but this possibility has not been sufficiently examined. Additionally, given evidence for cross-attraction between *D. frontalis* and *D. brevicomis* (Hofstetter et al., 2012; Pureswaran et al., 2008a) and between *D. frontalis* and *D. mesoamericanus* (Niño-Domínguez et al., 2015a), it would appear that for *D. frontalis* that bark beetle cross-attraction may be limited to congeners.

8.2 Exploitation of *D. frontalis* Pheromone by Predators and Parasitoids

Natural enemies have evolved attractive responses to the aggregation pheromones of their prey, presumably because these cues are highly reliable and detectable indicators of prey presence (Steidle and van Loon, 2002; Vet and Dicke, 1992). Apparency to predators is perhaps an unavoidable cost of the use of aggregation pheromones by aggressive bark beetle species during the colonization of trees (Raffa and Dahlsten, 1995; Raffa and Klepzig, 1989). The clerid beetle, *Thanasimus dubius* (Fabricius), is an abundant and highly visible predator of *D. frontalis* populations in the southeastern and eastern United States, and evidence indicates that it may be an important mortality agent (Moore, 1972; Reeve, 1997). It captures and consumes *D. frontalis* as they land on the bark of host trees undergoing mass attack, and its larvae, developing from eggs laid within bark crevices, move throughout the gallery system of *D. frontalis* and consume beetle brood (Thatcher and Pickard, 1966). There are obvious fitness benefits accrued by *T. dubius* in arriving precisely when *D. frontalis* are landing (Camors and Payne, 1973; Dixon and Payne, 1979a, 1980; Thatcher and Pickard, 1966) hence their attraction to the *D. frontalis* aggregation attractant achieves this purpose well. *Thanasimus dubius* apparently locate *D. frontalis*-attacked trees primarily by responding to frontalin and host monoterpenes (Vité and Williamson, 1970). In experimental trials the host component could be presented as either α-pinene or distilled pine resin (Billings, 1985; Billings and Cameron, 1984; Costa and Reeve, 2011; Staeben et al., 2015; Vité and Williamson, 1970). *Thanasimus dubius* may sometimes be observed congregating on pines where *D. frontalis* brood beetles are emerging (Clarke and Menard, 2006), presumably because the emergent *D. frontalis* contain their aggregation pheromone (Pureswaran et al., 2007) and are apparently "leaking" it in sufficient quantities during emergence to attract this predator. *Thanasimus dubius'* response is more strongly synergized by higher concentrations of host odours (eg, g/d levels from trap lures) in parallel with their *D. frontalis* prey (Billings, 1985, author's unpublished data). However, *T. dubius* appears to respond solely to the (−)-enantiomer of frontalin, whereas its prey *D. frontalis* responds to both enantiomers but produces predominantly (−) (ie, 85–95%); moreover *T. dubius* may entirely lack the capacity to sense the (+)-enantiomer (Payne et al., 1984). These authors suggested that *T. dubius* may be a semiochemical specialist on *D. frontalis*. This idea is supported by the generally stronger attraction by *T. dubius* to frontalin than pheromone components of its alternative *Ips* prey, as well as strong responses to frontalin by *T. dubius* populations located outside the range of *D. frontalis* or any other known frontalin-producing prey species (Aukema and Raffa, 2005; Reeve et al., 2009).

Investigations of predator–prey interactions in other bark beetle systems have suggested that disparities between the maximally attractive semiochemical blends for bark beetle prey and their predators may provide a "chemically mediated predator free space" that may have resulted from directional selection by predators on the pheromone composition of their prey (Aukema and Raffa, 2000; Aukema et al., 2000; Dahlsten et al., 2004). Such semiochemical-based predator escape may involve differing predator/prey responses to components of the prey's aggregation pheromone (Dahlsten et al., 2004; Raffa and Klepzig, 1989) as well as the host tree-produced pheromone synergists (Erbilgin and Raffa, 2001). In fact, *T. dubius* and their *D. frontalis* prey are not maximally attracted to the same semiochemical blend, suggesting a less-than-ideal "search image" is used by *T. dubius* in seeking its *D. frontalis* prey. Despite the strongly synergistic effect of aggregation pheromone component *endo*-brevicomin on attraction of *D. frontalis* to baited traps (Sullivan et al., 2007b; Vité et al., 1985) and thus its being a presumable indicator of localized concentrations of *D. frontalis*, *endo*-brevicomin does not appear to enhance attraction of *T. dubius* (Mizell et al., 1984; Richerson and Payne, 1979; Salom et al., 1992b), and neither does the *exo*-isomer which likewise can function as an attractive synergist for *D. frontalis* (Pureswaran et al., 2008a). A similar situation occurs with the aggregation pheromone component lanierone in *Ips pini* (Say) which is a potent attractive synergist for conspecifics but has no effect on attraction of predators including *T. dubius* (Aukema and Raffa, 2000). Lack of response by *D. frontalis*-seeking *T. dubius* to *endo*-brevicomin may be due to the fact that *endo*-brevicomin does not direct landings by *D. frontalis* to its point of release [whereas frontalin and host odours apparently have this effect; (Sullivan and Mori, 2009)], and furthermore since it is a "multifunctional" pheromone component it can have inhibitory effects on *D. frontalis* at high concentrations (Sullivan et al., 2011). As such, its presence might not increase the odds of a *T. dubius* encountering a prey insect at least at close range. A mismatch of attractive cues also occurs due to a preference by *D. frontalis* for the (+)-enantiomer of the synergistic host monoterpene α-pinene when combined with its aggregation pheromone, whereas no preference for enantiomers occurs in *T. dubius* (Staeben et al., 2015). As the host species of pine utilized by *D. frontalis* vary in the enantiomeric composition of α-pinene, presumably *D. frontalis* might enhance escape from predation in preferential attack of trees with relatively high (+)-α-pinene composition, such as *Pinus taeda* L. which is a preferred host for *D. frontalis* (Phillips et al., 1999). Another abundant predator apparently attracted to the pheromone components of *D. frontalis* is *Medetera bistriata* (Diptera: Dolichopodidae), which lay their eggs in the newly created beetle galleries such that their hatching larvae can feed on the early brood stages of their host (Dixon and Payne, 1979b).

In contrast to the predators, the guild of hymenopteran parasitoids of *D. frontalis* (which include at least eight common species predominantly in the families Braconidae and Pteromalidae) generally do not respond to the beetles' aggregation pheromone components (Dixon and Payne, 1979b; Payne, 1989). Pheromone release by the mass-attacking parent *D. frontalis* is finished well before the preferred host life stages (late-instar larvae and pupae within the bark) for the parasitoids are available (Berisford, 2011), hence the pheromones are not suitable cues. Rather, at least some of the parasitoid species appear to be attracted to as-yet unidentified olfactory cues (which likely include oxygenated monoterpenes) associated specifically with bark infested with their preferred host life stages (Sullivan et al., 1997, 2000).

9. EXPLOITATION OF SEMIOCHEMISTRY IN MANAGEMENT OF *D. FRONTALIS*

Not surprisingly, research on the semiochemical system of this serious pest has been driven largely by the desire to develop tools that can lessen its economic and environmental impact. With the first successful syntheses of attractant and attraction-inhibiting semiochemicals, efforts were initiated immediately at using each to manipulate movement of *D. frontalis* populations to disrupt their capacity to congregate on and kill trees (Vité, 1970; Vité and Francke, 1976). Despite the diversity of ways researchers have attempted to use attractants and inhibitors to prevent or reduce tree mortality from *D. frontalis*, no semiochemical-based technology is currently in common use for stand or tree protection. The research efforts are detailed in this section. Greater success can perhaps be claimed with regard of the use of attractants to monitor population fluctuations and make forecasts regarding outbreaks; this technology is in wide use and is a key component of state and federal efforts to address *D. frontalis* on an annual basis.

9.1 Population Monitoring and Prediction

Within the southeastern United States, attractant-baited traps are deployed by state and federal cooperators each spring (at bud-break of dogwood, *Cornus florida* L., which coincides approximately with *D. frontalis*' spring flight) and are used for making forecasts of the likelihood of *D. frontalis* outbreaks and their severity for the coming summer (Billings, 1988, 2011). Data on both the average daily catches of *D. frontalis* per trap as well as ratios of catches of *D. frontalis* to *T. dubius* are data used in a model which forecasts outbreaks with a demonstrated 60–80% reliability (Billings, 2011; Billings and Upton, 2010). These forecasts, which are completed by early summer, are then used by regional land managers in planning activities (eg, scheduling flights by reconnaissance aircraft) and in coordinating monetary and human resources necessary for beetle management activities expected during the coming months. Since the beginning of the trapping programme during the 1980s, the attractant has consisted of frontalin (4–8 mg/d) released with a high rate (ie, 1–10 g/d) of host odours. Initially, the host odour component consisted of turpentine distilled from *P. taeda*, but due to the loss of commercial sources of suitable turpentine during the 2000s, the host component was switched to a synthetic blend of α- and β-pinene. Recent efforts at enhancing the existing monitoring/forecasting programme have considered a range of modifications of methodology including revision of the composition of the trap lure (including the possible inclusion of *endo*-brevicomin to enhance sensitivity) and addition of further monitoring periods during the year (Billings, 2011). The beetle-produced (+)-enantiomer of *endo*-brevicomin and the racemate (at a doubled dose) appear to be similarly potent attractive synergists (Sullivan and Mori 2009), and thus the far less expensive racemate is recommended for inclusion in enhanced monitoring lures.

9.2 Direct Control with Semiochemicals

Development of semiochemical technology for direct control of *D. frontalis* during epidemics has focused on two spatial scales. (1) Stand level: suppression or reduction in the rate of growth of individual infestations and (2) Tree level: protection of individual, high value trees at risk of imminent mass attack due to their proximity to growing infestations (Strom and Clarke, 2011). Infestations are discovered by means of aerial surveys conducted systematically in the late spring through summer (Billings, 2011). Locations of trees with yellowing foliage are noted and then ground checked to determine whether the mortality agent is *D. frontalis* and whether the infestations are "active" (ie, continuing to accumulate freshly attacked trees) (Billings, 2011). Because foliage of mass-attacked, beetle-killed trees often requires weeks before a change in foliage colour is readily detected from the air (Clarke, 2001), spots are often large and growing quickly when first detected (Billings and Kibbe, 1978). Currently, two main techniques are recommended for infestation suppression: "cut-and-remove" and "cut-and-leave" and both involve felling the trees currently undergoing attack at the head of the infestation along with a buffer strip of apparently unattacked trees located in the path of the growing infestation (Clarke and Nowak, 2010; Schowalter, 2012). In "cut-and-remove", all infested trees and the buffer are removed from the site thereby eliminating both the natural source of the aggregating pheromone and the major source of beetles that would be aggregated by it. This method is of course ideal from the standpoint that removed trees can often be salvaged and brood beetles that would otherwise disperse from infested trees and potentially do harm in the surrounding forest are eliminated (Cronin et al., 1999; Gara, 1967). Cut-and-leave is used when tree removal is not practical or economical; trees occupied by beetles and the buffer trees are all felled opposite the direction of spot growth but left in place. The apparent mechanism underlying this approach is the elimination (or interruption) of the aggregation pheromone plume that is produced by freshly attacked trees at the spot head, as this plume is critical for sustaining spot growth (Gara, 1967). Cut trees are reported to rapidly lose attractiveness (Vité and Crozier, 1968), however, other reports indicate that downed timber can sustain mass attacks (Moser, 1987). Emergent beetles from felled brood trees tend to disperse rather than regenerate a new head at the same location (Cronin et al., 1999), however the need for retreatment is more common for cut-and-leave than cut-and-remove (Clarke, 2001; Clarke and Billings, 2003). Nonetheless, the major mechanism underlying the success of both methods appears to be interruption of the beetles' semiochemical communication.

Extensive efforts to develop treatments involving synthetic semiochemical devices to control infestations of *D. frontalis* and protect trees occurred during the 1970s through the 1990s, but despite these efforts no method is yet operational or in common practice as of this writing. Four general approaches have been pursued: (1) direct elimination of beetles by attracting them to trees where they would attack but be unable to reproduce, (2) broadcast of attractants within an infestation to interrupt orientation toward host trees undergoing mass attack [comparable to pheromone disruption techniques applied in management of lepidopteran pests (Bartell, 2008; Carde and Minks, 1995; Witzgall et al., 2010)], (3) deployment of attractants at points within brood-containing areas of the infestation to reduce beetle densities at the spot head sufficiently to prevent mass attack and switching, and (4) application of attraction inhibitors or repellants to trees in advance of the growing infestation to arrest infestation growth or protect individual selected, trees.

Approach (1) involves attaching devices releasing attractant (frontalin and α-pinene) to selected, healthy "trap trees" treated with herbicide (cacodylic acid, applied to axe wounds at the tree base) (Vité, 1970). Such trees are established in close proximity to brood trees within active or overwintering infestations, with the goal being to absorb emerging beetles onto the trap trees on which reproduction would be greatly reduced and brood production would become insufficient to support continued infestation growth. One notable advantage of this approach is that it allows initial ground check crews to treat trees and thereby take steps to arrest infestation growth immediately following infestation discovery and inspection. This "trap tree" method relies on several principles: (a) the lures should initiate attacks on the treated tree but, as attacks progress, pheromone released by the attacking beetles themselves should provide the "natural attractant" which would presumably be superior to that possible with synthetic lures and traps (Vité, 1970), (b) a higher number of "pitch-outs" should occur on the herbicide/attractant-treated trees which can cause direct mortality of parent adults (Copony and Morris, 1972), and (c) the offspring:parent ratio would be reduced below 1:1 causing local population decline (Copony and Morris, 1972; Coulson et al., 1973a). In one study, 60 of 65 treated infestations did not require retreatment (Copony and Morris, 1972). However, a large number of variables influence the method's potential for success (eg, beetle population densities, infestation size, time of year) (Coulson et al., 1973a,b), creating undesirable uncertainty and complexity with regard to practical implementation. Additionally, the need to poison large numbers of healthy trees makes the procedure unattractive to land managers (Wood, 1977).

Approach (2) was investigated in a single study in which rice soaked with frontalin and α-pinene was spread over 10 ha of pine forest enclosing an active infestation (Vité et al., 1976). Counter to the expectation that host-seeking beetles would become dispersed and disoriented (Bartell, 2008), the treatment instead increased beetle landings on pines already undergoing attack while decreasing landings on oak trees previously baited with an attractant. Thus the net effect appeared to be an increase in aggregation and attack on pines within the infestation's head.

Approach (3) involves deploying attractant release devices (ie, frontalin and α-pinene) onto trees within the interior of an active infestation that either contain brood (ie, larva through emerging adult stages) or are non-hosts. The goal is to attract brood adults emerging within the infestation [which are the beetles that make the greatest contribution to growth of an established infestation (Gara, 1967)] to trees that are unsuitable and located away from the head of the infestation and trees currently experiencing mass attack (Payne et al., 1985a,b; Richerson et al., 1980). In a three-replicate (and thus preliminary) study in two infestations, attractant-treatment of the infestation interior redistributed trap catches more evenly through the infestation, and it reduced landings on trees at the infestation head in a single trial (Payne and Richerson, 1979; Richerson et al., 1980). Additionally, no trees were mass attacked during the 3–6 days that the lures were in place, whereas infestation growth occurred both before and after treatment (Richerson et al., 1980). In an expanded study, reduction in infestation growth occurred invariably in 10 treated infestations in Georgia where beetles were at "endemic levels", whereas treatment of 15 infestations in Texas—where populations were "epidemic"—were not consistently effective (Payne et al., 1985a,b).

Approach (4) has received the most effort at development. Early studies with *D. frontalis* had indicated verbenone and brevicomin (both *endo-* and *exo-*isomers) could reduce responses by flying *D. frontalis* to attractant-baited traps (Payne et al., 1978a; Renwick and Vité, 1969; Vité and Renwick, 1971), which prompted attempts to use these compounds to protect trees at immediate risk (that is, trees located in the paths of growing infestations). In an attempt to protect individual trees, Richerson and Payne (1979) attached devices of either brevicomin (both *exo-* and *endo-*), verbenone, or their combination at 1 or 2 m intervals (four devices per height, encircling the bole) up to a height of 8–9 m. Paired, adjacent trees were left untreated as controls. None of the treatments prevented mortality, although the treatments which included brevicomin reduced landing of *D. frontalis* by 74% (verbenone alone, however, with a total release of approximately 80 mg/d/tree did not reduce landings) and there were few *D. frontalis* galleries in the brevicomin-treated trees (Payne and Richerson, 1979; Richerson and Payne, 1979). The brevicomin treatments however greatly increased the landings and attacks of the competing species *I. avulsus,* and this may have caused the mortality of the brevicomin-treated trees despite the reduced attacks by *D. frontalis*. The attraction of *I. avulsus* was unexpected given that in trapping studies *I. avulsus* has not demonstrated attraction to brevicomin or odours associated with *D. frontalis* pairs (Birch et al., 1980; Svihra, 1982). Since *I. avulsus* is less aggressive and cannot sustain infestation growth, it was proposed that, with reduced reproduction of *D. frontalis* both through semiochemical inhibition of attack and displacement by successfully competing *I. avulsus*, brevicomin/verbenone treatment would cause infestations to collapse (Payne and Richerson, 1985; Watterson et al., 1982). Thus brevicomin/verbenone treatment, although apparently not capable of saving trees actually receiving treatment, demonstrated potential to save other trees by accelerating spot collapse through semiochemical-induced "competitive replacement" (Payne and Richerson, 1985). This potential technology has not been developed further.

Deployment of releasers (soaked sponges enclosed in permeable plastic bags) of verbenone from all host pines undergoing mass attack as well as uninfested pines in a buffer up to 15 m width beyond the mass-attacked trees caused a fivefold reduction of additional attacked trees [relative to predictions of a spot growth model (Stephen and Lih, 1985)]. Furthermore, a ninefold reduction occurred when deployment of verbenone devices in the buffer was combined with cutting of infested trees at the spot head (Payne et al., 1992). In a different study, verbenone mixed with a liquid polymer that produced a controlled-release (average 160–200 mg/d/tree) was sprayed on infested and buffer trees (Payne and Billings, 1989). This treatment significantly reduced spot growth compared to model predictions, and combination of this verbenone application with removal of freshly attacked trees further improved treatment effectiveness. Infestations treated with releaser-packets of verbenone alone (with the numbers of packets adjusted to tree diametre) or combined with felling of freshly attacked trees were completely suppressed 69% and 86% of the time, respectively (compared to 90% success for cut-and-leave treatment) (Clarke et al., 1999). Given that verbenone has not been demonstrated to reduce landings on individual, infested trees (Richerson and Payne, 1979) nor alter the dispersion of beetles within infestations (Salom et al., 1995), the mechanism underlying the success of these "verbenone curtain" treatments is not clear, rendering it essentially impossible to make sufficiently informed attempts at method improvement.

On the other hand, brevicomin-incorporating treatments do not alter dispersion of flying beetles but rather appear to reduce *D. frontalis* landing rates on trees undergoing mass attack (Johnson and Coster, 1980; Payne et al., 1977). In the only study to examine infestation-wide effects of brevicomin, releasers of an endo/exo-brevicomin blend (36 total devices, 1.5 mg/d/device) were deployed from selected host trees within a grid that encompassed the active head of a growing *D. frontalis* infestation. The devices reduced landings of beetles on all host trees within the grid—treated or not—while having no effect on the densities of beetles flying in the infestation (Payne et al., 1977).

There are numerous additional *D. frontalis*-produced volatile compounds (most discussed above in the section on oxygenated monoterpenes) that have been shown to reduce beetle responses to attractant-baited traps, which suggests that these semiochemicals should be tested for potential in protecting trees or suppressing infestations (Sullivan, 2005; Sullivan et al., 2007a). One such *D. frontalis* attraction inhibitor, acetophenone, has demonstrated antiaggregant properties with other *Dendroctonus* spp. (Erbilgin et al., 2007a, 2008; Pureswaran and Borden, 2004) and has shown promise for inclusion in inhibitory semiochemical blends for managing the western and mountain pine beetles, *D. brevicomis and D. ponderosae* (Fettig et al., 2012a,b,c). Acetophenone is clearly worthy of further investigation as a potential management semiochemical for *D. frontalis*.

The phenylpropanoid 4-allylanisole which naturally occurs in relatively small amounts within resin of host pines of *D. frontalis* was found to repel walking *D. frontalis* when it was applied to the walking surface, and furthermore devices releasing >100 mg/d of 4-allylanisole significantly reduced *D. frontalis* attraction to traps baited with frontalin and α-pinene (Hayes et al., 1994; Strom et al., 1999). Despite promising initial trials in which strings of releasers suspended along the tree bole appeared to prevent attacks on at-risk trees (Hayes et al., 1994; Strom et al., 1995), three different methods of deployment of 4-allylanisole (polymer/semiochemical-filled paintballs, microencapsulated semiochemical in a sprayable form, and vials attached directly to trees) releasing approximately 1–5 g/d/tree failed to demonstrate efficacy in protecting trees challenged either with *D. frontalis* attractant lures or through weakening with application of a toxin (*N*-methyldithiocarbamate) (Strom et al., 2004).

Responses of *D. frontalis* to attractant-baited traps can be inhibited by certain volatiles associated with hardwoods and other non-host plants, as has been shown to occur within numerous other species of conifer-infesting bark beetles (Zhang and Schlyter, 2004). These "non-host volatiles" are believed to play an important role in mediating avoidance by coniferophagous bark beetles of habitats with few or no susceptible hosts and in preventing attack initiation on unsuitable trees (Byers et al., 2000; Zhang and Schlyter, 2003). Two such 6-carbon "green leaf alcohols", 1-hexanol and hexanal, both singly and in combination were found to significantly reduce *D. frontalis* responses to traps baited with frontalin and turpentine (Dickens et al., 1992). Likewise, a combination of non-host volatiles that included 1-hexanol, *cis*-3-hexen-1-ol, hexanal, and nonanal, in one of two tests significantly reduced catches of male *D. frontalis* in traps baited with frontalin and α-pinene (Sullivan et al., 2007a). Shepherd and Sullivan (2013) collected volatiles from leaves and bark of eight different species within six genera of sympatric non-hosts for *D. frontalis,* and they observed GC-EAD responses to 28 compounds. These included 11 compounds not typically present in host pines (benzaldehyde, benzyl alcohol, guaiacol, heptanal, hexanal, 1-hexanol, (Z)-3-hexen-1-ol, (E)-2-hexenal, methyl salicylate, nonanal, and salilicylaldehyde). Of these (E)-2-hexen-1-ol, and 1-hexanol had particularly low concentration thresholds of EAD response. The complete blend of 11 compounds significantly reduced trap responses of both sexes whereas subgroups within the blend (ie, the alcohols, aldehydes, green leaf volatiles, bark volatiles) did not. Data is still insufficient to assess the management potential of non-host volatiles for *D. frontalis*.

10. CONCLUDING REMARKS

There currently are no techniques adopted generally by foresters for managing *D. frontalis* with synthetic semiochemicals as biorational control agents. However, the fact that the mechanisms underlying the currently recommended procedures for direct control of *D. frontalis* (ie, cut-and-leave, cut-and-remove) involve disruption of this insect's semiochemical communication suggests that manipulation of beetle behaviour through deployment of synthetic semiochemicals may nonetheless have potential for use in pest management. Thus chemical ecology may still hold the key for new management technologies that could eliminate current methods that require cutting of trees. Simultaneously, it has become increasingly apparent that the semiochemical system of *D. frontalis* is substantially more complex than originally envisioned by Renwick and Vité (1969), and insufficient basic science on the ecological roles of compounds such as verbenone, *trans*-verbenol, and *endo*-brevicomin assures that well-informed decisions cannot be made with regard to development and enhancement of technology using these compounds. The list of naturally occurring volatile compounds that modify behaviours in *Dendroctonus* pine beetles is long and continues to increase (www.pherobase.com), and yet for many little is known about their ecology beyond what can be inferred from their ability to alter responses of beetles to traps.

REFERENCES

Anderson, N.H., Anderson, D.B., 1968. *Ips* bark beetle attacks and brood development on a lightning-struck pine in relation to its physiological decline. Fla. Entomol. 51, 23–30.

Andryszak, N.A., Payne, T.L., Billings, P.M., 1982. Effect of flight activity on laboratory response of the southern pine beetle to an attractant. J. Ga. Entomol. Soc. 17, 456–460.

Armendáriz-Toledano, F., Niño, A., Sullivan, B.T., Macías-Sámano, J., Víctor, J., Clarke, S.R., Zúñiga, G., 2014. Two species within *Dendroctonus frontalis* (Coleoptera: Curculionidae): evidence from morphological, karyological, molecular, and crossing studies. Ann. Entomol. Soc. Am. 107, 11–27.

Armendáriz-Toledano, F., Niño, A., Sullivan, B.T., Kirkendall, L.R., Zúñiga, G., 2015. A new species of bark beetle, *Dendroctonus mesoamericanus* sp. nov. (Curculionidae: Scolytinae), in southern Mexico and Central America. Ann. Entomol. Soc. Am. 108, 403–414.

Aukema, B.H., Raffa, K.F., 2000. Chemically mediated predator-free space: herbivores can synergize intraspecific communication without increasing risk of predation. J. Chem. Ecol. 26, 1923–1939.

Aukema, B.H., Raffa, K.F., 2005. Selective manipulation of predators using pheromones: responses to frontalin and ipsdienol pheromone components of bark beetles in the Great Lakes region. Agric. For. Entomol. 7, 193–200.

Aukema, B.H., Dahlsten, D.L., Raffa, K.F., 2000. Improved population monitoring of bark beetles and predators by incorporating disparate behavioral responses to semiochemicals. Environ. Entomol. 29, 618–629.

Ayres, M.P., Martinson, S.J., Friedenberg, N.A., 2011. Southern pine beetle ecology: populations within stands. In: Coulson, R.N., Klepzig, K.D. (Eds.), The Southern Pine Beetle II. In: General Technical Report SRS-140, USDA Forest Service Southern Research Station, Asheville, NC, pp. 75–89.

Bartell, R.J., 2008. Mechanisms of communication disruption by pheromone in the control of Lepidoptera: a review. Physiol. Entomol. 7, 353–364.

Bedard, W.D., Tilden, P.E., Lindahl Jr., K.Q., Wood, D.L., Rauch, P.A., 1980a. Effects of verbenone and *trans*-verbenol on the response of *Dendroctonus brevicomis* to natural and synthetic attractant in the field. J. Chem. Ecol. 6, 997–1013.

Bedard, W.D., Wood, D.L., Tilden, P.E., Lindahl Jr., K.Q., Silverstein, R.M., Rodin, J.O., 1980b. Field responses of the western pine beetle and one of its predators to host- and beetle-produced compounds. J. Chem. Ecol. 6, 625–641.

Bentz, B.J., Powell, J.A., Logan, J.A., 1996. Localized spatial and temporal attack dynamics of the mountain pine beetle in lodgepole pine. USDA Forest Service, Intermountain Research Station INT-RP-494, 10 pp.

Bennett, R.B., Borden, J.H., 1971. Flight arrestment of tethered *Dendroctonus pseudotsugae* and *Trypodendron lineatum* (Coleoptera: Scolytidae) in response to olfactory stimuli. Ann. Entomol. Soc. Am. 64, 1273–1286.

Berisford, C.W., 2011. Parasitoids of the southern pine beetle. In: Coulson, R.N., Klepzig, K.D. (Eds.), Southern Pine Beetle II. In: General Technical Report SRS-140 USDA Forest Service Southern Research Station, Asheville, NC, pp. 129–139.

Berisford, C.W., Payne, T.L., Berisford, Y.C., 1990. Geographical variation in response of southern pine beetle (Coleoptera: Scolytidae) to aggregating pheromones in laboratory bioassays. Environ. Entomol. 19, 1671–1674.
Billings, R.F., 1985. Southern pine bark beetles and associated insects: effects of rapidly-released host volatiles on response to aggregation pheromones. J. Appl. Entomol. 99, 483–491.
Billings, R.F., 1988. Forecasting southern pine beetle infestation trends with pheromone traps. In: Payne, T.L., Saarenmaa, H. (Eds.), Integrated Control of Scolytid Bark Beetles: Proceedings of Symposium. IUFRO Working Party and XVII International Congress of Entomology, Vancouver, BC, pp. 295–306.
Billings, R.F., 2011. Aerial detection, ground evaluation, and monitoring of the southern pine beetle: state perspectives. In: Coulson, R.N., Klepzig, K.D. (Eds.), Southern Pine Beetle II. In: General Technical Report SRS-140, USDA Forest Service Southern Research Station, Asheville, NC, pp. 245–261.
Billings, R.F., Cameron, R.S., 1984. Kairomonal responses of Coleoptera, *Monochamus titillator* (Cerambycidae), *Thanasimus dubius* (Cleridae), and *Temnochila virescens* (Trogositidae), to behavioral chemicals of southern pine bark beetles (Coleoptera: Scolytidae). Environ. Entomol. 13, 1542–1548.
Billings, R.F., Kibbe, C.A., 1978. Seasonal relationships between southern pine beetle brood development and loblolly pine foliage color in east Texas. Southwest. Entomol. 3, 89–96.
Billings, R.F., Upton, W.W., 2010. A methodology for assessing annual risk of southern pine beetle outbreaks across the southern region using pheromone traps. In: Pye, J.M., Rauscher, H.M., Sands, Y., Lee, D.C., Beatty, J.S. (Eds.), Advances in Threat Assessment and Their Application to Forest and Rangeland Management. U.S. Department of Agriculture, Forest Service, Pacific Northwest and Southern Research Stations Gen. Tech. Rep. PNW-GTR-802, Portland, OR, pp. 73–85.
Birch, M.C., Svihra, P., Paine, T.D., Miller, J.C., 1980. Influence of chemically mediated behavior on host tree colonization by four cohabitating species of bark beetle. J. Chem. Ecol. 6, 395–414.
Birt, A., 2011. Population dynamics of southern pine beetle in forest landscapes. In: Coulson, R.N., Klepzig, K.D. (Eds.), Southern Pine Beetle II. In: General Technical Report SRS-140, USDA Forest Service Southern Research Station, Asheville, NC, pp. 91–108.
Blanche, C.A., Hodges, J.D., Nebeker, T.E., 1985. Changes in bark beetle susceptibility indicators in a lightning-struck loblolly pine. Can. J. For. Res. 15, 397–399.
Blomquist, G.J., Figueroa-Teran, R., Aw, M., Song, M.M., Gorzalski, A., Abbott, N.L., Chang, E., Tittiger, C., 2010. Pheromone production in bark beetles. Insect Biochem. Mol. Biol. 40, 699–712.
Borden, J.H., 1997. Disruption of semiochemical-mediated aggregation in bark beetles. In: Cardé, R.T., Minks, A.K. (Eds.), Insect Pheromone Research: New Directions. Chapman and Hall, New York, pp. 421–438.
Borden, J.H., Hunt, D.W.A., Miller, D.R., Slesslor, K.N., 1986. Orientation in forest Coleoptera: an uncertain outcome of responses by individual beetles to variable stimuli. In: Payne, T.L., Birch, M.C., Kennedy, C.E.J. (Eds.), Mechanisms in Insect Olfaction. Clarendon Press, Oxford, pp. 97–109.
Borden, J.H., Chong, L.J., Lindgren, B.S., 1990. Redundancy in the semiochemical message required to induce attack on lodgepole pines by the mountain pine beetle, *Dendroctonus ponderosae* Hopkins (Coleoptera: Scolytidae). Can. Entomol. 122, 769–777.
Brand, J.M., Bracke, J.W., Markovetz, A.J., Wood, D.L., Browne, L.L., 1975. Production of verbenol pheromone by a bacterium isolated from bark beetles. Nature 254, 136–137.
Brand, J.M., Bracke, J.W., Britton, L.N., Markovetz, A.J., Barras, S.J., 1976. Bark beetle pheromones: production of verbenone by a mycangial fungus of *Dendroctonus frontalis*. J. Chem. Ecol. 2, 195–199.

Brand, J.M., Schultz, J., Barras, S.J., Edson, L.J., Payne, T.L., Hedden, R.L., 1977. Bark-beetle pheromones—enhancement of *Dendroctonus frontalis* (Coleoptera: Scolytidae) aggregation pheromone by yeast metabolites in laboratory bioassays. J. Chem. Ecol. 3, 657–666.

Bridges, J.R., 1982. Effects of juvenile hormone on pheromone synthesis in *Dendroctonus frontalis*. Environ. Entomol. 11, 417–420.

Browne, L.E., Wood, D.L., Bedard, W.D., Silverstein, R.M., West, J.R., 1979. Quantitative estimates of the western pine beetle attractive pheromone components, *exo*-brevicomin, frontalin, and myrcene in nature. J. Chem. Ecol. 5, 397–414.

Bunt, W.D., Coster, J.E., Johnson, P.C., 1980. Behavior of the southern pine beetle on the bark of host trees during mass attack. Ann. Entomol. Soc. Am. 73, 647–652.

Byers, J.A., 1983. Sex-specific responses to aggregation pheromone: regulation of colonization density in the bark beetle *Ips paraconfusus*. J. Chem. Ecol. 9, 129–142.

Byers, J.A., 1987. Interactions of pheromone component odor plumes of western pine beetle. J. Chem. Ecol. 13, 2143–2157.

Byers, J.A., 1989a. Behavioral mechanisms involved in reducing competition in bark beetles. Holarct. Ecol. 12, 466–476.

Byers, J.A., 1989b. Chemical ecology of bark beetles. Experientia 45, 271–283.

Byers, J.A., 1996. An encounter rate model of bark beetle populations searching at random for susceptible host trees. Ecol. Model. 91, 57–66.

Byers, J.A., Wood, D.L., Craig, J., Hendry, L.B., 1984. Attractive and inhibitory pheromones produced in the bark beetle, *Dendroctonus brevicomis*, during host colonization: regulation of inter- and intraspecific competition. J. Chem. Ecol. 10, 861–877.

Byers, J.A., Zhang, Q.H., Birgersson, G., 2000. Strategies of a bark beetle, Pityogenes bidentatus, in an olfactory landscape. Naturwissenschaften 87, 503–507.

Camors Jr., F.B., Payne, T.L., 1973. Sequence of arrival of entomophagous insects to trees infested with the southern pine beetle. Environ. Entomol. 2, 267–270.

Carde, R.T., Minks, A.K., 1995. Control of moth pests by mating disruption: successes and constraints. Ann. Rev. Entomol. 40, 559–585.

Chatelain, M.P., Schenk, J.A., 1984. Evaluation of frontalin and *exo*-brevicomin as kairomones to control mountain pine beetle (Coleoptera: Scolytidae) in lodgepole pine. Environ. Entomol. 13, 1666–1674.

Clarke, S.R., 2001. Review of the operational IPM program for the southern pine beetle. Integr. Pest Manag. Rev. 6, 293–301.

Clarke, S., 2012. Implications of population phases on the integrated pest management of the southern pine beetle, *Dendroctonus frontalis*. J. Integr. Pest Manage 3, 1–9.

Clarke, S.R., Billings, R.F., 2003. Analysis of the southern pine beetle suppression program on the national forests in Texas in the 1990s. So. J. Appl. For. 27, 122–129.

Clarke, S.R., Menard, R.D., 2006. Predation of an ambrosia beetle (Coleoptera: Platypodidae) by a checkered beetle (Coleoptera: Cleridae) congregating on pines containing brood adult southern pine beetles (Coleoptera: Curculionidae). J. Entomol. Sci. 41, 257.

Clarke, S.R., Nowak, J.T., 2010. Southern Pine Beetle. USDA Forest Service Forest Insect and Disease Leaflet 49. USDA Forest Service, Pacific Northwest Region, Portland, Oregon.

Clarke, S.R., Salom, S.M., Billings, R.F., Berisford, C.W., Upton, W.W., McClellan, Q.C., Dalusky, M.J., 1999. A *Scents*ible approach to controlling southern pine beetles. J. For. 97, 26–31.

Cook, S.P., Hain, F.P., 1988. Toxicity of host monoterpenes to *Dendroctonus frontalis* and *Ips calligraphus* (Coleoptera: Scolytidae). J. Entomol. Sci. 23, 287–292.

Copony, J.A., Morris, C.L., 1972. Southern pine beetle suppression with frontalure and cacodylic acid treatments. J. Econ. Entomol. 65, 754–757.

Costa, A., Reeve, J.D., 2011. Upwind flight response of the bark beetle predator *Thanasimus dubius* towards olfactory and visual cues in a wind tunnel. Agric. For. Entomol. 13, 283–290.

Coster, J.E., Vité, J.P., 1972. Effects of feeding and mating on pheromone release in the southern pine beetle. Ann. Entomol. Soc. Am. 65, 263–266.

Coster, J.E., Payne, T.L., Hart, E.R., Edson, L.J., 1977. Aggregation of the southern pine beetle in response to attractive host trees. Environ. Entomol. 6, 725–731.

Coster, J.E., Hicks Jr., R.R., Watterston, K.G., 1978. Directional spread of southern pine beetle (Coleoptera: Scolytidae) infestations in East Texas. J. Ga. Entomol. Soc. 13, 315–321.

Coulson, R.N., 1979. Population dynamics of bark beetles. Ann. Rev. Entomol. 24, 417–447.

Coulson, R.N., Oliveria, F.L., Payne, T.L., Houseweart, M.W., 1973a. Variables associated with use of frontalure and cacodylic acid in suppression of the southern pine beetle. 2. Brood reduction in trees treated with cacodylic acid. J. Econ. Entomol. 66, 897–899.

Coulson, R.N., Oliveria, F.L., Payne, T.L., Houseweart, M.W., 1973b. Variables associated with use of frontalure and cacodylic acid in supression of the southern pine beetle. 1. Factors influencing manipulation to prescribed trap trees. J. Econ. Entomol. 66, 893–896.

Coulson, R.N., Hennier, P.B., Flamm, R.O., Rykiel, E.J., Hu, L.C., Payne, T.L., 1983. The role of lightning in the epidemiology of the southern pine beetle. Z. Angew. Entomol. 96, 182–193.

Coulson, R.N., Flamm, R.O., Wagner, T.L., Rykiel, E.J., Sharpe, P.J.H., Payne, T.L., Lin, S.K., 1985. Population dynamics of initiation and growth of southern pine beetle infestations. In: Branham, S.J., Thatcher, R.C. (Eds.), Proceedings of the Integrated Pest Management Symposium. In: USDA Forest Service General Technical Report SO-56, pp. 136–151.

Coulson, R.N., Flamm, R.O., Pulley, P.E., Payne, T.L., Rykiel, E.J., Wagner, T.L., 1986. Response of the southern pine bark beetle guild (Coleoptera: Scolytidae) to host disturbance. Environ. Entomol. 15, 850–858.

Coulson, R.N., McFadden, B.A., Pulley, P.E., Lovelady, C.N., Fitzgerald, J.W., Jack, S.B., 1999. Heterogeneity of forest landscapes and the distribution and abundance of the southern pine beetle. For. Ecol. Manag. 114, 471–485.

Cronin, J.T., Turchin, P., Hayes, J.L., Steiner, C.A., 1999. Area-wide efficacy of a localized forest pest management practice. Environ. Entomol. 28, 496–504.

Dahlsten, D.L., Six, D.L., Rowney, D.L., Lawson, A.B., Erbilgin, N., Raffa, K.F., 2004. Attraction of *Ips pini* (Coleoptera: Scolytinae) and its predators to natural attractants and synthetic semiochemicals in Northern California: implications for population monitoring. Environ. Entomol. 33, 1554–1561.

Davis, T.S., Hofstetter, R.W., 2009. Effects of gallery density and species ratio on the fitness and fecundity of two sympatric bark beetles (Coleoptera: Curculionidae). Environ. Entomol. 38, 639–650.

Dickens, J.C., 1979. Electrophysiological investigations of olfaction in bark beetles. Bull. Swiss Entomol. Soc. 52, 203–216.

Dickens, J.C., Payne, T.L., 1977. Bark beetle olfaction: pheromone receptor system in *Dendroctonus frontalis*. J. Insect Physiol. 23, 481–489.

Dickens, J.C., Billings, R.F., Payne, T.L., 1992. Green leaf volatiles interrupt aggregation pheromone response in bark beetles infesting southern pines. Experientia 48, 523–524.

Dixon, J.C., Osgood, E.A., 1961. Southern pine beetle: a review of present knowledge. USDA Forest Service Southern Forest Experiment Station Research Note S0-128, 34 pp.

Dixon, W.N., Payne, T.L., 1979a. Aggregation of *Thanasimus* dubius on trees under mass-attack by the southern pine beetle. Environ. Entomol. 8, 178–181.

Dixon, W.N., Payne, T.L., 1979b. Attraction of entomophagous and associate insects of the southern pine beetle to beetle- and host tree-produced volatiles. J. Ga. Entomol. Soc. 15, 378–389.

Dixon, W.N., Payne, T.L., 1980. Sequence of arrival and spatial distribution of entomophagous and associate insects on southern pine beetle-infested trees. Texas Agricultural Experiment Station Bulletin Mp-1432, 27 pp.

Erbilgin, N., Raffa, K.F., 2000. Opposing effects of host monoterpenes on responses by two sympatric species of bark beetles to their aggregation pheromones. J. Chem. Ecol. 26, 2527–2548.

Erbilgin, N., Raffa, K.F., 2001. Modulation of predator attraction to pheromones of two prey species by stereochemistry of plant volatiles. Oecologia 127, 444–453.

Erbilgin, N., Powell, J.S., Raffa, K.F., 2003. Effect of varying monoterpene concentrations on the response of *Ips pini* (Coleoptera: Scolytidae) to its aggregation pheromone: implications for pest management and ecology of bark beetles. Agric. For. Entomol. 5, 269–274.

Erbilgin, N., Gillette, N.E., Mori, S.R., Stein, J.D., Owen, D.R., Wood, D.L., 2007a. Acetophenone as an anti-attractant for the western pine beetle, *Dendroctonus brevicomis* LeConte (Coleoptera: Scolytidae). J. Chem. Ecol. 33, 817–823.

Erbilgin, N., Krokene, P., Kvamme, T., Christiansen, E., 2007b. A host monoterpene influences *Ips typographus* (Coleoptera: Curculionidae, Scolytinae) responses to its aggregation pheromone. Agric. For. Entomol. 9, 135–140.

Erbilgin, N., Gillette, N.E., Owen, D.R., Mori, S.R., Nelson, A.S., Uzoh, F., Wood, D.L., 2008. Acetophenone superior to verbenone for reducing attraction of western pine beetle *Dendroctonus brevicomis* to its aggregation pheromone. Agric. For. Entomol. 10, 433–441.

Everaerts, C., Grégoire, J.C., Merlin, J., 2012. The toxicity of Norway spruce monoterpenes to two bark beetle species. In: Mattson, W.J., Levieux, J., Bernard-Dagan, C. (Eds.), Mechanisms of Woody Plant Defenses Against Insects: Search for Pattern. Springer, New York, pp. 335–344.

Fargo, W.S., Wagner, T.L., Coulson, R.N., 1985. Factors influencing the growth of multiple-tree infestations of *Dendroctonus frontalis*. Res. Popul. Ecol. 27, 25–38.

Fatzinger, C.W., 1985. Attraction of the black turpentine beetle (Coleoptera: Scolytidae) and other forest Coleoptera to turpentine-baited traps. Environ. Entomol. 14, 768–775.

Fatzinger, C.W., Siegfried, B.D., Wilkinson, R.C., Nation, J.L., 1987. *trans*-Verbenol, turpentine, and ethanol as trap baits for the black turpentine beetle, *Dendroctonus terebrans*, and other forest Coleoptera in North Florida. J. Entomol. Sci. 22, 201–209.

Fettig, C.J., Bulaon, B.M., Dabney, C.P., Hayes, C.J., McKelvey, S.R., 2012a. Verbenone Plus reduces levels of tree mortality attributed to mountain pine beetle infestations in whitebark pine, a tree species of concern. J. Biofert. Biopest 3, 123–128.

Fettig, C.J., McKelvey, S.R., Dabney, C.P., Huber, D.P., 2012b. Responses of *Dendroctonus brevicomis* (Coleoptera: Curculionidae) in behavioral assays: implications to development of a semiochemical-based tool for tree protection. J. Econ. Entomol. 105, 149–160.

Fettig, C.J., McKelvey, S.R., Dabney, C.P., Huber, D.P., Lait, C.G., Fowler, D.L., Borden, J.H., 2012c. Efficacy of "Verbenone Plus" for protecting ponderosa pine trees and stands from *Dendroctonus brevicomis* (Coleoptera: Curculionidae) attack in British Columbia and California. J. Econ. Entomol. 105, 1668–1680.

Flamm, R.O., Pulley, P.E., Coulson, R.N., 1993. Colonization of disturbed trees by the southern pine bark beetle guild (Coleoptera, Scolytidae). Environ. Entomol. 22, 62–70.

Flechtmann, C.A.H., Dalusky, M.J., Berisford, C.W., 1999. Bark and ambrosia beetle (Coleoptera: Scolytidae) responses to volatiles from aging loblolly pine billets. Environ. Entomol. 28, 638–648.

Francke, W., Vité, J.P., 1983. Oxygenated terpenes in pheromone systems of bark beetles. Zeits. Angew. Entomol. 96, 146–156.
Franklin, R.T., 1970. Southern pine beetle population behavior. J. Georgia Entomol. Soc. 5, 175–182.
Fronk, W.D., 1947. The southern pine beetle-its life history. Virginia Agricultural Experiment Station Technical Bulletin, vol. 108. Blacksburg, VA, 12 pp.
Gara, R.I., 1967. Studies on the attack behavior of the southern pine beetle. I. The spreading and collapse of outbreaks. Contrib. Boyce Thompson Inst. 23, 349–354.
Gara, R.I., Coster, J.E., 1968. Studies on the attack behavior of the southern pine beetle. III. Sequence of tree infestation within stands. Contrib. Boyce Thompson Inst. 24, 77–85.
Gara, R.I., Vité, J.P., Cramer, H.H., 1965. Manipulation of *Dendroctonus frontalis* by use of a population aggregation pheromone. Contrib. Boyce Thompson Inst. 23, 55–66.
Gara, R.I., Geiszler, D.R., Litke, W.R., 1984. Primary attraction of the mountain pine beetle to lodgepole pine in Oregon. Ann. Entomol. Soc. Am. 77, 333–334.
Gries, G., Nolte, R., Sanders, W., 1989. Computer simulated host selection in *Ips typographus*. Entomol. Exp. Appl. 53, 211–217.
Gries, G., Leufvén, A., Lafontaine, J., Pierce, H., Borden, J., Vanderwel, D., Oehlschlager, A., 1990a. New metabolites of α-pinene produced by the mountain pine beetle, *Dendroctonus ponderosae* (Coleoptera: Scolytidae). Insect Biochem. 20, 365–371.
Gries, G., Smirle, M.J., Leufven, A., Miller, D.R., Borden, J.H., Whitney, H.S., 1990b. Conversion of phenylalanine to toluene and 2-phenylethanol by the pine engraver *Ips pini* (Say) (Coleoptera, Scolytidae). Experientia 46, 329–331.
Grosman, D.M., 1996. Southern Pine Beetle, *Dendroctonus frontalis* Zimmermann (Coleoptera: Scolytidae): Quantitative Analysis of Chiral Semiochmicals, PhD Dissertation. Department of Entomology, Virginia Polytechnic Institute and State University, Blacksburg, Virginia, p.171.
Grosman, D.M., Salom, S.M., Ravlin, F.W., Young, R.W., 1997. Geographic and gender differences in semiochemicals in emerging adult southern pine beetle (Coleoptera: Scolytidae). Ann. Entomol. Soc. Am. 90, 438–446.
Hain, F.P., Duehl, A.J., Gardner, M.J., Payne, T.L., 2011. Natural history of the southern pine beetle. In: Coulson, R.N., Klepzig, K.D. (Eds.), Southern Pine Beetle II. In: General Technical Report SRS-140, USDA Forest Service Southern Research Station, Asheville, NC, pp. 13–24.
Harborne, J.B., 1988. Introduction to Ecological Biochemistry, third ed. Academic Press, London. 384 pp.
Hayes, J.L., Strom, B.L., Roton, L.M., Ingram, L.L., 1994. Repellent properties of the host compound 4-allylanisole to the southern pine beetle. J. Chem. Ecol. 20, 1595–1615.
Heikkenen, H.J., 1977. Southern pine beetle: a hypothesis regarding its primary attractant. J. Forestry 75, 412–413.
Hodges, J.D., Pickard, L.S., 1971. Lightning in the ecology of the southern pine beetle, *Dendroctonus frontalis* (Coleoptera: Scolytidae). Can. Entomol. 103, 44–51.
Hofstetter, R.W., Chen, Z., Gaylord, M.L., McMillin, J.D., Wagner, M.R., 2008. Synergistic effects of *alpha*-pinene and *exo*-brevicomin on pine bark beetles and associated insects in Arizona. J. Appl. Entomol. 132, 387–397.
Hofstetter, R.W., Gaylord, M.L., Martinson, S., Wagner, M.R., 2012. Attraction to monoterpenes and beetle-produced compounds by syntopic *Ips* and *Dendroctonus* bark beetles and their predators. Agric. For. Entomol. 14, 207–215.
Hopkins, A.D., 1909. Contributions toward a monograph of scolytid beeltes. I. Bark beetles of the genus Dendroctonus. U.S. Department of Agriculture, Bureau of Entomology Technical Series No. 17, Part 1, Washington, DC, 164 pp.
Hughes, P.R., 1973. *Dendroctonus*: production of pheromones and related compounds in response to host monoterpenes. Z. Angew. Entomol. 73, 294–312.

Hughes, P.R., 1976. Response of female southern pine beetles to the aggregation pheromone frontalin. Z. Angew. Entomol. 81, 463–466.
Hughes, P.R., Renwick, J.A.A., Vité, J.P., 1976. The identification and field bioassay of chemical attractants in the roundheaded pine beetle. Environ. Entomol. 5, 1165–1168.
Hunt, D.W.A., Borden, J.H., 1990. Conversion of verbenols to verbenone by yeasts isolated from *Dendroctonus ponderosae* (Coleoptera: Scolytidae). J. Chem. Ecol. 16, 1385–1397.
Hunt, D.W.A., Borden, J.H., Lindgren, B.S., Gries, G., 1989. The role of autoxidation of *alpha*-pinene in the production of pheromones of *Dendroctonus ponderosae* (Coleoptera: Scolytidae). Can. J. For. Res. 19, 1275–1282.
Hynum, B.G., Berryman, A.A., 1980. *Dendroctonus ponderosae* (Coleoptera: Scolytidae): pre-aggregation landing and gallery initiation on lodgepole pine. Can. Entomol. 112, 185–191.
Johnson, P.C., Coster, J.E., 1978. Probability of attack by southern pine beetle in relation to distance from an attractive host tree. For. Sci. 24, 574–580.
Johnson, P.C., Coster, J.E., 1980. Seasonal and behavioral chemical effects on dispersion of the southern pine beetle within infestations. In: Berryman, A.A., Safranyik, L. (Eds.), Proceedings of the Second IUFRO Conference on Dispersal of Forest Insects: Evaluation, Theory and Management Implications, Washington State University, Pullman, WA, pp. 173–193.
Kinn, D.N., 1986. Studies on the flight capabilities of *Dendroctonus frontalis* and *Ips calligraphus*: preliminary findings using tethered beetles. USDA Forest Service, Southern Forest Experiment Station Res. Note SO-324, 3 pp.
Kinzer, G.W., Fentiman Jr., A.F., Page Jr., T.F., Foltz, R.L., Vité, J.P., 1969. Bark beetle attractants: identification, synthesis and field bioassay of a new compound isolated from *Dendroctonus*. Nature 221, 477–478.
Kohnle, U., Mussong, M., Dubbel, V., Francke, W., 1987. Acetophenone in the aggregation of the beech bark beetle, *Taphrorychus bicolor* (Col., Scolytidae). J. Appl. Entomol. 103, 249–252.
Lanier, G.N., Burkholder, W.E., 1974. Pheromones in speciation of Coleoptera. In: Birch, M.C. (Ed.), Pheromones. North-Holland Publishing, Amsterdam, pp. 161–189.
Lanier, G.N., Wood, D.L., 1975. Specificity of response to pheromones in the genus *Ips* (Coleoptera: Scolytidae). J. Chem. Ecol. 1, 9–23.
Lanier, G.N., Birch, M.C., Schmitz, R.F., Furniss, M.M., 1972. Pheromones of *Ips pini* (Coleoptera: Scolytidae): variation in response among three populations. Can. Entomol. 104, 1917–1923.
Lanier, G.N., Hendrichs, J.P., Flores, J.E., 1988. Biosystematics of the *Dendroctonus frontalis* (Coleoptera: Scolytidae) complex. Ann. Entomol. Soc. Am. 81, 403–418.
Latty, T.M., Reid, M., 2010. Who goes first? Condition and danger dependent pioneering in a group-living bark beetle (*Dendroctonus ponderosae*). Behav. Ecol. Sociobiol. 64, 639–646.
Lindgren, B.S., Miller, D.R., 2002. Effect of verbenone on five species of bark beetles (Coleoptera: Scolytidae) in lodgepole pine forests. Environ. Entomol. 31, 759–765.
Lovelady, C.N., Pulley, P.E., Coulson, R.N., Flamm, R.O., 1991. Relation of lightning to herbivory by the southern pine bark beetle guild (Coleoptera: Scolytidae). Environ. Entomol. 20, 1279–1284.
Macias-Samano, J., Borden, J., Gries, R., Pierce Jr., H., Gries, G., King, G., 1998. Primary attraction of the fir engraver, *Scolytus ventralis*. J. Chem. Ecol. 24, 1049–1075.
Martinson, S.J., Ylioja, T., Sullivan, B.T., Billings, R.F., Ayres, M.P., 2013. Alternate attractors in the population dynamics of a tree-killing bark beetle. Popul. Ecol. 55, 95–106.
McCarty, F.A., Billings, P.M., Richerson, J.V., Payne, T.L., Edson, L.J., 1980. Response of the southern pine beetle to behavioral chemicals in the laboratory. J. Ga. Entomol. Soc. 15, 307–317.

Miller, M.C., 1983. Lightning Strike Simulation for Studying Southern Pine Bark and Engraver Beetle Attacks. USDA Forest Service, Southern Forest Experiment Station Research Note SO-296, 4 pp.

Miller, D.R., LaFontaine, J.P., 1991. *cis*-Verbenol: an aggregation pheromone for the mountain pine beetle *Dendroctonus ponderosae* (Coleoptera: Scolytidae). J. Entomol. Soc. Brit. Columb. 88, 34–38.

Miller, D.R., Rabaglia, R.J., 2009. Ethanol and (−)-α-pinene: attractant kairomones for bark and ambrosia beetles in the southeastern US. J. Chem. Ecol. 35, 435–448.

Miller, D.R., Madden, J.L., Borden, J.H., 1986. Primary attraction of *Ips latidens* (LeConte) and *Hylastes gracilis* LeConte (Coleoptera, Scolytidae) to high-girdled lodgepole pine, *Pinus contorta* var *latifolia* Engelmann. Can. Entomol. 118, 85–88.

Miller, D.R., Borden, J.H., Slessor, K.N., 1989. Interpopulation and intrapopulation variation of the pheromone ipsdienol produced by male pine engravers, *Ips pini* (Say) (Coleoptera, Scolytidae). J. Chem. Ecol. 15, 233–247.

Mirov, N.T., 1961. Composition of gum turpentines of pines. USDA Forest Service Pacific Southwest Forest and Range Experiment Station Technical Bulletin 1239, Washington, DC, 158 pp.

Mizell, R.F.I., Frazier, J.L., Nebeker, T.E., 1984. Response of the clerid predator *Thanasimus dubius* (F.) to bark beetle pheromones and tree volatiles in a wind tunnel. J. Chem. Ecol. 10, 177–187.

Moeck, H.A., 1978. Field test for primary attraction of the spruce beetle. Canadian Forest Service Bi-monthly Research Note 34. 8 pp.

Moeck, H.A., Simmons, C.S., 1991. Primary attraction of mountain pine beetle, *Dendroctonus ponderosae* Hopk. (Coleoptera: Scolytidae), to bolts of lodgepole pine. Can. Entomol. 123, 299–304.

Moeck, H.A., Wood, D.L., Lindahl Jr., K.Q., 1981. Host selection behavior of bark beetles (Coleoptera: Scolytidae) attacking *Pinus ponderosa*, with special emphasis on the western pine beetle, *Dendroctonus brevicomis*. J. Chem. Ecol. 7, 49–83.

Moore, G.E., 1972. Southern pine beetle mortality in North Carolina caused by parasites and predators. Environ. Entomol. 1, 58–65.

Moore, R.N., Golumbic, C., Fisher, G.S., 1956. Autoxidation of *alpha*-pinene. J. Am. Chem. Soc. 78, 1173–1176.

Moreno, B., Macías, J., Sullivan, B.T., Clarke, S.R., 2008. Field response of *Dendroctonus frontalis* (Coleoptera: Scolytinae) to synthetic semiochemicals in Chiapas. Mexico. *J. Econ. Entomol.* 101, 1821–1825.

Moser, J.C., 1987. Southern Pine Beetles Attack Felled Green Timber. USDA Forest Service, Southern Forest Experiment Station Res. Note SO-342, 8 pp.

Moser, J.C., Fitzgibbon, B.A., Klepzig, K.D., 2005. The mexican pine beetle, *Dendroctonus mexicanus*: first record in the United States and co-occurrence with the southern pine beetle—*Dendroctonus frontalis* (Coleoptera: Scolytidae or Curculionidae: Scolytinae). Entomol. News 116, 235–244.

Nebeker, T., 2011. Southern pine bark beetle guild. In: Coulson, R.N., Klepzig, K.D. (Eds.), Southern Pine Beetle II. In: General Technical Report SRS-140, USDA Forest Service Southern Research Station, Asheville, NC, pp. 199–209.

Niño-Domínguez, A., Sullivan, B.T., López-Urbina, J.H., Macías-Sámano, J., 2015a. Responses by *Dendroctonus frontalis* and *Dendroctonus mesoamericanus* (Coleoptera: Curculionidae) to semiochemical lures in Chiapas, Mexico: Possible roles of pheromones during joint host attacks. J. Econ. Entomol. 1–8. http://dx.doi.org/10.1093/jee/tov406, in press.

Niño-Domínguez, A., Sullivan, B.T., López-Urbina, J.H., Macías-Sámano, J.E., 2015b. Pheromone-mediated mate location and discrimination by two syntopic sibling species of *Dendroctonus* bark beetles in Chiapas. Mexico. *J. Chem. Ecol.* 41, 746–756.

Økland, B., Skarpaas, O., Kausrud, K., 2009. Threshold facilitations of interacting species. Popul. Ecol. 51, 513–523.

Økland, B., Erbilgin, N., Skarpaas, O., Christiansen, E., Långström, B., 2011. Inter-species interactions and ecosystem effects of non-indigenous invasive and native tree-killing bark beetles. Biol. Invasions 13, 1151–1164.

Paine, T.D., Birch, M.C., Svihra, P., 1981. Niche breadth and resource patitioning by four sympatric species of bark beetles (Coleoptera: Scolytidae). Oecologia 48, 1–6.

Payne, T.L., 1975. Bark beetle olfaction. III. Antennal olfactory responsiveness of *Dendroctonus frontalis* Zimmerman and *D. brevicomis* Le Conte (Coleoptera: Scolytidae) to aggregation pheromones and host tree terpene hydrocarbons. J. Chem. Ecol. 1, 233–242.

Payne, T.L., 1980. Life history and habits. In: Thatcher, R.C., Searcy, J.L., Coster, J.E., Hertel, G.D. (Eds.), The Southern Pine Beetle, pp. 31–54. USDA Forest Service Science and Education Administration Technical Bulletin 1631.

Payne, T.L., 1986. Olfaction and vision in host finding by a bark beetle. In: Payne, T.L., Birch, M.C., Kennedy, C.E. (Eds.), Mechanisms in Insect Olfaction. Clarendon Press, Oxford, UK, pp. 111–116.

Payne, T.L., 1989. Olfactory basis for insect enemies of allied species. In: Kulhavy, D.L., Miller, M.C. (Eds.), Potential for Biological Control of *Dendroctonus* and *Ips* Bark Beetles. University of Texas Press, Austin, TX, pp. 55–69.

Payne, T.L., Billings, R.F., 1989. Evaluation of (S)-verbenone applications for suppressing southern pine beetle (Coleoptera: Scolytidae) infestations. J. Econ. Entomol. 82, 1702–1708.

Payne, T.L., Richerson, J.V., 1979. Management implications of inhibitors for *Dendroctonus frontalis* (Col. Scolytidae). Bull. Soc. Entomol. Suisse. 52, 323–331.

Payne, T.L., Richerson, J.V., 1985. Pheromone-mediated competitive replacement between two bark beetle populations: influence on infestation suppression. Z. Angew. Entomol. 99, 131–138.

Payne, T.L., Coster, J.E., Johnson, P.C., 1977. Effects of slow-release formulation of synthetic *endo*- and *exo*-brevicomin on southern pine beetle flight and landing behavior. J. Chem. Ecol. 3, 133–141.

Payne, T.L., Coster, J.E., Richerson, J.V., Edson, L.J., Hart, E.R., 1978a. Field response of the southern pine beetle to behavioral chemicals. Environ. Entomol. 7, 578–582.

Payne, T.L., Coster, J.E., Richerson, J.V., Hart, E.R., Hedden, R.L., Edson, L.J., 1978b. Reducing variation in field tests of behavioral chemicals for the southern pine beetle. J. Ga. Entomol. Soc. 13, 85–90.

Payne, T.L., Richerson, J.V., Dickens, J.C., West, J.R., Mori, K., Berisford, C.W., Hedden, R.L., Vité, J.P., Blum, M.S., 1982. Southern pine beetle: olfactory receptor and behavior discrimination of enantiomers of the attractant pheromone frontalin. J. Chem. Ecol. 8, 873–881.

Payne, T.L., Dickens, J.C., Richerson, J.V., 1984. Insect predator-prey coevolution via en antiomeric specificity in a kairomone-pheromone system. J. Chem. Ecol. 10, 487–492.

Payne, T.L., Kudon, L.H., Berisford, C.W., O'Donnel, B.P., Walsh, D.K., 1985a. Effects of frontalure in suppressing southern pine beetle spot growth under endemic and epidemic population levels. In: Branham, S.J., Thatcher, R.C. (Eds.), Integrated Pest Management Research Symposium: The Proceedings. In: General Technical Report SO-56. USDA Forest Service Southern Experiment Station, New Orleans, LA, pp. 281–287.

Payne, T.L., Kudon, L.H., Walsh, K.D., Berisford, C.W., 1985b. Influence of infestation density on suppression of *D. frontalis* infestations with attractant. Z. Angew. Entomol. 99, 39–43.

Payne, T.L., Billings, R.F., Delorme, J.D., Andryszak, N.A., Bartels, J., Francke, W., Vité, J.P., 1987. Kairomonal-pheromonal system in the black turpentine beetle, *Dendroctonus terebrans* (Ol.). J. Appl. Entomol. 103, 15–22.

Payne, T.L., Andryszak, N.A., Wieser, H., Dixon, E.A., Ibrahim, N., Coers, J., 1988. Antennnal olfactory and behavioral response of southern pine beetle, *Dendroctonus frontalis*, to analogs of its aggregation pheromone frontalin. J. Chem. Ecol. 14, 1217–1225.

Payne, T.L., Billings, R.F., Berisford, C.W., Salom, S.M., Grosman, D.M., Dalusky, M.J., Upton, W.W., 1992. Disruption of *Dendroctonus frontalis* (Col., Scolytidae) infestations with an inhibitor pheromone. J. Appl. Entomol. 114, 341–347.

Phillips, T.W., Wilkening, A.J., Atkinson, T.H., Nation, J.L., Wilkinson, R.C., Foltz, J.L., 1988. Synergism of turpentine and ethanol as attractants for certain pine-infesting beetles (Coleoptera). Environ. Entomol. 17, 456–462.

Phillips, T.W., Nation, J.L., Wilkinson, R.C., Foltz, J.L., 1989. Secondary attraction and field activity of beetle-produced volatiles in *Dendroctonus terebrans*. J. Chem. Ecol. 15, 1513–1533.

Phillips, M.A., Savage, T.J., Croteau, R., 1999. Monoterpene synthases of loblolly pine (*Pinus taeda*) produce pinene isomers and enantiomers. Arch. Biochem. Biophys. 372, 197–204.

Pitman, G.B., Vité, J.P., 1969. Aggregation behavior of *Dendroctonus ponderosae* (Coleoptera: Scolytidae) in response to chemical messengers. Can. Entomol. 101, 143–149.

Pitman, G.B., Vité, J.P., Kinzer, G.W., Fentiman Jr., A.F., 1968. Bark beetle attractants: *trans*-verbenol isolated from *Dendroctonus*. Nature 218, 168–169.

Pitman, G.B., Vité, J.P., Kinzer, G.W., Fentiman Jr., A.F., 1969. Specificity of population-aggregating pheromones in *Dendroctonus*. J. Insect Physiol. 15, 363–366.

Plummer, E.L., Stewart, T.E., Byrne, K., Pearce, G.T., Silverstein, R.M., 1976. Determination of the enantiomeric composition of several insect pheromone alcohols. J. Chem. Ecol. 2, 307–331.

Powell, J., Tams, J., Bentz, B., Logan, J., 1998. Theoretical analysis of "switching" in a localized model for mountain pine beetle mass attack. J. Theor. Biol. 194, 49–63.

Pureswaran, D.S., Borden, J.H., 2004. New repellent semiochemicals for three species of *Dendroctonus* (Coleoptera: Scolytidae). Chemoecology 14, 67–75.

Pureswaran, D.S., Borden, J.H., 2005. Primary attraction and kairomonal host discrimination in three species of *Dendroctonus* (Coleoptera: Scolytidae). Agric. For. Entomol. 7, 219–230.

Pureswaran, D.S., Sullivan, B.T., 2012. Semiochemical emission from individual galleries of the southern pine beetle, (Coleoptera: Curculionidae: Scolytinae), attacking standing trees. J. Econ. Entomol. 105, 140–148.

Pureswaran, D.S., Gries, R., Borden, J.H., Pierce Jr., H.D., 2000. Dynamics of pheromone production and communication in the mountain pine beetle, *Dendroctonus ponderosae* Hopkins, and the pine engraver, *Ips pini* (Say) (Coleoptera: Scolytidae). Chemoecology 10, 153–168.

Pureswaran, D.S., Sullivan, B.T., Ayres, M.P., 2006. Fitness consequences of pheromone production and host selection strategies in a tree-killing bark beetle (Coleoptera: Curculionidae: Scolytinae). Oecologia 148, 720–728.

Pureswaran, D.S., Sullivan, B.T., Ayres, M.P., 2007. High individual variation in pheromone production by tree-killing bark beetles (Coleoptera: Curculionidae: Scolytinae). Naturwissenschaften 95, 33–44.

Pureswaran, D.S., Hofstetter, R.W., Sullivan, B.T., 2008a. Attraction of the southern pine beetle, *Dendroctonus frontalis*, to pheromone components of the western pine beetle, *Dendroctonus brevicomis* (Coleoptera:Curculionidae:Scolytinae), in an allopatric zone. Environ. Entomol. 37, 70–78.

Pureswaran, D.S., Sullivan, B.T., Ayres, M.P., 2008b. High individual variation in pheromone production by tree-killing bark beetles (Coleoptera:Curculionidae:Scolytinae). Naturwissenschaften 95, 33–44.

Pye, J.M., Holmes, T.P., Prestemon, J.P., Wear, D.N., 2011. Economic impacts of the southern pine beetle. In: Southern Pine Beetle II. General Technical Report SRS-140, US Forest Service Southern Research Station, Asheville, NC, pp. 213–234.

Raffa, K.F., Berryman, A.A., 1980. Flight responses and host selection by bark beetles. In: Proceedings of the Second IUFRO Conference on Dispersal of Forest Insects: Evaluation, Theory, and Management Implications. Conference Office, Cooperative Extension Service, Washington State University, Pullman Washington, pp. 213–233.

Raffa, K.F., Dahlsten, D.L., 1995. Differential responses among natural enemies and prey to bark beetle pheromones. Oecologia 102, 17–23.

Raffa, K.F., Klepzig, K.D., 1989. Chiral escape of bark beetles from predators responding to a bark beetle pheromone. Oecologia 80, 566–569.

Raffa, K.F., Phillips, T.W., Salom, S.M., 1993. Strategies and mechanisms of host colonization by bark beetles. In: Schowalter, R.D., Filip, G.M. (Eds.), Beetle-Pathogen Interactions in Conifer Forests. Academic Press, London, pp. 103–128.

Redlich, H., Bruns, W., Francke, W., Schurig, V., Payne, T.L., Vité, J.P., 1987. Chiral building units from carbohydrates. XIII. Identification of the absolute configuration of *endo*-brevicomin from *Dendroctonus frontalis* and synthesis of both enantiomers from D-ribose. Tetrahedron 43, 2029–2034.

Reeve, J.D., 1997. Predation and bark beetle dynamics. Oecologia 112, 48–54.

Reeve, J.D., Strom, B.L., Rieske, L.K., Ayres, B.D., Costa, A., 2009. Geographic variation in prey preference in bark beetle predators. Ecol. Entomol. 34, 183–192.

Renwick, J.A.A., 1967. Identification of two oxygenated terpenes from the bark beetles *Dendroctonus frontalis* and *Dendroctonus brevicomis*. Contrib. Boyce Thompson Inst. 23, 355–360.

Renwick, J.A.A., 1970. Chemical aspects of bark beetle aggregation. Contrib. Boyce Thompson Inst. 24, 337–341.

Renwick, J.A.A., Hughes, P.R., 1975. Oxidation of unsaturated cyclic hydrocarbons by *Dendroctonus frontalis*. Insect Biochem. 5, 459–463.

Renwick, J.A.A., Vité, J.P., 1968. Isolation of the population aggregating pheromone of the southern pine beetle. Contrib. Boyce Thompson Inst. 24, 65–68.

Renwick, J.A.A., Vité, J.P., 1969. Bark beetle attractants: mechanism of colonization by *Dendroctonus frontalis*. Nature 224, 1222–1223.

Renwick, J.A.A., Vité, J.P., 1970. Systems of chemical communication in *Dendroctonus*. Contrib. Boyce Thompson Inst. 24, 283–292.

Renwick, J.A.A., Hughes, P.R., Ty, T.D., 1973. Oxidation products of pinene in the bark beetle *Dendroctonus frontalis*. J. Insect Physiol. 19, 1735–1740.

Renwick, J.A.A., Hughes, P.R., Vité, J.P., 1975. The aggregation pheromone system of a *Dendroctonus* bark beetle in Guatemala. J. Insect Physiol. 21, 1097–1100.

Renwick, J.A.A., Hughes, P.R., Pitman, G.B., Vité, J.P., 1976a. Oxidation products of terpenes identified from *Dendroctonus* and *Ips* bark beetles. J. Insect Physiol. 22, 725–727.

Renwick, J.A.A., Pitman, G.B., Vité, J.P., 1976b. 2-phenylethanol isolated from bark beetles. Naturwissenschaften 63, 198.

Richerson, J.V., Payne, T.L., 1979. Effects of bark beetle inhibitors on landing and attack behavior of the southern pine beetle and beetle associates. Environ. Entomol. 8, 360–364.

Richerson, J.V., McCarthy, F.A., Payne, T.L., 1980. Disruption of southern pine beetle infestations with frontalure. Environ. Entomol. 9, 90–93.

Rudinsky, J.A., 1973a. Multiple functions of the Douglas fir beetle pheromone 3-methyl-2-cyclohexen-1-one. Environ. Entomol. 2, 579–585.

Rudinsky, J.A., 1973b. Multiple functions of the southern pine beetle pheromone verbenone. Environ. Entomol. 2, 511–514.

Rudinsky, J.A., Michael, R.R., 1974. Sound production in Scolytidae: 'rivalry' behaviour of male *Dendroctonus* beetles. J. Insect Physiol. 20, 1219–1230.
Rudinsky, J.A., Morgan, M.E., Libbey, L.M., Putnam, T.B., 1974. Antiaggregative-rivalry pheromone of the mountain pine beetle, and a new arrestant of the southern pine beetle. Environ. Entomol. 3, 90–97.
Ryker, L.C., 1988. Acoustic studies of *Dendroctonus* bark beetles. Fla. Entomol. 71, 447–461.
Rykiel Jr., E.J., Coulson, R.N., Sharpe, P.J., Allen, T.F., Flamm, R.O., 1988. Disturbance propagation by bark beetles as an episodic landscape phenomenon. Landscape Ecol. 1, 129–139.
Saint-Germain, M., Buddle, C.M., Drapeau, P., 2007. Primary attraction and random landing in host-selection by wood-feeding insects: a matter of scale? Agric. For. Entomol. 9, 227–235.
Salom, S.M., Ascoli-Christensen, A., Birgersson, G., Payne, T.L., Berisford, C.W., 1992a. Electroantennogram responses of the southern pine beetle parasitoid *Coeloides pissodis* (Ashmead) (Hym., Braconidae) to potential semiochemicals. J. Appl. Entomol. 114, 472–479.
Salom, S.M., Billings, R.F., Upton, W.W., Dalusky, M.J., Grosman, D.M., Payne, T.L., Berisford, C.W., Shaver, T.N., 1992b. Effect of verbenone enantiomers and racemic *endo*-Brevicomin on response of *Dendroctonus frontalis* (Coleoptera, Scolytidae) to attractant-baited traps. Can. J. For. Res. 22, 925–931.
Salom, S.M., Grosman, D.M., McClellan, Q.C., Payne, T.L., 1995. Effect of an inhibitor-based suppression tactic on abundance and distribution of the southern pine beetle (Coleoptera: Scolytidae) and its natural enemies. J. Econ. Entomol. 88, 1703–1716.
Schlyter, F., Byers, J.A., Löfqvist, J., 1987a. Attraction to pheromone sources of different quantity, quality, and spacing: density-regulation mechanisms in bark beetle *Ips typographus*. J. Chem. Ecol. 13, 1503–1523.
Schlyter, F., Lofqvist, J., Byers, J.A., 1987b. Behavioural sequence in the attraction of the bark beetle *Ips typographus* to pheromone sources. Physiol. Entomol. 12, 185–196.
Schowalter, T., 2012. Ecology and management of bark beetles (Coleoptera: Curculionidae: Scolytinae) in southern pine forests. J. Integr. Pest Manage 3, 1–7.
Schowalter, T.D., Pope, D.N., Coulson, R.N., Fargo, W.S., 1981. Patterns of southern pine beetle (*Dendroctonus frontalis* Zimm.) infestation enlargement. For. Sci. 27, 837–849.
Seybold, S.J., 1993. Role of chirality in olfactory-directed behavior—aggregation of pine engraver beetles in the genus *Ips* (Coleoptera, Scolytidae). J. Chem. Ecol. 19, 1809–1831.
Seybold, S.J., Vanderwel, D., 2003. Biosynthesis and endocrine regulation of pheromone production in the Coleoptera. In: Blomquist, G.J., Vogt, R.G. (Eds.), Insect Pheromone Biochemistry and Molecular Biology. Elsevier Academic Press, London, pp. 137–200.
Seybold, S.J., Huber, D.P.W., Lee, J.C., Graves, A.D., Bohlmann, J., 2006. Pine monoterpenes and pine bark beetles: a marriage of convenience for defense and chemical communication. Phytochem. Rev. 5, 143–178.
Shepherd, W.P., Sullivan, B.T., 2013. Southern pine beetle, *Dendroctonus frontalis*, antennal and behavioral responses to nonhost leaf and bark volatiles. J. Chem. Ecol. 39, 481–493.
Siegfried, B.D., Fatzinger, C.W., Wilkinson, R.C., Nation, J.L., 1986. In-flight responses of the black turpentine beetle (Coleoptera: Scolytidae) to individual monoterpenes, turpentine, and paraquat-treated slash pines. Environ. Entomol. 15, 710–714.
Skillen, E.L., Berisford, C.W., Camaan, M.A., Reardon, R.C., 1997. Semiochemicals of Forest and Shade Tree Insects in North America and Management Applications. USDA Forest Service Forest Health Technology Enterprise Team Publication FHTET-96-15.
Smith, M.T., Payne, T.L., Birch, M.C., 1990. Olfactory-based behavioral interactions among five species in the southern pine bark beetle group. J. Chem. Ecol. 16, 3317–3331.

Smith, M.T., Salom, S.M., Payne, T.L., 1993. The southern pine bark beetle guild: an historical review of the research on the semiochemical-based communication system of the five principal species. Va. Agric. Expt. Stn. Bull. 93–94, 106.
Staeben, J.C., Sullivan, B.T., Nowak, J.T., Gandhi, K.J., 2015. Enantiospecific responses of southern pine beetle (*Dendroctonus frontalis*) and its clerid predator, *Thanasimus dubius*, to α-pinene. Chemoecology 25, 73–83.
Steidle, J.L., van Loon, J.J., 2002. Chemoecology of parasitoid and predator oviposition behaviour. In: Hilker, M., Meiners, T. (Eds.), Chemoecology of Insect Eggs and Egg Deposition. Blackwell, London, pp. 291–317.
Stephen, F.M., 2011. Southern pine beetle competitors. In: Coulson, R.N., Klepzig, K.D. (Eds.), Southern Pine Beetle II. In: General Technical Report SRS-140, USDA Forest Service Southern Research Station, Asheville, NC, pp. 183–198.
Stephen, F.M., Lih, M.P., 1985. A *Dendroctonus frontalis* infestation growth model: organization, refinement, and utilization. In: Branham, S.J., Thatcher, R.C. (Eds.), Proceedings, Integrated Pest Management Research Symposium. In: General Technical Report SO-56, USDA Forest Service Southern Forest Experiment Station, New Orleans. pp. 186–194.
Stewart, T.E., Plummer, E.L., McCanless, L.L., West, J.R., Silverstein, R.M., 1977. Determination of enantiomer composition of several bicyclic ketal insect pheromone components. J. Chem. Ecol. 3, 27–43.
Strom, B.L., Clarke, S.R., 2011. Use of semiochemicals for southern pine beetle infestation management and resource protection. In: Coulson, R.N., Klepzig, K.D. (Eds.), Southern Pine Beetle II. In: General Technical Report SRS-140, USDA Forest Service Southern Research Station, Asheville, NC, pp. 381–397.
Strom, B.L., Goyer, R.A., Hayes, J.L., 1995. Naturally occuring compound can protect pines from the southern pine beetle. Louisiana Agric. 38, 5–7.
Strom, B.L., Roton, L.M., Goyer, R.A., Meeker, J.R., 1999. Visual and semiochemical disruption of host finding in the southern pine beetle. Ecol. Appl. 9, 1028–1038.
Strom, B.L., Clarke, S.R., Shea, P.J., 2004. Efficacy of 4-allylanisole-based products for protecting individual loblolly pines from *Dendroctonus frontalis* Zimmermann (Coleoptera: Scolytidae). Can. J. For. Res. 34, 659–665.
Sullivan, B.T., 1997. The chemical ecology of host habitat location by larval parasitoids of the southern pine beetle, *Dendroctonus frontalis* Zimmermann: olfactory cues and their possible sources. PhD Thesis, Department of Entomology. University of Georgia, Athens, GA. p. 204.
Sullivan, B.T., 2005. Electrophysiological and behavioral responses of *Dendroctonus frontalis* (Coleoptera: Curculionidae) to volatiles isolated from conspecifics. J. Econ. Entomol. 98, 2067–2078.
Sullivan, B.T., Mori, K., 2009. Spatial displacement of release point can enhance activity of an attractant pheromone synergist of a bark beetle. J. Chem. Ecol. 35, 1222–1233.
Sullivan, B.T., Berisford, C.W., Dalusky, M.J., 1997. Field response of southern pine beetle parasitoids to some natural attractants. J. Chem. Ecol. 23, 837–856.
Sullivan, B.T., Pettersson, E.M., Seltmann, K.C., Berisford, C.W., 2000. Attraction of the bark beetle parasitoid *Roptrocerus xylophagorum* (Hymenoptera: Pteromalidae) to host-associated olfactory cues. Environ. Entomol. 29, 1138–1151.
Sullivan, B.T., Dalusky, M.J., Wakarchuk, D., Berisford, C.W., 2007a. Field evaluations of potential aggregation inhibitors for the southern pine beetle, *Dendroctonus frontalis* (Coleoptera: Curculionidae). J. Entomol. Sci. 42, 139–149.
Sullivan, B.T., Shepherd, W.P., Pureswaran, D.S., Tashiro, T., Mori, K., 2007b. Evidence that (+)-*endo*-brevicomin is a male-produced component of the southern pine beetle aggregation pheromone. J. Chem. Ecol. 33, 1510–1527.

Sullivan, B.T., Dalusky, M.J., Mori, K., Brownie, C., 2011. Variable responses by southern pine beetle, *Dendroctonus frontalis* Zimmermann, to the pheromone component *endo*-brevicomin: influence of enantiomeric composition, release rate, and proximity to infestations. J. Chem. Ecol. 37, 403–411.

Sullivan, B.T., Niño, A., Moreno, B., Brownie, C., Macías-Sámano, J., Clarke, S.R., Kirkendall, L.R., Zúñiga, G., 2012. Biochemical evidence that *Dendroctonus frontalis* consists of two sibling species in Belize and Chiapas. Mexico. *Ann. Entomol. Soc. Am.* 105, 817–831.

Svihra, P., 1982. Influence of opposite sex on attraction produced by pioneer sex of four bark beetles species cohabiting pine in the southern United States. J. Chem. Ecol. 8, 373–378.

Svihra, P., Paine, T.D., Birch, M.C., 1980. Interspecific olfactory communications in southern pine beetles. Naturwissenschaften 67, 518–520.

Symonds, M.R.E., Elgar, M.A., 2004. Species overlap, speciation and the evolution of aggregation pheromones in bark beetles. Ecol. Lett. 7, 202–212.

Tchakerian, M.D., Coulson, R.N., 2011. Ecological impacts of southern pine beetle. In: Coulson, R.N., Klepzig, K.D. (Eds.), Southern Pine Beetle II. In: General Technical Report SRS-140, USDA Forest Service Southern Research Station, Asheville, NC, pp. 223–234.

Thatcher, R.C., 1960. Bark Beetles Affecting Southern Pines: A Review of Current Knowledge. USDA Forest Service Southern Forest Experiment Station Occ. Pap. 180, Pineville, LA, 25 pp.

Thatcher, R.C., Pickard, L.S., 1964. Seasonal variations in activity of the southern pine beetle in east Texas. J. Econ. Entomol. 57, 840–842.

Thatcher, R.C., Pickard, L.S., 1966. The clerid beetle, *Thanasimus dubius*, as a predator of the southern pine beetle. J. Econ. Entomol. 59, 955–957.

Turchin, P., Thoeny, W.T., 1993. Quantifying dispersal of southern pine beetles with mark-recapture experiments and a diffusion model. Ecol. Appl. 3, 187–198.

Vet, L.E.M., Dicke, M., 1992. Ecology of infochemical use by natural enemies in a tritrophic context. Ann. Rev. Entomol. 37, 141–172.

Vité, J.P., 1970. Pest management systems using synthetic pheromones. Contrib. Boyce Thompson Inst. 24, 343–350.

Vité, J.P., Crozier, R.G., 1968. Studies on the attack behavior of the southern pine beetle. IV. Influence of host condition on aggregation pattern. Contrib. Boyce Thompson Inst. 24, 87–93.

Vité, J.P., Francke, W., 1976. The aggregation pheromones of bark beetles: progress and problems. Naturwissenschaften 63, 550–555.

Vité, J.P., Pitman, G.B., 1968. Bark beetle aggregation: effects of feeding on the release of pheromones in *Dendroctonus* and *Ips*. Nature 218, 169–170.

Vité, J.P., Renwick, J.A.A., 1968. Insect and host factors in the aggregation of the southern pine beetle. Contrib. Boyce Thomps. Inst. 24, 61–63.

Vité, J.P., Renwick, J.A.A., 1971. Inhibition of *Dendroctonus frontalis* response to frontalin by isomers of brevicomin. Naturwissenschaften 58, 418–419.

Vité, J.P., Williamson, D.L., 1970. *Thanasimus dubius*: prey perception. J. Insect Phys. 16, 233–239.

Vité, J.P., Bakke, A., Renwick, J.A.A., 1972. Pheromones in *Ips* (Coleoptera: Scolytidae): occurence and production. Can. Entomol. 104, 1967–1975.

Vité, J.P., Islas, S.F., Renwick, J.A.A., Hughes, P.R., Kliefoth, R.A., 1974. Biochemical and biological variation of southern pine beetle populations in North and Central America. Z. Angew. Entomol. 75, 422–435.

Vité, J.P., Hughes, P.R., Renwick, J.A.A., 1976. Southern pine beetle: effect of aerial pheromone saturation on orientation. Naturwissenschaften 63, 44.

Vité, J.P., Billings, R.F., Ware, C.W., Mori, K., 1985. Southern pine beetle: enhancement or inhibition of aggregation response mediated by enantiomers of *endo*-brevicomin. Naturwissenschaften 72, 99–100.
Watterson, G.P., Payne, T.L., Richerson, J.V., 1982. The effects of verbenone and brevicomin on the within-tree populations of *Dendroctonus frontalis*. J. Ga. Entomol. Soc. 17, 118–126.
Werner, R.A., 1972. Aggregation behaviour of the beetle *Ips grandicollis* in response to host-produced attractants. J. Insect Phys. 18, 423–437.
Witzgall, P., Kirsch, P., Cork, A., 2010. Sex pheromones and their impact on pest management. J. Chem. Ecol. 36, 80–100.
Wood, D.L., 1977. Manipulation of forest insect pests. In: Shorey, H.H., McKelvey, J.J. (Eds.), Chemical Control of Insect Behavior. John Wiley & Sons, New York, pp. 369–384.
Wood, D.L., 1982a. The role of pheromones, kairomones, an allomones in the host selection and colonization behavior of bark beetles. Ann. Rev. Entomol. 27, 411–446.
Wood, S.L., 1982b. The bark and ambrosia beetles of North and Central America (Coleoptera: Scolytidae), a taxonomic monograph. Great Basin Nat. Mem. 6, 1359.
Wood, D.L., Bedard, W.D., 1977. The role of pheromones in the population dynamics of the western pine beetle. Proc. XV Int. Congr. Entomol. 15, 643–652.
Xu, L., Lou, Q., Cheng, C., Lu, M., Sun, J., 2015. Gut-associated bacteria of *Dendroctonus valens* and their involvement in verbenone production. Microb. Ecol. 70, 1012–1023.
Zhang, Q.H., Schlyter, F., 2003. Redundancy, synergism, and active inhibitory range of non-host volatiles in reducing pheromone attraction in European spruce bark beetle *Ips typographus*. Oikos 101, 299–310.
Zhang, Q.H., Schlyter, F., 2004. Olfactory recognition and behavioural avoidance of angiosperm nonhost volatiles by conifer-inhabiting bark beetles. Agric. For. Entomol. 6, 1–19.
Zhang, Q.H., Schlyter, F., Liu, G.T., Sheng, M.L., Birgersson, G., 2007. Electrophysiological and behavioral responses of *Ips duplicatus* to aggregation pheromone in inner mongolia, China: amitinol as a potential pheromone component. J. Chem. Ecol. 33, 1303–1315.

CHAPTER FIVE

The Evolution of Aggregation Pheromone Diversity in Bark Beetles

M.R.E. Symonds*, C.W. Gitau-Clarke†
*Centre for Integrative Ecology, School of Life and Environmental Sciences, Deakin University, Burwood, Vic, Australia
†Department of Economic Development, Jobs, Transport and Resources, Rutherglen, Vic, Australia

Contents

Abstract

Bark beetles of the subfamily Scolytinae employ aggregation pheromones to attract conspecifics to their host trees (predominantly conifers) for the purposes of mating and resource exploitation. These aggregation pheromones consist of blends of chemical components whose composition is typically species specific. Across tribes and genera within the Scolytinae there are clear phylogenetic patterns in these compounds.

Advances in Insect Physiology, Volume 50
ISSN 0065-2806
http://dx.doi.org/10.1016/bs.aiip.2015.12.003

However, within these clades differences in pheromone composition may be as great between closely related species as between more distantly related species.

Here we review our understanding of the evolution of aggregation pheromone diversity in this group including discussion of the mode of evolution, and the ecological, genetic and biosynthetic mechanisms that may explain this diversity. We also collate information from the literature on the aggregation pheromone composition for 65 species of bark beetles across 17 genera, including intraspecific variation. We relate this information to evolutionary relationships to ascertain the extent to which pheromone composition reflects phylogeny. As with previous studies, we show that the mode of pheromone evolution can be characterised by substantial saltational changes in chemical composition at speciation events. We also demonstrate that some individual components are strongly phylogenetically conserved within clades, while others are phylogenetically labile. This suggests that differences in aggregation pheromone composition may arise through a combination of relatively rapid 'switching on and switching off' of certain compounds (eg, ipsdienol and ipsenol in the Ipini), while other compounds remain more stable once their use has evolved. We discuss these findings in the context of recent theory on the evolution of species-specific attractive pheromones.

1. BARK BEETLES AS PARADIGMS FOR PHEROMONE RESEARCH

For almost 60 years since the first description of a pheromone (Karlson and Lüscher, 1959) the field of chemical ecology has seen enormous research effort into the compounds used in communication between species, with thousands of semiochemical compounds now described (El-Sayed, 2014). There have been taxonomic biases in this research and, after Lepidoptera, the Coleoptera is the second-most well-studied group of animals, representing nearly one-fifth of all pheromone publications (Symonds and Elgar, 2008). The Coleoptera as a whole present a diverse array of chemical compounds used for a range of behavioural functions (Francke and Dettner, 2005). Within the Coleoptera, the attractant pheromones of the Scolytinae (bark beetles) were one of the earliest kinds of pheromones to be studied intensively across multiple species (eg, Kinzer et al., 1969; Silverstein et al., 1966; Silverstein et al., 1968; Vité and Renwick, 1971), and to-date almost 70 bark beetle species have had the chemical composition of these pheromones described (see reviews by Blomquist et al., 2010; Borden, 1985; Byers, 1989; Wood, 1982).

The strong interest in bark beetle chemical communication undoubtedly arose from the role that many species play as significant silvicultural pests, and the desire to develop pheromone traps with synthetic lures as a means to reduce population densities below damaging levels (El-Sayed et al., 2006; Schiebe et al., 2011). Serendipitously, this research focus has provided us with an ideal group with which to examine broader issues regarding pheromone evolution and our understanding of the ecological drivers and genetic mechanisms that underlie diversity in pheromone composition among closely related species (Symonds and Elgar, 2004a, 2008). It was realised fairly early on that bark beetle aggregation pheromones frequently comprised two or three behaviourally significant compounds and that the exact composition could be species specific (Vité et al., 1972), but understanding how these differences between species arose, and the evolutionary processes that explain them, is often relegated in the literature to speculation rather than systematic comparative analysis.

2. BARK BEETLE AGGREGATION PHEROMONES: FUNCTION AND CHEMICAL DIVERSITY

Scolytine bark beetles are largely characterised by their tendency to aggregate in mass attacks on their host trees (predominantly species of the Pinaceae), where they bore into the phloem and construct galleries in which they mate and oviposit. Aggregation pheromones are a communication tool among bark beetles. Aggregation is initiated by the production of volatile compounds after feeding by the attacking pioneers, and excretion of these compounds in faecal material (frass). Aggregation arising from pheromone-mediated communication results in mate selection, mass attacks on healthy or dying trees, and assists in overcoming tree defence mechanisms. Beetles are generally attracted to sources of pheromones produced by conspecifics and avoid pheromones produced by individuals of other species (Borden, 1982; Hofstetter et al., 2012). For example, among the sibling bark beetles species *Pityogenes bidentatus* and *Pityogenes quadridens*, the aggregation pheromone components from males of each species not only attract conspecifics but also aid individuals in avoiding interspecific mating interference, competition for food and spatial resources within the bark phloem (Byers et al., 2013).

Aggregation pheromones are technically distinguished from sex pheromones (although see Section 3.1) because they attract both sexes. Within the subfamily Scolytinae, there is variation in which sex initiates the aggregation process. Thus, in *Dendroctonus* species, the females initiate attacks on trees and construct galleries in the phloem and are later joined by males, while in *Ips* species the opposite is the case (Raffa, 2001). Note that in many species individuals may produce antiaggregation pheromones that assist in dispersing beetles to new trees and thus reducing intraspecific competition. Several of these antiaggregants may be attractants at lower concentration but become repellent at higher concentrations, for example, *endo*-brevicomin in *Dendroctonus frontalis* (Sullivan et al., 2007). Hence these dual action pheromones may act effectively as colonisation density regulators (Raffa, 2001).

Aggregation pheromone compounds are typically terpenes and terpenoids and for many years it was assumed that they were solely derived from host-produced compounds like myrcene and α-pinene (Borden, 1985; Byers, 1981; Hendry et al., 1980). However, more recent evidence indicates that these compounds can also be synthesised *de novo* by the bark beetles (Seybold and Tittiger, 2003; Seybold et al., 2000; Song et al., 2013). There are also likely to be differences in the biosynthetic pathways utilised to produce specific aggregation pheromone components. Even apparently structurally closely related compounds (like frontalin and *exo*-brevicomin, as produced by *Dendroctonus* species) may be derived via very different biosynthetic pathways (Barkawi et al., 2003; Vanderwel et al., 1992). Changes in enzymes can therefore profoundly alter pheromone composition, or at least the potential for synthesis (Blomquist et al., 2010). If the changes occur to prevailing metabolic pathways, this can create a diverse suite of pheromone components (Blomquist et al., 2010; Symonds and Elgar, 2008). For instance, in the pine engraver *Ips pini* pheromone composition is variable, and these differences can be attributed to a single or few genetic loci (Domingue and Teale, 2007; Song et al., 2013).

There have been several comprehensive reviews of the chemical structures of scolytine aggregation pheromone components (eg, Blomquist et al., 2010, Francke and Dettner, 2005; Seybold et al., 2000), and it is not our aim to recapitulate those here. However, a brief summary of the taxonomic spread of commonly used components is useful in the context of the subsequent discussion.

In Table 1, we provide a summary of known attractive aggregation pheromone components across bark beetle species. It is clear that there are distinct taxonomic patterns to the use of certain compounds. If we look at the two major genera of bark beetles on which work has focussed, *Dendroctonus* and *Ips*, each genus has certain 'palettes' of chemicals which typify their aggregation pheromones. In *Ips* (and the closely related *Orthotomicus*) the pheromone chemistry is dominated by the use of ipsenol, ipsdienol and *cis*-verbenol (derived from the (−)-enantiomer of α-pinene)

Table 1 Aggregation Pheromone Components of Bark Beetle Species

Species	Ie	Id	cV	tV	Fr	exoB	endoB	Other Components	Reference(s)
Corthylini									
Gnathotrichus materiarius								Sulcatol	Flechtmann and Berisford (2003)
Gnathotrichus retusus								Sulcatol	Borden et al. (1980)
Gnathotrichus sulcatus								Sulcatol	Byrne et al. (1974)
Dryocoetini									
Dryocoetes affaber						X	X		Camacho et al. (1994)
Dryocoetes autographus						X	X		Kohnle (1985)
Dryocoetes confusus						X	X		Camacho et al. (1993)
Xylocleptes bispinus	X								Klimetzek et al. (1989b)
Hylesinini									
Dendroctonus adjunctus					X	X			Hughes et al. (1976)
Dendroctonus armandi				X		X		Seudenol, MCOL	Chen et al. (2015)
Dendroctonus brevicomis					X	X			Wood et al. (1976)
Dendroctonus frontalis (Arizona)				X	X	X	X	Myrtenol, verbenone	Pureswaran et al. (2008)
Dendroctonus frontalis (California)				X	X		X	Myrtenol, verbenone	Rudinsky et al. (1974)
Dendroctonus frontalis (Texas)				X	X		X		Vité et al. (1985)

Continued

Table 1 Aggregation Pheromone Components of Bark Beetle Species—cont'd

Species	Ie	Id	cV	tV	Fr	exoB	endoB	Other Components	Reference(s)
Dendroctonus jeffreyi					X	X		1-Heptanol	Paine et al. (1999)
Dendroctonus mesoamericanus					X		X		Niño-Domínguez et al. (2015)
Dendroctonus ponderosae (British Columbia 1)				X	X	X			Borden et al. (1987)
Dendroctonus ponderosae (British Columbia 2)			X	X		X			Miller and Lafontaine (1991)
Dendroctonus ponderosae (Oregon)				X		X		3-Caren-10-ol	Libbey et al. (1985)
Dendroctonus pseudotsugae (British Columbia 1)					X			Seudenol, MCOL	Lindgren et al. (1992)
Dendroctonus pseudotsugae (British Columbia 1)				X	X				Pureswaran et al. (2004)
Dendroctonus rufipennis					X			Seudenol, MCOL, verbenene	Gries et al. (1992) and Ross et al. (2005)
Dendroctonus simplex					X			Seudenol	Francke et al. (1995) and Baker et al. (1977)
Dendroctonus terebrans (Northern Florida)				X	X	X		Myrtenol	Payne et al. (1987)
Dendrocotnus terebrans (Southern Florida)				X	X			Myrtenol	Phillips et al. (1989b)

Dendroctonus valens				X	X				Myrtenol	Zhang and Sun (2006)
Hylesinus pruinosus								X		Shepherd et al. (2010)
Pteleobius vittatus									Pityol, 2-methyl-3-buten-2-ol, vittatol	Klimetzek et al. (1989a)
Ipini										
Ips acuminatus	X	X	X							Bakke (1978)
Ips amitinus	X	X							Amitinol	Francke et al. (1980)
Ips avulsus		X							Lanierone	Birgersson et al. (2012)
Ips calligraphus		X	X							Birgersson et al. (2012)
Ips cembrae	X	X							3-Methyl-3-buten-1-ol	Zhang et al. (2000)
Ips duplicatus (China)		X							Amitinol, E-myrcenol	Zhang et al. (2007b)
Ips duplicatus (Sweden)		X							E-myrcenol	Byers et al. (1990b)
Ips grandicollis	X		X							Birgersson et al. (2012)
Ips nitidus		X	X						2-Methyl-3-buten-2-ol	Schlyter et al. (2015)
Ips paraconfusus	X	X	X							Byers et al. (1979) and Byers and Birgersson (2012)
Ips perroti	X	X								Ayres et al. (2001)
Ips perturbatus	X		X	X						Graves (2008)

Continued

Table 1 Aggregation Pheromone Components of Bark Beetle Species—cont'd

Species	Ie	Id	cV	tV	Fr	exoB	endoB	Other Components	Reference(s)
Ips pini		X						Lanierone	Birgersson et al. (2012)
Ips schmutzenhoferi	X		X						Kohnle et al. (1988)
Ips sexdentatus		X							Francke et al. (1986)
Ips shangrila		X	X						Schlyter et al. (2015)
Ips subelongatus (China)	X								Zhang et al. (2007a)
Ips subelongatus (Jilin, Liaoning Provinces, China)	X	X							Chen et al. (2016)
Ips typographus			X					2-Methyl-3-buten-2-ol	Schlyter et al. (1987)
Orthotomicus caelatus	X	X							Phillips et al. (1989a)
Orthotomicus erosus		X						2-Methyl-3-buten-2-ol	Giesen et al. (1984)
Orthotomicus latidens	X								Miller et al. (1991)
Orthotomicus mannsfeldi	X	X						Amitinol	Kohnle et al. (1993)
Orthotomicus tridentatus	X	X						2-Methyl-3-buten-2-ol	Birgersson (unpublished) *in* Schlyter et al. (2015)
Pityogenes chalcographus								Chalcogran, methyl decadienoate	Byers et al. (1990a)
Pityogenes hopkinsi			X					Chalcogran, ethyl dodecanoate	Birgersson et al. (2000)
Pityogenes knechteli	X	X						Hexanol	Savoie et al. (1998)

Pityogenes quadridens				Chalcogran, grandlure	Byers et al. (2013)
Pityokteines curvidens	X				Harring (1978)
Pityokteines elegans	X	X			Macías-Sámano et al. (1997)
Pityokteines minutus	X				Macías-Sámano et al. (1997)
Pityokteines vorontzovi	X	X			Harring (1978)
Pseudips orientalis	X		X		Zhang et al. (2011)
Pityophthorini					
Pityophthorus carmeli				Pityol, conophthorin	Dallara et al. (2000)
Pityophthorus pityographus				Pityol, grandisol	Francke et al. (1987)
Pityophthorus pubescens				Pityol	López et al. (2011)
Pityophthorus setosus				Pityol	Dallara et al. (2000)
Polygraphini					
Polygraphus poligraphus				Terpinene-4-ol	Schurig et al. (1985)
Polygraphus rufipennis				3-Methyl-3-buten-1-ol	Bowers et al. (1991)
Scolytini					
Scolytus amygdali				4-Methyl-3-hexanol, 4-methyl-3-heptanol	Zada et al. (2004)

Continued

Table 1 Aggregation Pheromone Components of Bark Beetle Species—cont'd

Species	Ie	Id	cV	tV	Fr	exoB	endoB	Other Components	Reference(s)
Scolytus multistriatus								4-Methyl-3-hexanol, α-cubenene, α-multistriatin	Lanier et al. (1977)
Scolytus scolytus								4-Methyl-3-heptanol, 4-methyl-3-heptanone, α-cubenene	Blight et al. (1983)
Taphrorychini									
Taphrorychus bicolor								Acetophenone, bicolorin	Kohnle et al. (1987)
Tomicini									
Tomicus minor				X					Lanne et al. (1987)
Xyloterini									
Trypodendron domesticus								3-Hydroxy-3-methyl-butan-2-one	Francke and Heemann (1974)
Trypodendron lineatum								Lineatin	MacConnell et al. (1977)

Common components are *Ie*, ipsenol; *Id*, ipsdienol; *cV*, *cis*-Verbenol; *tV*, *trans*-Verbenol; *Fr*, frontalin; *exoB*, *exo*-brevicomin; *endoB*, *endo*-brevicomin.

(Cognato et al. 1997). All species in these genera (and indeed, in the tribe Ipini, with the exception of *Pityogenes chalcographus*—Byers et al., 2013) use at least one of these components in their aggregation pheromone blend. Additional compounds used by several species but exclusively found within the Ipini include amitinol (itself closely related to ipsdienol), and 2-methyl-3-buten-2-ol (Francke et al., 1980; Giesen et al., 1984). Several compounds within the Ipini are more specific—for example, lanierone is utilised only by *I. pini* and *I. avulsus* (Birgersson et al., 2012), while E-myrcenol is solely used by *I. duplicatus* (Byers et al., 1990b).

Within *Dendroctonus* the predominant compounds are frontalin, *trans*-verbenol (derived from (+)-enantiomer of α-pinene) and *exo*- and *endo*-brevicomin (Renwick and Vité, 1970), with all species utilising at least one of these compounds in their aggregation pheromone blends (see Table 1 and references therein). Other less common compounds in this genus include myrtenol, and the structurally related compounds 3-methyl-2-cyclohexen-1-ol (seudenol) and 1-methyl-2-cyclohexen-1-ol (MCOL) which are expressed within the clade that includes *D. armandi, D. pseudotsugae, D. rufipennis* and *D. simplex* (Baker et al., 1977; Chen et al., 2015; Lindgren et al., 1992). As within the Ipini, certain compounds in *Dendroctonus* are utilised more rarely, such as the compounds 1-heptanol, 3-caren-10-ol and verbenene which have been identified only in single species (*D. jeffreyi, D. ponderosae* and *D. rufipennis*, respectively) (Gries et al., 1992; Libbey et al., 1985; Paine et al., 1999). In addition to these major clades, certain compounds are exclusively found, and characterise, the more minor genera. For example, sulcatol is exclusively used by *Gnathotrichus* species (Flechtmann and Berisford, 2003), pityol by *Pityophthorus* (Dallara et al., 2000) and chalcogran by *Pityogenes* (Harring, 1978).

While there are clear taxonomic (and hence, we assume phylogenetic) trends in the use of the chemical compounds highlighted above, nevertheless, diversity in pheromone composition even between closely related species can still be generated in a couple of ways. First, by using varying combinations of the most common components (for example, the sibling species *Ips grandicollis* and *I. calligraphus* utilise ipsenol plus *cis*-verbenol, and ipsdienol plus *cis*-verbenol, respectively—Birgersson et al., 2012). Second, additional diversity and species specificity may be achieved by the addition of less common compounds. Therefore, despite the phylogenetic pattern to aggregation pheromone diversity within the Scolytinae as a whole, there is still capacity to generate substantial differences between closely related species, and this may influence our understanding of how the evolution of differences in pheromone blends has evolved.

3. THE MODE OF EVOLUTION OF BARK BEETLE AGGREGATION PHEROMONES

3.1 Saltational Evolution of Bark Beetle Pheromone Composition

As intraspecific signals, pheromones are subject to selection dependent on their function and their environment, and therefore can vary substantially between species (Wyatt, 2014). The term 'aggregation' pheromone rather generically describes the behavioural result of the emission of this signal in bark beetles. But in fact aggregation pheromones may elicit aggregations that serve a variety of roles such as protection, reproduction, resource exploitation or some combination of these (Borden, 1985). Delineating between these functions is not straightforward. For example, in *Ips*, male pioneer beetles initiate attacks by producing the aggregation pheromone which attracts females with whom they mate, but also other males who participate in the feeding attack (Raffa, 2001). This promotes speculation as to whether the original function of the pheromone was to attract a female, with aggregation by eavesdropping males being a secondary consideration, or whether the aggregation was the primary benefit, with mating with females being a secondary inevitable consequence. The situation is more complex still in *Dendroctonus* where females are the pioneers, but only produce part of the aggregation pheromone blend (eg, frontalin in many species), while males produce other components like *exo*-brevicomin which promote the aggregation when they arrive.

Either way the role that aggregation pheromones play in mate attraction and location is highly likely to influence the way in which these pheromone blends evolve. Generally speaking, the composition of pheromones that serve as sexual attractants tends to be highly species specific (Smadja and Butlin, 2009). But the question of how this specificity (and consequent diversity) arises is one that is still at the early stages of our understanding (Steiger et al., 2010; Symonds and Elgar, 2008). With a gradual mode of evolution, speciation is associated with small changes in the pheromone composition: perhaps the addition or removal of single components from the pheromone blend, or even just changes in the ratios of components (eg, Roelofs and Brown, 1982). However, for signals involved in species recognition, a likely driver of divergence lies in the need to avoid mismating and hybridisation with closely related species (Gröning and Hochkirch, 2008). In such circumstances there should be strong selection for rapid divergence in the signal through reproductive character displacement and reinforcement (Marshall et al., 2002; Noor, 1999). Consequently, we may predict larger 'saltational' changes in pheromone composition between closely related species (Baker, 2002; Butlin and Trickett, 1997). By contrast, pheromones that do not play a role in species recognition or isolation should exhibit the gradual mode of evolution, as seen for example, in *Drosophila* aggregation pheromones (Symonds and Wertheim, 2005) where interspecific attraction is actively beneficial to improve exploitation of an oviposition resource.

In terms of bark beetles, previous literature that had discussed the evolution of aggregation pheromone composition had suggested different patterns, with Cane et al. (1990) implicating close links between pheromone composition and phylogeny among *Ips* species, but Cognato et al. (1997) suggested high levels of lability among pheromone components across the phylogeny, indicative of the capacity for saltational changes.

Symonds and Elgar (2004a) utilised the extensive literature on bark beetle aggregation pheromones, coupled with phylogenetic information to elucidate the mode of evolution among 34 species of *Dendroctonus* and *Ips* (including *Orthotomicus*, which at that time had only recently been separated into a separate genus). Their analysis showed strong evidence for saltational evolution in aggregation pheromone composition within each genus (although with some phylogenetic patterns in individual components both between and within genera). Indeed, in *Dendroctonus*, they found that the greatest level of difference in pheromone composition was between sibling species, and in fact there was as many, or greater, differences in pheromone composition within *Dendroctonus* than there were between the genera.

3.2 Ecological Drivers of Aggregation Pheromone Diversity

What caused the observed diversity in bark beetle pheromones to evolve? While the answer to this question involves mechanistic process at the molecular level, such as changes in gene expression, biosynthetic pathways and receiver structure (Andersson et al., 2013; Domingue and Teale, 2007; Seybold et al., 2000), the ultimate selective drivers are likely to be ecological (Symonds and Elgar, 2008). The most commonly proposed candidate for drivers of pheromone divergence is overlap with closely related species, both geographical (sympatry) and at the microhabitat or niche level (syntopy). With such species we predict 'chemical niche differentiation' (Zimmermann et al., 2009), in order to prevent mixed messages when species overlap. These different compounds may even serve as inhibitors of the sympatric species (eg,

McElfresh and Millar, 2001). This mechanism has often been mooted, or implied, to explain pheromone composition differences both within and between species of bark beetle (eg, Ayres et al., 2001; Birgersson et al., 2012; Raffa, 2001; Schlyter et al., 1987). However, experimental and comparative tests of this hypothesis remain scanty even among insects as a whole. Higgie et al.'s (2000) experimental breeding in sympatry of two *Drosophila* species selecting for divergence in cuticular hydrocarbon profiles (used as sexual signals) remains one of the few really good examples of this effect in action.

Testing the prediction that reproductive characters should diverge to a greater extent in sympatry (Coyne and Orr, 1989), Symonds and Elgar (2004b) assessed the role that species overlap (both geographical and host tree) may have played in generating differences in bark beetle aggregation pheromone composition. They found no evidence within genera (*Dendroctonus* and *Ips*) for any greater difference in pheromone composition with syntopic vs allotropic species, suggesting no consistent effect of interspecific competition in driving divergence in pheromone profiles.

The lack of greater divergence in sympatry may reflect the dual roles of aggregation pheromones (ie, not just species recognition for mating) and that there may be advantages to heterospecific aggregations in order to overcome host defences. Heterospecific attraction has been reported in quite a number of bark beetle studies (eg, Cane et al., 1990; Davis and Hofstetter, 2009; Pureswaran et al., 2004). For example, the sympatric distribution of *D. ponderosae* and *I. pini*, their utilisation of a common food source and mating sites, and the involvement of general metabolic pathways in the conversion of nutrients would account for the production of common volatile compounds serving as aggregants, resulting in heterospecific attraction (Hofstetter et al., 2012). Within *Ips*, the sympatric species *I. pini*, *I. perroti*, and *I. grandicollis* appear to actively benefit from coattraction and cross-species aggregation. However, the aggregation pheromones serve multiple functions such as attracting conspecifics, acquiring mates, preventing intersection with heterospecific galleries and ensuring resource partitioning along the bole (Ayres et al., 2001). All three species exploit this same short-lived resource, so pheromone plumes from any of the species indicates suitable host material, and individuals that can orient to heterospecific pheromones have an increased probability of finding food resources (Ayres et al., 2001). The attraction of *Ips* for heterospecific pheromones could result from a directional selection to exploit these benefits. Males, the colonising sex within *Ips* species are more strongly attracted to heterospecific signals than females because they benefit from attacking trees en masse (Ayres et al., 2001). Females would not normally benefit from this attraction, and the detriment for the females is that they would suffer the negative effects of interspecific competition in the larval stage. Thus, if heterospecific attraction has been favoured through natural selection, then it should be strongest among sympatric species, where it may counteract selection for reproductive isolation (Hofstetter et al., 2012). Individuals may sometimes benefit from joining aggregations of other species because of predator swamping, improved success in attacking live trees and location of suitable, recently dead trees. These benefits should be greatest for males (which locate and colonise host trees before signalling females, eg, in *Ips* species) and generally, males tend to be more attracted by heterospecific signals than females (Ayres et al., 2001).

These patterns of heterospecific attraction are also reflected in Symonds and Elgar's (2004b) finding that species show somewhat greater similarity in pheromone profile when they share host trees, which may reflect the derivation of aggregation pheromone components from the host tree compounds (Renwick et al., 1976a,b). Given that species utilising the same hosts also tend to be more phylogenetically closely related (Symonds and Elgar, 2004b), then this suggests that divergence between closely related species is not driven by host shifting at speciation.

A final cause of evolutionary change in pheromone composition may be that caused by the effects of predators or parasitoids that eavesdrop on their host's signal. Variation in aggregation pheromone composition seen between local populations of *I. pini* has been attributed to predation by clerid beetles (Erbilgin and Raffa, 2001; Raffa et al., 2007) with the exact ratio of enantiomers of ipsdienol reflecting the mixture to which predators respond least, and the use of lanierone (which the predatory beetles cannot detect) applying added specificity. To-date, though there has be no wider comparative study of the potential for predator and parasitoid eavesdropping to drive evolutionary change and divergence in bark beetle pheromones.

4. A REANALYSIS OF BARK BEETLE AGGREGATION PHEROMONE EVOLUTION

4.1 Rationale

We have reanalysed the composition of bark beetle aggregation pheromones in relation to their phylogenetic relationships with the aim of further elucidating patterns regarding the mode of pheromone evolution, but also the degree to which individual pheromone components are phylogenetically conserved. Our first aim was to assess whether, across the Scolytinae as a whole, and within two major clades (*Dendroctonus* and the tribe Ipini), there are phylogenetic patterns in pheromone composition, and further ascertain the evidence that evolution of these pheromone blends is best described by a saltational model of evolution. Our second aim was to assess, through the analysis of phylogenetic signal in key aggregation pheromone components, which compounds are most variable (and hence most evolutionary labile) and therefore likely to be more responsible for the patterns of diversity observed.

Since the previous analysis of the mode of evolution of these pheromones (Symonds and Elgar, 2004a), there have been a number of advances in research on bark beetle pheromones. These include the description of chemical composition in new species (eg, Niño-Domínguez et al., 2015; Schlyter et al., 2015; Zhang et al., 2007a, 2011) and greater understanding of phylogenetic relationships among the Scolytinae (Jordal and Cognato, 2012). These enable us to greatly extend the taxonomic scope of our analysis. Further, we can now apply newer techniques, used in others studies of pheromone evolution (Symonds and Wertheim, 2005; Symonds et al., 2009; Van Wilgenburg et al., 2011), to more systematically analyse patterns. Finally, we take the opportunity to refine some of the characterisation of the aggregation pheromone components to stricter criteria than were applied in the Symonds and Elgar's (2004a) analysis. The present analysis therefore represents the most extensive, systematic and accurate comparative analysis of bark beetle aggregation pheromone evolution to-date.

4.2 Study Species and Collation of Pheromonal and Phylogenetic Information

We collated information from the literature on the aggregation pheromone composition of 65 scolytine bark beetle species, across 17 genera (see Table 1), almost doubling the representation of species compared to Symonds and Elgar's (2004a) analysis. We recorded the presence or absence of chemical compounds as active components of the aggregation pheromone blend for each species. In other words, we considered only compounds that are both produced by, and responded to behaviourally by, those species. In cases where there were multiple publications investigating the aggregation pheromone composition of a species we deferred to the more recent publications if it was obvious that these were simply adding additional compounds to already established blends; however, in several species (*D. frontalis, D. ponderosae, D. pseudotsugae, D. terebrans, I. duplicatus, I. subelongatus*) clear population level variation in pheromone composition has been reported. Because such differences may represent the beginnings of speciational processes, or even cryptic species (eg, the newly described

Dendroctonus mesoamericanus was previously considered to be a separate population of *D. frontalis*—Armendáriz-Toledano et al., 2015; Sullivan et al., 2012), we consider these pheromone 'races' as separate species in our analysis—making our total number of taxa being compared equal to 73. In total we identified 39 chemical compounds as aggregation pheromone components across these species.

Differences in chirality of compounds may provide an additional vital level of specificity and diversity among species (Byers, 1989; Vité et al., 1978). For example, some species of *Ips* use primarily the (*S*)-(+)-enantiomer of ipsdienol (eg, *I. paraconfusus, I. subelongatus*), others predominantly use the (*R*)-(−)-enantiomer (eg, *I. pini*), while others use both in varying ratios (eg, *I. avulsus, I. calligraphus*). However, the enantiomeric identity and composition of these compounds is inconsistently reported in the literature, making broader comparison difficult. We therefore did not distinguish between enantiomers of these compounds (primarily ipsdienol and ipsenol), and simply considered presence or absence of the compound more generically. Consequently, our analysis is likely to underestimate between species variation in pheromone composition, particularly within the Ipini.

We considered only compounds that are known attractive components of the aggregation pheromone blend. In Symonds and Elgar (2004a), compounds that had antiaggregation properties were also included. This was justified on the basis that some attractive compounds become repellent at higher concentration, something which may help regulate the optimal colonisation density (Raffa, 2001). However, other compounds (eg, verbenone in many species—Borden, 1997; Lindgren and Miller, 2002) may serve solely as antiattractants. In this latter case, since these are not involved specifically in attracting conspecifics at any stage, we did not include them in this analysis (thus explaining several of the differences in pheromone composition between the present analysis and Symonds and Elgar's (2004a) earlier analysis).

We subsequently mapped the chemical components onto a composite phylogeny of bark beetle species derived from recently published molecular phylogenies of the scolytine subfamily. Specifically, the relationships between genera were derived from Jordal and Cognato's (2012) molecular phylogeny of the Scolytinae. Interspecific relationships within the tribe Ipini (comprising the genera *Ips, Orthotomicus, Pityogenes, Pityokteines* and *Pseudips*) were resolved based on Cognato's (2013) most parsimonious tree, with additional resolution of *Ips* species provided by Zhang et al. (2011). Relationships within *Dendroctonus* were taken from Reeve et al. (2012). No published phylogeny for *Pityophthorus* exists, but we took the parsimonious view that the two European species (*P. pityographus, P. pubescens*) and the two Californian species

(*P. carmeli* and *P. setosus*) in our analysis represented two pairs of sibling species. We resolved branch lengths on the phylogeny by making the final phylogeny ultrametric, using the Grafen (1989) branch length algorithm, where the height of each node in the phylogeny is equal to the number of daughter taxa

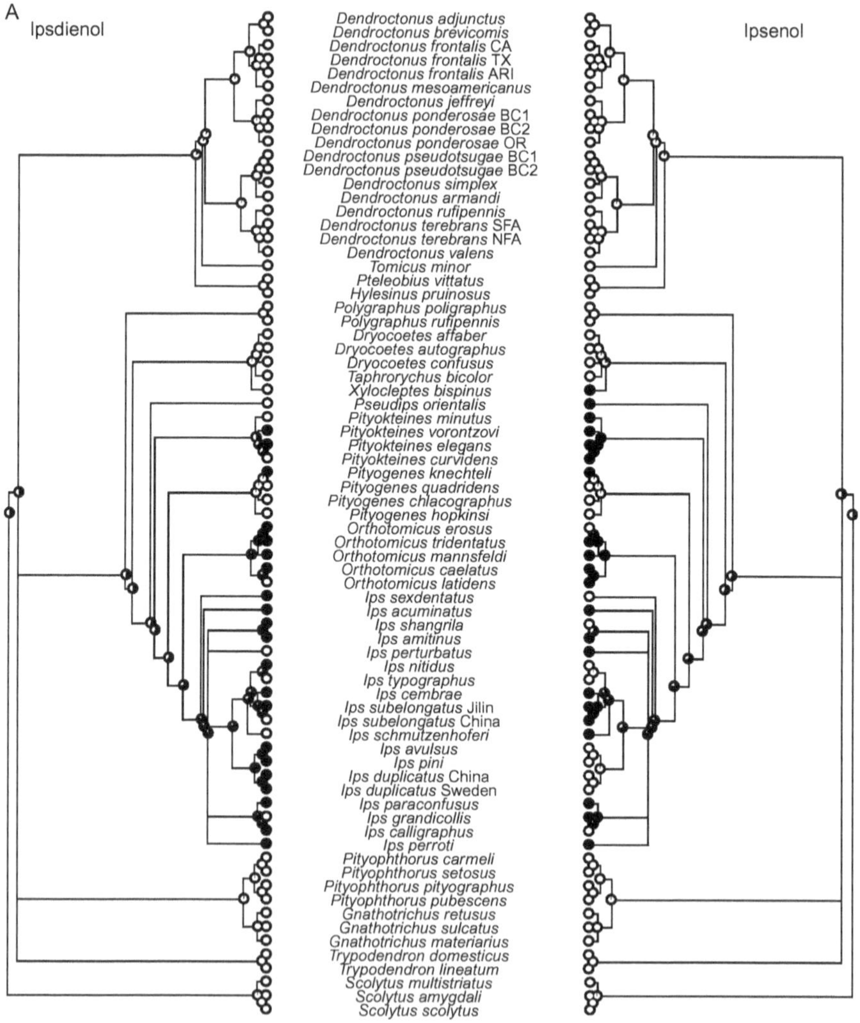

Fig. 1 Phylogeny of bark beetle species used in the analysis. Common pheromone components and ancestral state reconstructions (based on maximum likelihood estimation using Mesquite: Maddison and Maddison, 2015) are mapped on as follows: (A) ipsdienol and ipsenol, (B) *cis*-verbenol and *trans*-verbenol, (C) *exo*-brevicomin and *endo*-brevicomin, (D) frontalin. Shading on tips represents presence (shaded) or absence of the component as part of the aggregation pheromone. Shading on nodes represents proportional likelihood of ancestral state reconstructions.

Fig. 1—Cont'd

deriving from that node. The phylogeny with the most common chemical characters mapped on is shown in Fig. 1A–D.

4.3 Assessing the Mode of Pheromone Evolution and Phylogenetic Signal in Pheromone Chemistry

As with earlier analyses (Symonds and Elgar, 2004a; Symonds and Wertheim, 2005; Symonds et al., 2009; Van Wilgenburg et al., 2011) we first assessed the relationship between amount of phenotypic (pheromonal) differences and phylogenetic distance. We therefore constructed a pairwise

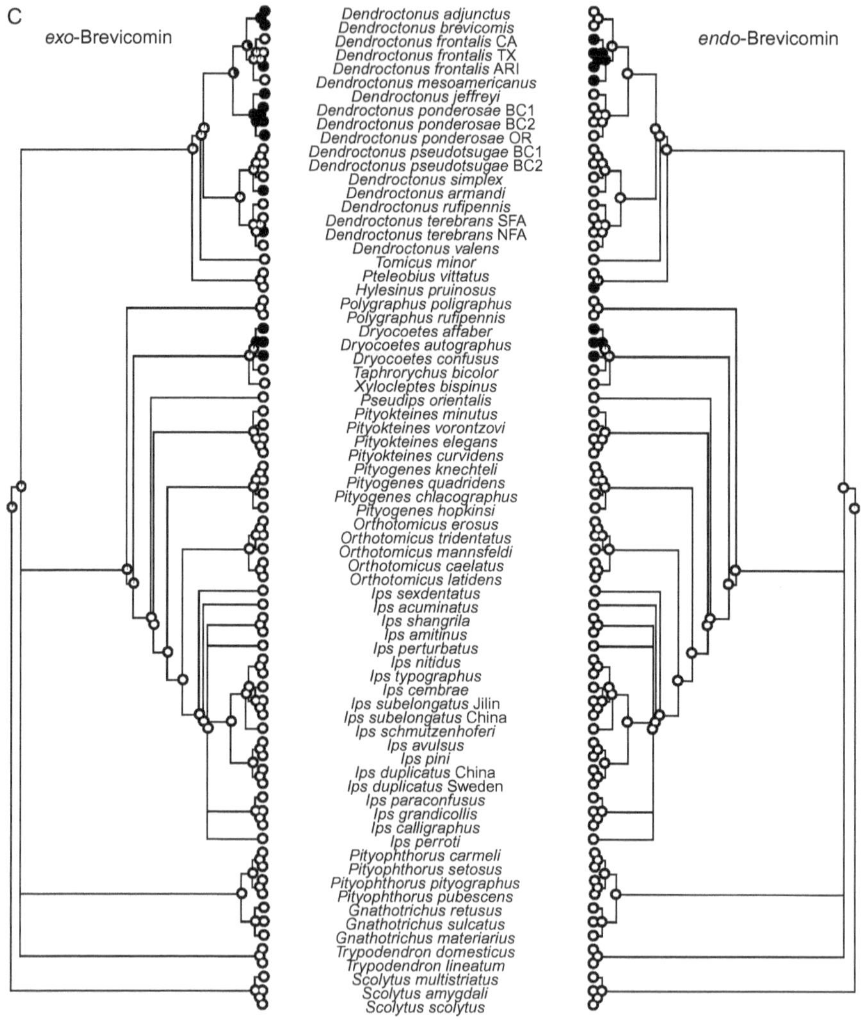

Fig. 1—Cont'd

distance matrix of phylogenetic distances by summing the branch lengths between each and every pair of species. Likewise we constructed a second matrix detailing the binary squared Euclidean differences in pheromone composition between the pairs (the total number of chemical components that are absent in one species but present in the other, and vice versa). The correlation between amount of pheromonal differences and phylogenetic distance between species was then assessed using Mantel tests using the package GenAlEx (Peakall and Smouse, 2006, 2012). This involved randomly

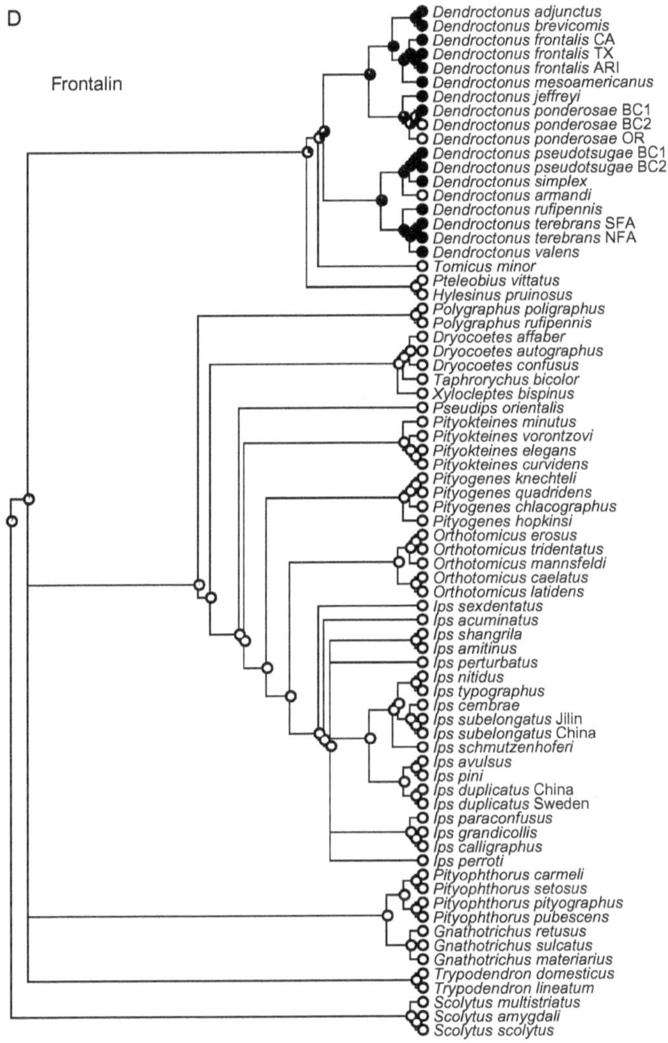

Fig. 1—Cont'd

perturbing the rows and columns of the distance matrix 999 times to derive a null distribution of the correlation coefficient, r, against which our observed correlation could be compared.

To more closely determine the mode of pheromone evolution requires examination of the relationship between the amount of phenotypic change along individual branches of the phylogeny (Pagel, 1999). Specifically, we would expect under a gradual mode of evolution that greater amounts of phenotypic change would be observable on longer branches of the phylogeny, with the least amount of change on the shorter branches in the phylogeny. To evaluate this we first used a maximum likelihood approach to reconstruct ancestral character histories for each chemical component, as implemented in the package Mesquite (Maddison and Maddison, 2015). For each node in the phylogeny a probability of the component being present in the ancestor of the two daughter species can be assessed. We considered the component to be present when this probability was greater than 0.5. From this gains or losses of the component on each branch could be estimated and summed across components to derive a total summed amount of chemical change along each branch length. This total was then correlated with branch length to test the mode of evolution.

Finally, we assessed the degree of phylogenetic signal in individual pheromone chemical components (in other words, the extent to which the distribution of these components across species is related to phylogeny). For all components found in three or more species/populations, we calculated phylogenetical signal. These components include the compounds expressed by eight or more species/populations: ipsenol, ipsdienol, frontalin, *cis*-verbenol, *trans*-verbenol, *exo*-brevicomin and *endo*-brevicomin, whose distributions on the phylogeny are shown in Fig. 1A–D.

The degree of phylogenetic signal was estimated using the D metric of Fritz and Purvis (2010), which was developed specifically for the analysis of binary characters, and implemented through the R package *caper* (Orme et al., 2012). D compares the number of changes in a trait over the phylogeny with that expected under two different scenarios. The first of these is where there is strong phylogenetic dependence of the trait—and the number of changes equals that predicted by a Brownian motion model of evolution that produces the same number of tip species with each character state (see Fritz and Purvis, 2010 for details of the calculation). The second model is one where the amount of change is that expected by effectively randomly perturbing the presence of the trait across the tips of the phylogeny. A D value of 1 indicates that the trait has evolved in a manner that is indistinguishable from the latter (random) model (ie, no phylogenetic signal), while a D value of 0 indicates the trait is highly correlated with phylogeny, as predicted using the Brownian motion model. Statsitically significant differences from $D=1$ (no phylogenetic signal) and $D=0$ (strong phylogenetic signal) can be obtained by randomly perturbing the species across the tips of the phylogeny or randomly simulating the evolution of trait by Brownian motion 1000 times to derive a null distribution of D values under these two scenarios.

We carried out all analysis using three sets of species in order to assess possible differences in patterns at different taxonomic scales and within different clades. First, we analysed the complete set of 73 species and populations. Second, we examined the separate patterns within the genus *Dendroctonus* alone (18 species and populations in total) and within the tribe Ipini alone (33 species and populations).

4.4 Patterns of Aggregation Pheromone Evolution in Bark Beetles

Given the overlying phylogenetic patterns in the expression of particular components (see Section 2) it is perhaps not surprising to see that across the whole subfamily there is a strongly significant positive correlation between pheromonal differences and phylogenetic distance between species (Mantel test $r=0.435$, $p=0.001$; Fig. 2A). Note that even with this pattern though there is on average 2.4 component differences even between the most closely related sibling species, suggesting that substantial differences in pheromone composition can be generated within a short period of evolutionary time. Likewise, the test of mode of pheromone evolution indicates no significant correlation between branch length and the amount of pheromonal change ($r=0.076$, $n=137$, $p=0.377$), indicating that substantial changes in pheromone composition take place just as readily over short branches, and that the differences in pheromone blends between species have not slowly accumulated via gradual divergent evolution. While the branch lengths used here do not directly estimate time, they are likely to provide reasonable nonparametric approximation. For example, changes within *Dendroctonus* are substantial, despite the genus being recently evolved (c. 2Mya—Jordal and Cognato, 2012).

Within the two smaller clades (the genus *Dendroctonus* and the tribe Ipini), different relationships between pheromonal difference and phylogenetic distance occur. Within *Dendroctonus* the pattern is similar to that across Scolytinae as a whole, with a positive correlation between difference and distance (Mantel test: $r=0.221$, $p=0.009$; Fig. 2B). Again there is also a degree of difference between even closely related species (mean number of pheromone component difference between sibling species $=2.8$). Within the Ipini, however, there is no relationship, with closely related species being just as different in their pheromone composition as more distantly related species in the tribe (Mantel test: $r=0.042$, $p=0.319$; Fig. 2C). However, the mean number of pheromone component differences across

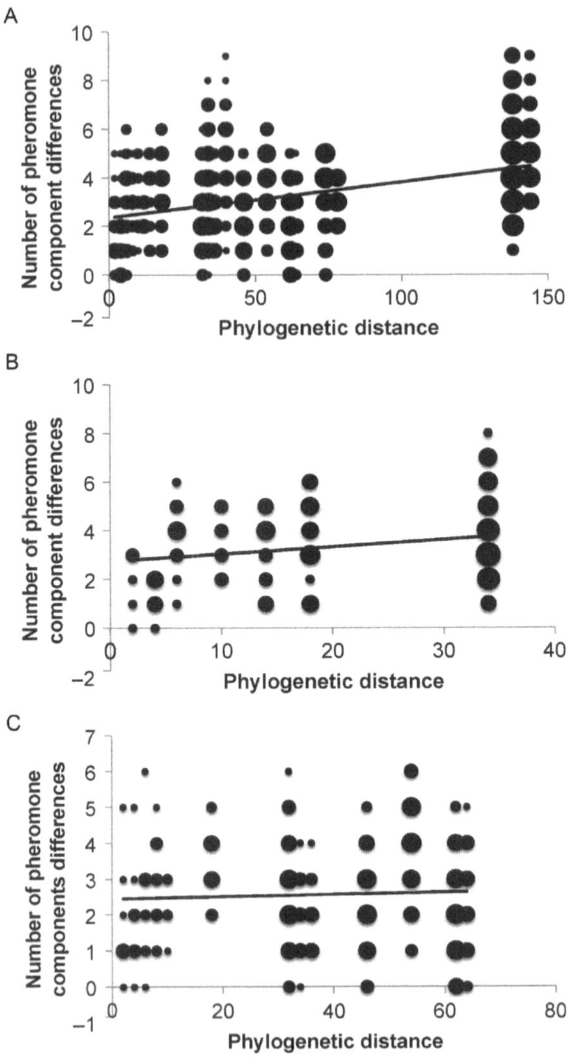

Fig. 2 Relationship between pheromonal differences and phylogenetic distance between species, with best-fit regression line. Size of each point in the bubble plot is proportional to the log-transformed number of data points contributing to it. Plots are shown for comparisons between (A) species across all of the Scolytinae, (B) species within *Dendroctonus*, (C) species within the Ipini.

all species in the tribe (2.6) is fewer than that seen between the mostly closely related species in *Dendroctonus*.

In respect to the individual pheromone components that are commonly used by bark beetle species, patterns again differ according to the taxonomic

Table 2 Values of the *D* Metric Indicating Levels of Phylogenetic Signal in Key Bark Beetle Pheromone Components at Three Different Taxonomic Scales

Compound	Across Scolytinae	Within *Dendroctonus*	Within Ipini
Common components			
Ipsenol	0.235[a]		0.653[b]
Ipsdienol	0.272[a]		1.129[b]
cis-Verbenol	0.348[a]		0.166[a]
trans-Verbenol	0.114[a]	0.533	
Frontalin	−0.342[a]	0.493	
exo-Brevicomin	0.147[a]	0.570	
endo-Brevicomin	−0.127[a]	−0.932[a]	
Rarer components			
2-Methyl-3-buten-2-ol	0.243[a]		−0.010[a]
Amitinol	1.055[b]		1.256[b]
Myrtenol	0.066	0.177[a]	
Seudenol	0.054[a]	0.050[a]	
MCOL	0.381[a]	0.318	
Pityol	−0.441[a]		
Sulcatol	−0.999[a]		
Chalcogran	−0.172[a]		

[a]Value of D significantly different from $D=1$.
[b]Value of D significantly different from $D=0$.
$D=1$ is equivalent to no signal, while $D=0$ is a strong signal equivalent to that expected under a gradual Brownian motion model of evolution.

scale examined (see Table 2). As expected given the clear phylogenetic patterns to the general use of these compounds across the Scolytinae, all seven of the major compounds show significant phylogenetic signal generally. However, when examined within clades completely different patterns emerge, with some compounds exhibiting strong phylogenetic signal, and others having much less signal, or indeed being completely phylogenetically labile. Thus, within the Ipini, both ipsdienol and ipsenol have a phylogenetic distribution significantly different from that predicted by a gradual model of evolution, and indeed their distribution is indistinguishable from that which would be observed if the distribution of these chemicals across the Ipini phylogeny was random. By contrast, *cis*-verbenol remains a highly phylogenetic conserved compound in this tribe. In *Dendroctonus*, *endo*-brevicomin is the single compound which preserves a strong phylogenetic signal with *exo*-brevicomin, frontalin and *trans*-verbenol displaying weaker signals (although the observed values of D are not significantly different from either 0 or 1, indicating that the signal in these components cannot be statistically distinguished from either no phylogenetic signal or a Brownian motion model of evolution signal).

5. WHAT ARE THE BROAD PATTERNS UNDERLYING BARK BEETLE PHEROMONE DIVERSITY?

5.1 Diversity with a Limited Palette

Different clades of bark beetles have specific chemical palettes from which they generate their aggregation pheromone signals. These are likely due to fundamental differences in both synthetic pathways and the utilisation of host tree compounds (Blomquist et al., 2010). However, even given these phylogenetic constraints to the composition of aggregation pheromones at the broader scale of the whole subfamily, the analysis here supports Symonds and Elgar's (2004a) characterisation of saltational changes in aggregation pheromone composition at speciation events.

Within the Ipini closely related species are as different as more distantly related species. Within *Dendroctonus* (and across Scolytinae as a whole), although there is a positive correlation between phylogenetic distance and pheromonal differences the intercepts of the relationships shown in Fig. 2, are all above two component differences, indicating, on average that even very closely related species show considerable differences. The positive correlation in *Dendroctonus* vs Ipini therefore more likely reflects the larger capacity for pheromone diversity within *Dendroctonus*. In this genus 12 functionally active compounds have been identified from 12 species. By comparison in the Ipini, a similar number of compounds (14) have been identified from 31 species: a broader taxonomic range, and a more ancient lineage (Jordal and Cognato, 2012). Additionally, on average members of *Dendroctonus* have more complex aggregation pheromone composition (mean number of components $= 3.17 \pm 0.26$ s.e.) than members of the Ipini (mean $= 2.18 \pm 0.12$ s.e.). This may reflect the more complex nature of aggregation pheromone production in *Dendroctonus* where the final blend is often a combination of both male- and female-produced volatiles.

Symonds and Elgar (2004a) demonstrated that, given a limited palette of chemical components, one would predict an asymptotic relationship between pheromonal difference and phylogenetic distance. The level of this asymptote would depend on the size of the palette, as well as the number of components utilised by individual species. This capacity for greater diversity and complexity likely explains why differences in pheromone composition become greater still with increasing phylogenetic distance in *Dendroctonus*, despite the fact that even closely related species may have substantial differences (and would also explain the same pattern across Scolytinae as a whole). Note also that our analysis incorporates a number of within species (between population) comparisons in *Dendroctonus* (*D. frontalis, D. terebrans, D. pseudotsugae, D. ponderosae*), and so underestimates differences between closely related but separate species (mean number of pheromone differences between populations within species = 2).

5.2 Phylogenetic Conservatism vs Lability in Compound Use

In an earlier analysis of pheromone evolution in eight *Ips* species, Cognato et al. (1997) suggested a high degree of homoplasy (ie, the degree of parallel or convergent evolution) in certain pheromone components but not others. These results suggest that the capacity to change pheromone composition rapidly is profound and that the evolutionary history of pheromone use within bark beetles has been characterised by the 'switching on' and 'switching off' of components with relative ease.

Our analysis confirms Cognato et al.'s (1997) findings and indicates that some compounds are highly plastic within clades but that others are far more phylogenetically constrained. This means that rapid evolution in aggregation pheromone composition may be achieved by changing some compounds, but not others. Of the major components perhaps the best examples of these lie in the comparison of use of ipsenol, ipsdienol and *cis*-verbenol within the Ipini. Here the former two monoterpene alcohols are highly variable across the phylogeny, and the latter is phylogenetically conserved (eg, present in the *I. grandicollis* subgroup, but absent in the *I. pini* subgroup and the genera *Orthotomicus* and *Pityokteines*). It is difficult to identify the reasons why some compounds within these clades are more prone to be phylogenetically conserved, but others are not. Potentially it may reflect the relative ease or difficulty of synthesis. Both ipsenol and ipsdienol (and the similarly phylogenetically labile amitinol) are most commonly derived from myrcene, while *cis*-verbenol derives from (−)-α-pinene. But *trans*-verbenol is similarly derived from (+)-α-pinene, and yet is more phylogenetically labile within *Dendroctonus*. So it is unclear from a biosynthetic perspective why *cis*-verbenol shows strong phylogenetic signal and *trans*-verbenol does not. It would be interesting, however, to investigate in more detail, through relationship to genetic control and biosynthetic pathways, why some compounds appear more variable in the phylogeny than others.

6. MECHANISMS FOR EVOLUTIONARY CHANGE IN AGGREGATION PHEROMONE COMPOSITION

From an evolutionary perspective, the phylogenetic patterns in compound use raise interesting corollaries with a recently proposed mechanism for species-specific pheromone evolution. One of the key puzzles is that even if rapid divergence in pheromone composition between closely related species is demonstrated, we still need to explain how these changes are able to be detected and recognised. In any communication system, the sender must produce a signal that the receiver identifies and appropriately responds to. If a pheromone system responds to directional selection, there must therefore also be a genetic mechanism by which signal preference and production remains associated if the system is to be sustained (Greenfield, 2002). The mechanism of genetic coordination and evolution in pheromone production is based on pleiotropic effects of a gene or genes controlling the neural network underlying both the signal and response, and selection acts correspondingly on the traits. One proposition that explains the evolution and maintenance of such communication among conspecifics is genetic coupling where a linkage or sharing of common components for signal production and recognition takes place (eg, Marcillac et al., 2005); however, theoretical models of pheromone evolution do not require this (Phelan, 1992), and the evidence from bark beetles also suggests this is not the case (Hager and Teale, 1996). Rather, it simply requires the presence of individuals in the population who have the appropriate genetic mutation that makes them capable of responding to the novel blend (Roelofs et al., 2002).

Recently a possible solution to this conundrum has been identified (Niehuis et al., 2013). In jewel wasps, *Nasonia vitripennis,* the male-produced sex pheromone consists of three components: 4(*R*),5(*R*)-5-hydroxy-4-decanolide, 4(*R*),5(*S*)-5-hydroxy-4-decanolide, and 4-methylquinazoline. Closely related ancestral species produce only the latter two compounds, indicating that at some stage, *N. vitripennis* has added 4(*R*),5(*R*)-5-hydroxy-4-decanolide as an active component in its blend. Interestingly, in behavioural experiments, the related *N. giraulti* does not show any reduced attraction to the three-compound blend, even though its own specific sex pheromone comprises only the ancestral compounds. Consequently, it may be that new pheromone compositions arise initially through modification of blends through addition of components that may not initially be selected for, but not selected against either—the preference for these new components arising later (Niehuis et al., 2013).

One implication of this model is that the capacity to produce compounds is present in species prior to those compounds being a functional component of the pheromone blend. However, it may also be that the capacity to respond to components could evolve prior to the capability of producing those components. There is evidence for both scenarios in bark beetles. For example, many species of *Ips* produce *trans*-verbenol (eg, Birgersson et al., 2012), but as yet only *Ips perturbatus* shows evidence of responding to this compound as an aggregation pheromone component (Graves, 2008). Likewise, *I. duplicatus* and *I. grandicollis* both produce trace amounts of ipsenol and ipsdienol, respectively, despite neither of these compounds being functionally attractive in those species (Birgersson et al., 2012; Zhang et al., 2007b). From the receiver perspective, there is also evidence that heterospecifics can positively respond to aggregation pheromone components that aren't produced by themselves—as is seen in the attraction of *D. frontalis* to *exo*-brevicomin produced by *D. brevicomis*. Interestingly, though, Arizona populations of *D. frontalis* do produce and respond to *exo*-brevicomin (Pureswaran et al., 2008).

A second implication of the model by Niehuis et al. (2013) is that we would therefore expect to see 'core' chemical components shared by several species that show strong phylogenetic conservatism as being fundamental components of the pheromone composition of those species, whilst other components should be exclusive to individual species and show no phylogenetic pattern, which fits with the patterns that we have seen in bark beetles.

In conclusion, the capacity to produce and respond to individual components would seem to be considerably flexible among bark beetles, and this may explain the capacity to shift preferences for particular compounds. When coupled to the possibility that such changes are under the control of single genetic loci (Domingue et al., 2006) and are heritable (Hager and Teale, 1996), the ability for rapid evolutionary change is apparent.

7. OTHER CONSIDERATIONS

7.1 Host Tree Synergistic Kairomones

Here we have focussed discussion exclusively on the evolution of beetle-produced compounds, but many behavioural responses to odours from aggregation pheromone components are synergistic. Synergy can be an evolutionary adaptation that facilitates interactions between bark beetles making their relationships more specific and cost-effective (Landolt and Phillips, 1997; Visser, 1986). Synergistic effects on attraction often occur when host-produced kairomones are released in combination with insect-produced pheromones (Billings et al., 1976; Borden, 1985; Miller and Borden, 1990). For instance, in *I. pini*, intermediate concentrations of α-pinene were shown to have a synergistic effect but inhibitory at high concentrations, whereas low concentrations of α-pinene have no effect on attraction to its pheromone (Erbilgin et al., 2003). The presence of host plant kairomones also influences the production and release of aggregation pheromones as seen in *Ips paraconfusus* and *I. pini*, where ipsenol, ipsdienol, and amitinol are synthesised *de novo* when the bark beetles are fed a diet containing host monoterpenes (Seybold et al., 1995). Clearly though, the host tree component of the attractive signal for bark beetles is likely under very different selective pressures to the components produced by the bark beetles themselves. It is difficult to formulate an argument that selection of species-specific character divergence could drive changes in host tree chemistry. However, it would be interesting to examine further the idea that host switching (as has been suggested particularly among generalist species—Kelley and Farrell, 1998) might enable further divergence in attractive blend composition.

7.2 Chemical Similarity

Many of the aggregation pheromone components utilised by bark beetles are closely derived from each other and hence share large parts of their synthetic pathways (eg, sulcatol, pityol and vittatol; or verbenol, verbenone and verbenene; Francke and Dettner, 2005). Consequently, we might expect that certain changes in aggregation pheromone composition would be more likely to occur than others. For example, given the production of ipsenol and ipsdienol by *Ips* species, the production of amitinol, generated by a simple allylic rearrangement of ipsdienol, may not necessarily be as large a step as the evolution of use of the cyclic compound lanierone (a compound whose synthetic pathway remains unknown).

Symonds and Elgar (2004a) attempted to control for this by grouping aggregation pheromone components into structurally related groupings. After they did so, they still found significant differences in pheromone composition between closely related species, indicating that variation was not solely being achieved by producing 'variations on a theme'. However, the difficulty in such comparisons is that structural similarities may not necessarily accurately reflect the ease of synthesis, or indeed the sharing of biosynthetic pathways. As Blomquist et al. (2010) have pointed out, for example, frontalin and brevicomin, while structurally similar (and thus grouped together by Symonds and Elgar), are derived from completely different synthetic pathways.

Nevertheless, the relative ease by which a new component can be produced (and utilised) in relation to existing biosynthetic pathways could be a key element in our understanding of how new components are added to existing blends, and also why some components are evolutionary conserved and others more labile (see Section 5.2). Ultimately this does not influence the fact that substantial differences in signal can be achieved rapidly, but it does influence our interpretation of how such change occurs.

Although our understanding of the biosynthesis of aggregation pheromone components has substantially improved in recent years, there are still many uncertainties including the extent to which *de novo* synthesis overrides synthesis from host tree compounds. Additionally, a fundamental compound such as frontalin has three putative synthetic pathways for its precursor 6-methylhept-6-en-2-one (Blomquist et al., 2010). Consequently we are still some way from identifying good estimates of the 'chemical distances' between compounds that would enable us to get a better understanding of how chemical diversity can be generated.

8. CONCLUDING REMARKS

The evolution of aggregation pheromone diversity in bark beetles is characterised by broad-scale phylogenetic patterns in the expression of components, but remarkable variation within clades given limited palettes of chemicals and the relatively low number of components used by each species individually (mean number of components per species $= 2.25 \pm 0.12$ s.e.). The capacity to switch on and switch off the use of key components suggests many species retain the genetic capacity to both produce and respond to a wide range of components as and when required for adaptive reasons. The existence of populations within species that do utilise additional components in comparison with the 'established' pheromone composition for that species (eg, *exo*-brevicomin in Arizona populations of *D. frontalis*—Pureswaran et al. 2008; ipsdienol in Jilin and Liaoning provinces populations of *Ips subelongatus*—Chen et al., 2016) provides further indication of this capacity for species to switch expression of components and hence provides insights into how saltational change may be facilitated.

This capacity for rapid change in pheromone composition may have interesting consequences for how we might manage pest species. Artificial pheromones are used to both monitor populations and protect plants, and so understanding the composition of the attractive blend is essential (Witzgall et al., 2010). This is particularly so in bark beetles and their pathogenic associates (Gitau et al., 2013). However, as with pest moth species (Groot et al., 2014; Tabata et al., 2007), the capacity for individuals within a species to produce and respond to different blends suggests that the use of synthetic lures needs to be carefully monitored on both geographic and temporal scales, as rapid evolutionary change may effectively render a population 'lure resistant'. Human imposed artificial control is merely one addition to the numerous selective forces which have already driven diversity in bark beetle aggregation pheromones.

REFERENCES

Andersson, M.N., Grosse-Wilde, E., Keeling, C.I., Bengtsson, J.M., Yuen, M.M.S., Li, M., Hillbur, Y., Bohlmann, J., Hansson, B.S., Schlyter, F., 2013. Antennal transcriptome analysis of the chemosensory gene families in the tree killing bark beetles, *Ips typographus* and *Dendroctonus ponderosae* (Coleoptera: Curculionidae: Scolytinae). BMC Genomics 14, 198.

Armendáriz-Toledano, F., Niño, A., Sullivan, B.T., Kirkendall, L.R., Zúñiga, G., 2015. A new species of bark beetle, *Dendroctonus mesoamericanus* sp. nov. (Curculionidae, Scolytinae) in southern Mexico and central America. Ann. Entomol. Soc. Am. 108, 403–414.

Ayres, B.D., Ayres, M.P., Abrahamson, M.D., Teale, S.A., 2001. Resource partitioning and overlap in three sympatric species of *Ips* bark beetles Coleoptera: Scolytidae. Oecologia 128, 443–453.

Baker, T.C., 2002. Mechanism for saltational shifts in pheromone communication systems. Proc. Natl. Acad. Sci. U.S.A. 99, 13368–13370.

Baker, B.H., Hostetler, B.B., Furniss, M.M., 1977. Response of eastern larch beetle (Coleoptera: Scolytidae) in Alaska to its natural attractant and to douglas-fir beetle pheromones. Can. Entomol. 109, 289–294.

Bakke, A., 1978. Aggregation pheromone components of the bark beetle *Ips acuminatus*. Oikos 31, 184–188.

Barkawi, L.S., Franke, W., Blomquist, G.J., Seybold, S.J., 2003. Frontalin: de novo biosynthesis of an aggregation pheromone component by *Dendroctonus* spp. bark beetles (Coleoptera: Scolytidae). Insect Biochem. Mol. Biol. 33, 773–788.

Billings, R.F., Gara, R.I., Hrutfiord, B.F., 1976. Influence of ponderosa pine resin volatiles on the response of *Dendroctonus ponderosae* to synthetic trans-verbenol. Environ. Entomol. 5, 171–179.

Birgersson, G., Dalusky, M.J., Berisford, C.W., 2000. Identification of an aggregation pheromone for *Pityogenes hopkinsi* (Coleptera: Scolytidae). Can. Entomol. 132, 951–963.

Birgersson, G., Dalusky, M.J., Espelie, K.E., Berisford, C.W., 2012. Pheromone production, attraction, and interspecific inhibition among four species of *Ips* bark beetles in the southwestern USA. Psyche Entomol. J. 1–14. ID 532652.

Blight, M.M., Henderson, N.C., Wadhams, L.J., 1983. The identification of 4-methyl-3-heptanone from *Scolytus scolytus* (F.) and *S. multistriatus* (Marsham). Absolute configuration, laboratory bioassay and electrophysiological studies on *S. scolytus*. Insect Biochem. 13, 27–38.

Blomquist, G.J., Figueroa-Teran, R., Aw, M., Song, M., Gorzalski, A., Abbott, N.L., Tittiger, C., 2010. Pheromone production in bark beetles. Insect Biochem. Mol. Biol. 40, 699–712.

Borden, J.H., 1982. Aggregation pheromones. In: Mitton, J.B., Sturgeon, K.B. (Eds.), Bark Beetles in North American Conifers. University of Texas Press, Austin, Texas, pp. 74–139.

Borden, J.H., 1985. Aggregation pheromones. In: Kerkut, G.A. (Ed.), In: Comprehensive Insect Physiology Biochemistry and Pharmacology, vol. 9. Pergamon, New York, pp. 257–285.

Borden, J.H., 1997. Disruption of semiochemical-mediated aggregation in bark beetles. In: Cardé, R.T., Minks, A.K. (Eds.), Pheromone Research: New Directions. Chapman and Hall, New York, pp. 421–438.

Borden, J.H., Handley, J.R., McLean, J.A., Silverstein, R.M., Chong, L., Slessor, K.N., Johnston, B.D., Schuler, H.R., 1980. Enantiomer-based specificity in pheromone communication by two sympatric *Gnathotrichus* species (Coleoptera: Scolytidae). J. Chem. Ecol. 6, 445–456.

Borden, J.H., Ryker, L.C., Chong, L.J., Pierce Jr., H.D., Johnston, B.D., Oehlschlager, A.C., 1987. Response of the mountain pine beetle, *Dendroctonus ponderosae* Hoplins (Coleoptera: Scolytidae), to five semiochemicals in British Columbia lodgepole pine forests. Can. J. For. Res. 17, 118–128.

Bowers, W.W., Gries, G., Borden, J.H., Pierce Jr., H.D., 1991. 3-Methyl-3-buten-1-ol: an aggregation pheromone of the four-eyed spruce bark beetle, *Polygraphus rufipennis* (Kirby) (Coleoptera: Scolytidae). J. Chem. Ecol. 17, 1989–2002.

Butlin, R.K., Trickett, A.J., 1997. Can population genetic simulations help to interpret pheromone evolution? In: Carde, R.T., Minks, A.K. (Eds.), Insect Pheromone Research. Chapman and Hall, New York, pp. 548–562.

Byers, J.A., 1981. Pheromone biosynthesis in the bark beetle, *Ips paraconfusus*, during feeding or exposure to vapours of host plant precursors. Insect Biochem. 11, 563–569.

Byers, J.A., 1989. Chemical ecology of bark beetles. Experientia 45, 271–283.

Byers, J.A., Birgersson, G., 2012. Host-tree monoterpenes and biosynthesis of aggregation pheromones in the bark beetle *Ips paraconfusus*. Psyche J. Entomol. 10, 1–6.

Byers, J.A., Wood, D.L., Browne, L.E., Fish, R.H., Piatek, B., Hendry, L.B., 1979. Relationship between a host plant compound, myrcene and pheromone production in the bark beetle, *Ips paraconfusus*. J. Insect Physiol. 25, 477–482.

Byers, J.A., Birgersson, G., Löfqvist, J., Appelgren, M., Bergström, G., 1990a. Isolation of pheromone synergists of bark beetle, *Pityogenes chalcographus*, from complex insect-plant odors by fractionation and subtractive-combination bioassay. J. Chem. Ecol. 16, 861–876.

Byers, J.A., Schlyter, F., Birgersson, G., Francke, W., 1990b. E-Myrcenol in *Ips duplicatus*: an aggregation pheromone component new for bark beetles. Experientia 46, 1209–1211.

Byers, J.A., Birgersson, G., Francke, W., 2013. Aggregation pheromones of bark beetles, *Pityogenes quadridens* and *P. bidentatus*, colonizing Scotch pine: olfactory avoidance of interspecific mating and competition. Chemoecology 23, 251–261.

Byrne, K.J., Swigar, A.A., Silverstein, R.M., Borden, J.H., Stokkink, E., 1974. Sulcatol: population aggregation pheromone in the scolytid beetle, *Gnathotrichus sulcatus*. J. Insect Physiol. 20, 1895–1900.

Camacho, A.D., Pierce Jr., H.D., Borden, J.H., 1993. Geometrical and optical isomerism of pheromones in two sympatric *Dryocoetes* species (Coleoptera: Scolytidae), mediates species specificity and response level. J. Chem. Ecol. 19, 2169–2182.

Camacho, A.D., Pierce Jr., H.D., Borden, J.H., 1994. Aggregation pheromones in *Dryocoetes affaber* (Mann.) (Coleoptea: Scolytidae): stereoisomerism and species specificity. J. Chem. Ecol. 20, 111–124.

Cane, J.H., Wood, D.L., Fox, J., 1990. Ancestral semiochemical attraction persists for sympatric populations of sibling *Ips* bark beetles (Coleoptera: Scolytidae). J. Chem. Ecol. 16, 993–1013.

Chen, G., Song, Y., Wang, P., Chen, J., Zhang, Z., Wang, S., Huang, X., Zhang, Q.-H., 2015. Semiochemistry of *Dendroctonus armandi* Tsai and Li (Coleoptera: Curculionidae: Scolytinae): both female-produced aggregation pheromone and host tree kairomone are critically important. Chemoecology 25, 135–145.

Chen, D.-F., Li, Y.-J., Zhang, Q.-H., Zhang, S.-F., Wang, H.-B., Zhang, Z., Zhao, L.-L., Kong, X.-B., 2016. Population divergence of aggregation pheromone responses in *Ips subelongatus* in northeastern China. Insect Sci.in press.

Cognato, A.I., 2013. Molecular phylogeny and taxonomic review of Premnobiini Browne, 1962 (Coleoptera: Curculionidae: Scolytinae). Front. Ecol. Evol. 1, 1.

Cognato, A.I., Seybold, S.J., Wood, D.L., Teale, S.A., 1997. A cladistic analysis of pheromone evolution in *Ips* bark beetles (Coleoptera: Scolytidae). Evolution 51, 313–318.

Coyne, J.A., Orr, H.A., 1989. Patterns of speciation in *Drosophila*. Evolution 43, 362–381.

Dallara, P.L., Seybold, S.J., Meyer, H., Tolasch, T., Francke, W., Wood, D.L., 2000. Semiochemicals from three species of *Pityophthorus* (Coleoptera: Scolytidae): identification and field response. Can. Entomol. 132, 889–906.

Davis, T.S., Hofstetter, R.W., 2009. Effects of gallery density and species ratio on the fitness and fecundity of two sympatric bark beetles Coleoptera: Curculionidae. Environ. Entomol. 38, 639–650.

Domingue, M.J., Teale, S.A., 2007. The genetic architecture of pheromone production between populations distant from the hybrid zone of the pine engraver, *Ips pini*. Chemoecology 17, 255–262.

Domingue, M.J., Starmer, W.T., Teale, S.A., 2006. Genetic control of the enantiomeric composition of ipsdienol in the pine engraver, *Ips pini*. J. Chem. Ecol. 32, 1005–1026.

El-Sayed, A.M., 2014. The Pherobase: Database of Pheromones and Semiochemicals. http://www.pherobase.com.

El-Sayed, A.M., Suckling, D.M., Wearing, C.H., Byers, J.A., 2006. Potential of mass trapping for long-term pest management and eradication of invasive species. J. Econ. Entomol. 99, 1550–1564.

Erbilgin, N., Raffa, K.F., 2001. Kairomonal range of generalist predators in specialized habitats. Entomol. Exp. Appl. 99, 205–210.

Erbilgin, N., Powell, J.S., Raffa, K.F., 2003. Effect of varying monoterpene concentrations on the response of *Ips pini* Coleoptera: Scolytidae to its aggregation pheromone: implications for pest management and ecology of bark beetles. Agric. For. Entomol. 5, 269–274.

Flechtmann, C.A.H., Berisford, C.W., 2003. Identification of sulcatol, a potential pheromone of the ambrosia beetle *Gnathotrichus materiarius* (Col., Scolytidae). J. Appl. Entomol. 127, 189–194.

Francke, W., Dettner, K., 2005. Chemical signalling in beetles. Top. Curr. Chem. 240, 85–166.
Francke, W., Heemann, V., 1974. Tests on the attractiveness of 3-hydroxy-3-methylbutan-2-one to *Xylosterus domesticus* L. and *X. lineatus* Oliv. (Coleoptera: Scolytidae). Z. Angew. Entomol. 75, 67–72.
Francke, W., Saurwein, P., Vité, J.P., Klimetzek, D., 1980. The pheromone bouquet of *Ips amitinus*. Naturwissenschaften 67, 147–148.
Francke, W., Pan, M.-L., Bartels, J., König, W.A., Vité, J.P., Krawieltizki, S., Kohnle, U., 1986. The odour bouquet of three pine engraver beetles (*Ips* spp.). J. Appl. Entomol. 101, 453–461.
Francke, W., Pan, M.-L., Konig, W.A., Mori, K., Paupoomchareon, P., Heuer, H., Vité, J.P., 1987. Identification of 'pityol' and 'grandisol' as pheromone components of the bark beetle, *Pityophthorus pityographus*. Naturwissenschaften 74, 343–345.
Francke, W., Bartels, J., Meyer, H., Schroder, F., Kohnle, U., Baader, E., Vité, J.P., 1995. Semiochemicals from bark beetles: new results, remarks, and reflections. J. Chem. Ecol. 21, 1043–1063.
Fritz, S.A., Purvis, A., 2010. Selectivity in mammalian extinction risk and threat types: a new measure of phylogenetic signal strength in binary traits. Conserv. Biol. 24, 1042–1051.
Giesen, H., Kohnle, U., Vité, J.P., Pan, M.-L., Francke, W., 1984. The aggregation pheromone of the Mediterranean pine bark-beetle *Ips* (*Orthotomicus*) *erosus*. Z. Angew. Entomol. 98, 95–97.
Gitau, C.W., Bashford, R., Carnegie, A.J., Gurr, G.M., 2013. A review of semiochemicals associated with bark beetle (Coleoptera: Curculionidae: Scolytinae) pests of coniferous trees: a focus on beetle interactions with other pests and their associates. For. Ecol. Manag. 297, 1–14.
Grafen, A., 1989. The phylogenetic regression. Philos. Trans. R. Soc. Lond. B 326, 119–157.
Graves, A.D., 2008. The Chemical Ecology of the Northern Spruce Engraver, *Ips perturbatus* (Eichhoff) (Coleoptera: Scolytidae), and Associated Insects in Spruce Forests of Alaska. PhD dissertation, University of Minnesota.
Greenfield, M.D., 2002. Signalers and Receivers—Mechanisms and Evolution of Arthropod Communication. Oxford University Press, New York.
Gries, G., Borden, J.H., Gries, R., Lafontaine, J.P., Dixon, E.A., Wieser, H., Whitehead, A.T., 1992. 4-Methylene-6,6-dimethylbicyclo[3.1.1]hept-2-ene (verbenene): new aggregation pheromone of the scolytid beetle *Dendroctonus rufipennis*. Naturwissenschaften 79, 367–368.
Gröning, J., Hochkirch, A., 2008. Reproductive interference between animal species. Q. Rev. Biol. 83, 257–282.
Groot, A.T., Schöfl, G., Inglis, O., Donnerhacke, S., Claßen, A., Schmaltz, A., Santangelo, R.G., Emerson, J., Gould, F., Schal, C., Heckel, D.G., 2014. Within-population variability in a moth sex pheromone blend: genetic basis and behavioural consequences. Proc. R. Soc. Lond. B 281, 20133054.
Hager, B.J., Teale, S.A., 1996. Genetic correlation of pheromone production and response in the pine engraver beetle, *Ips pini*. Heredity 77, 100–107.
Harring, C.M., 1978. Aggregation pheromones of the European fir engraver beetles *Pityokteines curvidens*, *P. spinidens* and *P. vorontzovi* and the role of juvenile hormone in pheromone biosynthesis. Z. Angew. Entomol. 85, 281–317.
Hendry, L.B., Piatek, B., Browne, L.E., Wood, D.L., Byers, J.A., Fish, R.H., Hicks, R.A., 1980. In vivo conversion of labelled host chemical to pheromones in the bark beetle *Ips paraconfusus*. Nature 284, 485.
Higgie, M., Chenoweth, S., Blows, M.W., 2000. Natural selection and the reinforcement of mate recognition. Science 290, 519–521.

Hofstetter, R.W., Gaylord, M.L., Martinson, S., Wagner, M.R., 2012. Attraction to monoterpenes and beetle-produced compounds by syntopic *Ips* and *Dendroctonus* bark beetles and their predators. Agric. For. Entomol. 14, 207–215.

Hughes, P.R., Renwick, J.A.A., Vité, J.P., 1976. The identification and field bioassay of chemical attractants in the roundheaded pine beetle. Environ. Entomol. 5, 1165–1168.

Jordal, B.H., Cognato, A.I., 2012. Molecular phylogeny of bark and ambrosia beetles reveals multiple origins of fungus farming during periods of global warming. BMC Evol. Biol. 12, 133.

Karlson, P., Lüscher, M., 1959. 'Pheromones': a new term for a class of biologically active substances. Nature 183, 55–56.

Kelley, S.T., Farrell, B.D., 1998. Is specialization a dead end? The phylogeny of host use in *Dendroctonus* bark beetles (Scolytidae). Evolution 52, 1731–1743.

Kinzer, G.W., Fentiman Jr., A.F., Page Jr., T.F., Foltz, R.L., Vité, J.P., Pitman, G.B., 1969. Bark beetle attractants: identification, synthesis and field bioassay of a new compound isolated from *Dendroctonus*. Nature 221, 477–478.

Klimetzek, D., Bartels, J., Francke, W., 1989a. The pheromone system of the elm bark beetle *Pteleobius vittatus* (F.) (Col., Scolytidae). J. Appl. Entomol. 107, 518–523.

Klimetzek, D., Kohler, J., Krohn, S., Francke, W., 1989b. The pheromone system of the clematis bark beetle, *Xylocleptes bispinus* Duft. (Col., Scolytidae). J. Appl. Entomol. 107, 304–309.

Kohnle, U., 1985. Studies on the pheromone systems of secondary bark-beetles (Col., Scolytidae). Z. Angew. Entomol. 100, 197–218.

Kohnle, U., Mussong, M., Dubbel, V., Francke, W., 1987. Acetophenone in the aggregation of the beech bark beetle, *Taphrorychus bicolor* (Col., Scolytidae). J. Appl. Entomol. 103, 249–252.

Kohnle, U., Schmutzenhofer, H., Bartels, J., Francke, W., 1988. Oxygenated terpenes in the chemical communication system of the bark beetle, *Ips schmutzenhoferi* (Col., Scolytidae), a species recently described for the Southeastern Himalaya. J. Appl. Entomol. 106, 46–51.

Kohnle, U., Pajares, J.A., Bartels, J., Meyer, H., Francke, W., 1993. Chemical communication in the Europaen pine engraver, *Ips mannsfeldi* (Wachtl) (Coleoptera Scolytidae). J. Appl. Entomol. 115, 1–7.

Landolt, P.J., Phillips, W.T., 1997. Host plant influences on sex pheromone behavioral phytophagous insects. Annu. Rev. Entomol. 42, 371–391.

Lanier, G.N., Gore, W.E., Pearce, G.T., Peacock, J.W., Silverstein, R.M., 1977. Response of the European elm bark beetle, *Scolytus multistriatus* (Coleoptera: Scolytidae), to isomers and components of its pheromone. J. Chem. Ecol. 3, 1–8.

Lanne, B.S., Schlyter, F., Byers, J.A., Löfqvist, J., Leufvén, A., Bergström, G., Van Der Pers, J.N.C., Unelius, R., Baeckström, P., Norin, T., 1987. Differences in attraction to semiochemicals present in sympatric pine shoot beetles, *Tomicus minor* and *T. piniperda*. J. Chem. Ecol. 13, 1045–1067.

Libbey, L.M., Ryker, L.C., Yandell, K.L., 1985. Laboratory and field studies of volatiles released by *Dendroctonus ponderosae* Hopkins (Coleoptera, Scolytidae). Z. Angew. Entomol. 100, 381–392.

Lindgren, B.S., Miller, D.R., 2002. Effect of verbenone on five species of bark beetles (Coleoptera: Scolytidae) in lodgepole pine forests. Environ. Entomol. 31, 759–765.

Lindgren, B.S., Gries, G., Pierce Jr., H.D., Mori, K., 1992. *Dendroctonus pseudotsugae* Hopkins (Coleoptera: Scolytidae): production of and response to enatiomers of 1-methylcyclohex-2-en-1-ol. J. Chem. Ecol. 18, 1201–1208.

López, S., Quero, C., Iturrondobeitia, J.C., Guerrero, A., Goldarazena, A., 2011. Evidence for (E)-pityol as an aggregation pheromone of *Pityophthorus pubescens* (Coleoptera: Curculionidae: Scolytinae). Can. Entomol. 143, 447–454.

MacConnell, J.G., Borden, J.H., Silverstein, R.H., Stokkink, E., 1977. Isolation and tentative identification of lineatin, a pheromone from the frass of *Trypodendron lineatum* (Coleoptera: Scolytidae). J. Chem. Ecol. 3, 549–561.
Macías-Sámano, J.E., Borden, J.H., Pierce Jr., H.D., Gries, R., Gries, G., 1997. Aggregation pheromone of *Pityokteines elegans*. J. Chem. Ecol. 23, 1333–1347.
Maddison, W.P., Maddison, D.R., 2015. Mesquite: A Modular System for Evolutionary Analysis. Version 304, http://mesquiteprojectorg.
Marcillac, F., Grosjean, Y., Ferveur, J.-F., 2005. A single mutation alters production and discrimination of *Drosophila* sex pheromones. Proc. R. Soc. Lond. B 272, 303–309.
Marshall, J.L., Arnold, M.L., Howard, D.J., 2002. Reinforcement: the road not taken. Trends Ecol. Evol. 17, 558–563.
McElfresh, J.S., Millar, J.G., 2001. Geographic variation in the pheromone system of the saturniid moth *Hemileuca eglanterina*. Ecology 82, 3505–3518.
Miller, D.R., Borden, J.H., 1990. Beta-phellandrene: kairomone for pine engraver, *Ips pini* (Say) (Coleoptera: Scolytidae). J. Chem. Ecol. 16, 2519–2531.
Miller, D.R., Lafontaine, J.P., 1991. cis-Verbenol: an aggregation pheromone for the mountain pine beetle, *Dendroctonus ponderosae* Hopkins (Coleoptera: Scolytidae). J. Entomol. Soc. British Columbia 88, 34–38.
Miller, D.R., Borden, J.H., King, G.G.S., Slessor, K.N., 1991. Ipsenol: an aggregation pheromone for *Ips latidens* (LeConte) (Coleoptem: Scolytidae). J. Chem. Ecol. 17, 1517–1527.
Niehuis, O., Buellesbach, J., Gibson, J.D., Pothmann, D., Hanner, C., Mutti, N.S., Judson, A.K., Gadau, J., Ruther, J., Schmitt, T., 2013. Behavioural and genetic analyses of *Nasonia* shed light on the evolution of sex pheromones. Nature 494, 345–348.
Niño-Domínguez, A., Sullivan, B., López-Urbina, J., Macías-Sámano, J., 2015. Pheromone-mediated mate location and discrimination by two syntopic sibling species of *Dendroctonus* bark beetles in Chiapas, Mexico. J. Chem. Ecol. 41, 746–756.
Noor, M.A.F., 1999. Reinforcement and other consequences of sympatry. Heredity 83, 503–508.
Orme, D., Freckleton, R., Thomas, G., Petzoldt, T., Fritz, S., Isaac, N., 2012. Caper: Comparative Analyses of Phylogenetics and Evolution in R. R Package Version 0.5, http://CRAN.R-project.org/package=caper.
Pagel, M., 1999. Inferring the historical patterns of biological evolution. Nature 401, 877–884.
Paine, T.D., Millar, J.G., Hanlon, C.C., Hwang, J.S., 1999. Identification of semiochemicals associated with Jeffrey pine beetle, *Dendroctonus jeffreyi*. J. Chem. Ecol. 25, 433–453.
Payne, T.L., Billings, R.F., Delorme, J.D., Andryszak, N.A., Bartels, J., Francke, W., Vité, J.P., 1987. Kairomonal-pheromonal system in the black turpentine beetle *Dendroctonus terebrans* (Ol.). J. Appl. Entomol. 103, 15–22.
Peakall, R., Smouse, P.E., 2006. GENALEX 6: genetic analysis in Excel. Population genetic software for teaching and research. Mol. Ecol. Notes 6, 288–295.
Peakall, R., Smouse, P.E., 2012. GenAlEx 6.5: genetic analysis in Excel. Population genetic software for teaching and research-an update. Bioinformatics 28, 2537–2539.
Phelan, P.L., 1992. Evolution of sex pheromones and the role of asymmetric tracking. In: Roitberg, B.D., Isman, M.B. (Eds.), Insect Chemical Ecology: An Evolutionary Approach. Chapman and Hall, New York, pp. 265–314.
Phillips, T.W., Atkinson, T.H., Foltz, J.L., 1989a. Pheromone-based aggregation in *Orthotomicus caelatus* (Eichhoff) (Coleoptera: Scolytidae). Can. Entomol. 121, 933–940.
Phillips, T.W., Nation, J.L., Wilkinson, R.C., Foltz, J.L., 1989b. Secondary attraction and field activity of beetle-produced volatiles in *Dendroctonus terebrans*. J. Chem. Ecol. 15, 1513–1533.
Pureswaran, D.S., Gries, R., Borden, J.H., 2004. Antennal responses of four species of tree-killing bark beetles (Coleoptera: Scolytidae) to volatiles collected from beetles, and their host and nonhost conifers. Chemoecology 14, 59–66.

Pureswaran, D.S., Sullivan, B.T., Ayres, M.P., 2008. High individual variation in pheromone production by tree-killing bark beetles (Coleoptera: Curculionidae: Scolytinae). Naturwissenschaften 95, 33–44.

Raffa, K.F., 2001. Mixed messages across multiple trophic levels: the ecology of bark beetle chemical communication systems. Chemoecology 11, 49–65.

Raffa, K.F., Hobson, K.R., LaFontaine, S., Aukema, B.H., 2007. Can chemical communication be cryptic? Adaptations by herbivores to natural enemies exploiting prey semiochemistry. Oecologia 153, 1009–1019.

Reeve, J.D., Anderson, F.E., Kelley, S.T., 2012. Ancestral state reconstruction for *Dendroctonus* bark beetles: evolution of a tree killer. Environ. Entomol. 41, 723–730.

Renwick, J.A.A., Vité, J.P., 1970. Systems of chemical communication in *Dendroctonus*. Contrib. Boyce Thompson Inst. Plant Res. 24, 282–292.

Renwick, J.A.A., Hughes, P.R., Krull, I.S., 1976a. Selective production of cis and trans verbenol from (−) and (+) α pinene by a bark beetle. Science 191, 199–201.

Renwick, J.A.A., Pitman, G.B., Vité, J.P., 1976b. 2-Phenylethanol isolated from bark beetles. Naturwissenschaften 63, 198.

Roelofs, W.L., Brown, R.L., 1982. Pheromones and evolutionary relationships of Tortricidae. Annu. Rev. Ecol. Syst. 13, 395–422.

Roelofs, W.L., Liu, W., Hao, G., Jiao, H., Rooney, A.P., Linn Jr., C.E., 2002. Evolution of moth sex pheromones via ancestral genes. Proc. Natl. Acad. Sci. U.S.A. 99, 13621–13626.

Ross, D.W., Daterman, G.E., Munson, A.S., 2005. Spruce beetle (Coleoptera: Scolytidae) response to traps baited with selected semiochemicals in Utah. West. N. Am. Naturalist. 65, 123–126.

Rudinsky, J.A., Morgan, M.E., Libbey, L.M., Putnam, T.B., 1974. Additional components of the douglas-fir beetle (Col., Scolytidae) aggregative pheromone and their possible utility in pest control. Z. Angew. Entomol. 76, 65–77.

Savoie, A., Borden, J.H., Pierce Jr., H.D., Gries, R., Gries, G., 1998. Aggregation pheromone of *Pityogenes knechteli* and semiochemical-based interactions with three other bark beetles. J. Chem. Ecol. 24, 321–337.

Schiebe, C., Blaženec, M., Jakuš, R., Unelius, C.R., Schlyter, F., 2011. Semiochemical diversity diverts bark beetle attacks from Norway spruce edges. J. Appl. Entomol. 135, 726–737.

Schlyter, F., Birgersson, G., Byers, J.A., Löfqvist, J., Bergström, G., 1987. Field response of the spruce bark beetle, *Ips typographus*, to aggregation pheromone candidates. J. Chem. Ecol. 13, 701–716.

Schlyter, F., Jakus, R., Han, F.-Z., Ma, J.-H., Kalinova, B., Mezei, P., Sun, J.-H., Ujhelyiová, L., Zhang, Q.-H., 2015. Reproductive isolation of *Ips nitidus* and *I. shangrila* in mountain forests of western China: responses to chiral and achiral candidate pheromone components. J. Chem. Ecol. 41, 678–688.

Schurig, V., Leyrer, U., Kohnle, U., 1985. Enantiomer composition and absolute configuration of terpinene-4-ol from the bark beetle, *Polygraphus poligraphus*. Naturwissenschaften 72, 211.

Seybold, S.J., Tittiger, C., 2003. Biochemistry and molecular biology of de novo isoprenoid pheromone production in the Scolytidae. Annu. Rev. Entomol. 48, 425–453.

Seybold, S.J., Quilici, D.R., Tillman, J.A., Vanderwel, D., Wood, D.L., Blomquist, G.J., 1995. De novo biosynthesis of the aggregation pheromone components ipsenol and ipsdienol by the pine engraver beetles *Ips paraconfusus* Lanier and *Ips pini* Say Coleoptera: Scolytidae. Proc. Natl. Acad. Sci. U.S.A. 92, 8393–8397.

Seybold, S.J., Bohlmann, J., Raffa, K.F., 2000. Biosynthesis of coniferophagous bark beetle pheromones and conifer isoprenoids: evolutionary perspective and synthesis. Can. Entomol. 132, 697–753.

Shepherd, W.P., Sullivan, B.T., Hoosier, B.M., Barrett, J.P., Bauman, T.A., 2010. Characterization of an aggregation pheromone in *Hylesinus pruinosus* (Coleoptera: Curculionidae: Scolytinae). Environ. Entomol. 39, 633–641.

Silverstein, R.M., Rodin, J.O., Wood, D.L., 1966. Sex attractants in frass produced by male *Ips confusus* in ponderosa pine. Science 154, 509–510.

Silverstein, R.M., Brownlee, R.G., Bellas, T.G., Wood, D.L., Browne, L.E., 1968. Brevicomin: principal sex attractant in the frass of the female western pine beetle. Science 159, 889–891.

Smadja, C., Butlin, R.K., 2009. On the scent of speciation: the chemosensory system and its role in premating isolation. Heredity 102, 77–97.

Song, M., Kim, A.C., Gorzalski, A.J., MacLean, M., Young, S., Ginzel, M.D., Tittiger, C., 2013. Functional characterization of myrcene hydroxylases from two geographically distinct *Ips pini* populations. Insect Biochem. Mol. Biol. 43, 336–343.

Steiger, S., Schmitt, T., Schaefer, H.M., 2010. The origin and dynamic evolution of chemical information transfer. Proc. R. Soc. Lond. B 278, 970–979.

Sullivan, B.T., Shepherd, W.P., Pureswaran, D.S., Tashiro, T., Mori, K., 2007. Evidence that (+)-*endo*-brevicomin is a male-produced component of the southern pine beetle aggregation pheromone. J. Chem. Ecol. 33, 1510–1527.

Sullivan, B.T., Niño, A., Moreno, B., Brownie, C., Macías-Sámano, J., Clarke, S.R., Kirkendall, L.R., Zúñiga, G., 2012. Biochemical evidence that *Dendroctonus frontalis* consists of two sibling species in Belize and Chiapas, Mexico. Ann. Entomol. Soc. Am. 105, 817–831.

Symonds, M.R.E., Elgar, M.A., 2004a. The mode of pheromone evolution: evidence from bark beetles. Proc. R. Soc. Lond. B 271, 839–846.

Symonds, M.R.E., Elgar, M.A., 2004b. Species overlap, speciation and the evolution of aggregation pheromones in bark beetles. Ecol. Lett. 7, 202–212.

Symonds, M.R.E., Elgar, M.A., 2008. The evolution of pheromone diversity. Trends Ecol. Evol. 23, 220–228.

Symonds, M.R.E., Wertheim, B., 2005. The mode of evolution of aggregation pheromones in *Drosophila* species. J. Evol. Biol. 18, 1253–1263.

Symonds, M.R.E., Moussalli, A., Elgar, M.A., 2009. The evolution of sex pheromones in an ecologically diverse genus of flies. Biol. J. Linn. Soc. 97, 594–603.

Tabata, J., Noguchi, H., Kainoh, Y., Mochizuki, F., Sugie, H., 2007. Sex pheromone production and perception in the mating disruption-resitant strain of the smaller tea leafroller moth, *Adoxophyes honmai*. Entomol. Exp. Appl. 122, 145–153.

Van Wilgenburg, E., Symonds, M.R.E., Elgar, M.A., 2011. Evolution of cuticular hydrocarbon diversity in ants. J. Evol. Biol. 24, 1188–1198.

Vanderwel, D., Gries, G., Singh, S.M., Borden, J.H., Oehlschlager, A.C., 1992. (E)- and (Z)-6-nonen-2-one: biosynthetic precursors of endo- and exo-brevicomin in two bark beetles (Coleoptera: Scolytidae). J. Chem. Ecol. 18, 1389–1404.

Visser, J.H., 1986. Host odour perception in phytophagous insects. Annu. Rev. Entomol. 31, 121–144.

Vité, J.P., Renwick, J.A.A., 1971. Population aggregating pheromone in the bark beetle *Ips grandicollis*. J. Insect Physiol. 17, 1699–1704.

Vité, J.P., Bakke, A., Renwick, J.A.A., 1972. Pheromones in *Ips* Coleoptera: Scolytidae: occurrence and production. Can. Entomol. 104, 1967–1975.

Vité, J.P., Ohloff, G., Billings, R.F., 1978. Pheromonal chirality and integrity of aggregation response in southern species of the bark beetle *Ips* sp. Nature 272, 817–818.

Vité, J.P., Billings, R.F., Ware, C.W., Mori, K., 1985. Southern pine beetle: enhancement or inhibition of aggregation response mediated by enantiomers of endo-brevicomion. Naturwissenschaften 72, 99–100.

Witzgall, P., Kirsch, P., Cork, A., 2010. Sex pheromones and their impact on pest management. J. Chem. Ecol. 36, 80–100.
Wood, D.L., 1982. The role of pheromones, kairomones and allomones in the host selection and colonization of bark beetles. Annu. Rev. Entomol. 27, 411–446.
Wood, D.L., Browne, L.E., Ewing, B., Lindahl, K., Bedard, W.D., Tilden, P.E., Mori, K., Pitman, G.B., Hughes, P.R., 1976. Western pine beetle: specificity among enantiomers of male and female components of an attractant pheromone. Science 192, 896–898.
Wyatt, T.D., 2014. Pheromones and Animal Behavior: Chemical Signals and Signatures. Cambridge University Press, Cambridge.
Zada, A., Ben-Yehuda, S., Dunkelblum, E., Harel, M., Assael, F., Mendel, Z., 2004. Synthesis and biological activity of the four stereoisomers of 4-methyl-3-heptanol: main components of the aggregation pheromone of *Scolytus amygdali*. J. Chem. Ecol. 30, 631–641.
Zhang, L.W., Sun, J.H., 2006. Electrophysiological and behavioral responses of *Dendroctonus valens* (Coleoptera: Curculionidae: Scolytinae) to candidate pheromone components identified in hindgut extracts. Environ. Entomol. 35, 1232–1237.
Zhang, Q.-H., Birgersson, G., Schlyter, F., Chen, G.F., 2000. Pheromone components in the larch bark beetle *Ips cembrae* from China: quantitative variation among attack phases and individuals. J. Chem. Ecol. 26, 841–858.
Zhang, Q.-H., Schlyter, F., Chen, G., Wang, Y., 2007a. Electrophysiological and behavioural responses of *Ips subelongatus* to semiochemicals from its hosts, non-hosts, and conspecifics in China. J. Chem. Ecol. 33, 391–404.
Zhang, Q.-H., Schlyter, F., Liu, G.-T., Sheng, M.-L., Birgersson, G., 2007b. Electrophysiological and behavioral responses of *Ips duplicatus* to aggregation pheromone in Inner Mongolia, China: amitinol as a potential pheromone component. J. Chem. Ecol. 33, 1303–1315.
Zhang, Q.-H., Ma, J.-H., Zhao, F.-Y., Song, L.-W., Sun, J.-H., Cognato, A.I., 2011. Aggregation pheromone of the oriental spruce engraver *Pseudips orientalis*. Agric. For. Entomol. 13, 67–75.
Zimmermann, Y., Ramírez, S.R., Eltz, T., 2009. Chemical niche differentiation among sympatric species of orchid bees. Ecology 90, 2994–3008.

CHAPTER SIX

Pheromone Production in Pine Bark Beetles

C. Tittiger, G.J. Blomquist
University of Nevada, Reno, NV, United States

Contents

Abstract

The first aggregation pheromone components from a pine bark beetle were identified in 1966 as a mixture of ipsdienol, ipsenol and verbenol from *Ips paraconfus*. Since then, a number of additional components have been identified as both aggregation and antiaggregation pheromones, with many of them being monoterpenoids. The structural similarity between the major pheromone components of bark beetles and the monoterpenes found in the host trees, along with the association of monoterpenoid production with plant tissue, led to the paradigm that most if not all bark beetle pheromone components were derived from host tree precursors, often with only a simple hydroxylation to produce the pheromone. In the 1990s there was a paradigm shift as evidence for de novo biosynthesis of pheromone components began to accumulate, and it is

Advances in Insect Physiology, Volume 50
ISSN 0065-2806
http://dx.doi.org/10.1016/bs.aiip.2016.02.002

now recognized that most, but not necessarily all, pine bark beetle monoterpenoid aggregation pheromone components are primarily biosynthesized de novo. The bark beetle aggregation pheromones are released from the frass, which is consistent with the isoprenoid aggregation pheromones, including ipsdienol, ipsenol and frontalin, being produced in midgut tissue. It appears that *exo*-brevicomin is produced de novo in fat body tissue, and that verbenol, verbenone and verbenene are produced from dietary α-pinene in fat body tissue. Combined biochemical, molecular and functional genomic studies in *Ips pini* yielded the discovery and characterization of the enzymes that convert mevalonate pathway intermediates to pheromone components, including a novel bifunctional geranyl diphosphate synthase/myrcene synthase, a cytochrome P450 that hydroxylates myrcene to ipsdienol, and an oxidoreductase that interconverts ipsdienol and ipsdienone, presumably to help achieve the appropriate stereochemistry of ipsdienol for pheromonal activity. Furthermore, the regulation of these genes and their corresponding enzymes proved complex and diverse in different species. Mevalonate pathway genes in pheromone-producing male *I. pini* exhibit much higher basal levels than in females, and feeding induces their expression. In *I. duplicatus* and *I. pini*, juvenile hormone III (JH III) induces pheromone production in the absence of feeding, whereas in *I. paraconfusus* and *I. confusus*, topically applied JH III does not induce pheromone production. In all four species, feeding induces pheromone production. Many of the details of pheromone production, including the site of synthesis, pathways and knowledge of the enzymes involved, are known for *Ips*, and recent work has done much to unravel the processes of pheromone production in *Dendroctonus*. Recent genomic studies in *D. ponderosae* are rapidly increasing our understanding of pheromone production in this genus. This chapter presents a historical development of what is known about pheromone production in bark beetles, emphasizes the genomic and postgenomic work in *I. pini* and *D. ponderosae* and points out areas where research is needed to obtain a more complete understanding of pheromone production.

1. INTRODUCTION

Bark beetles use aggregation pheromones to coordinate mass attacks on host pine. Feeding induces production of an aggregation pheromone in the pioneer sex that attracts both sexes to 'mass attack' the tree. For *Ips* spp., the male is the pioneer sex and the first to bore into the bark and feed on phloem. *Dendroctonus* spp. attacks on host trees begin with the arrival of one or a few pioneer females and both sexes can contribute to the aggregation pheromone complex. The tree responds systemically by producing a defensive resin containing toxic mono-, sesqui- and diterpenoid chemicals to pitch the beetles out (Phillips and Croteau, 1999; Steele et al., 1995). The combination of toxins and increased resin flow volume is usually sufficient for a healthy tree to kill the beetles. However, under stressful conditions (eg, drought) or outbreak situations as is now occurring in Western North America, resin production may not be sufficient to stop the infestation (Kurz et al., 2008; Rudinsky, 1962). Mass attacks result in extensive gallery construction in the phloem and the introduction of beetle-associated fungi. Both factors reduce water and nutrient flow and contribute to tree mortality (Seybold et al., 2000). Egg laying, hatching, development through larval instars, pupation and eclosion to teneral adults all occur beneath the bark of the brood tree. Some bark beetle species are very aggressive and can attack and overcome live trees, whereas other species concentrate on recently dead branches or dying trees.

2. ORIGIN OF PHEROMONES: PLANT PRECURSORS VS DE NOVO

The first bark beetle aggregation pheromone was identified in *Ips paraconfusus* as a mixture of ipsdienol, ipsenol and *cis*-verbenol (Silverstein et al., 1966). This was also the first multicomponent insect pheromone described; multicomponent pheromones are now the norm. In the genus *Ips*, the pheromone components ipdienol and ipsenol are extensively used, with varying ratios of both the *S*-(+) and *R*-(−) enantiomers of both ipsdienol and ipsenol (Borden, 1985). The IUPAC name for ipsdienol, 2-methyl-6-methylene-2,7-octadien-4-ol, led to the name ipsdienol rather than ipstrienol, even though there are three double bonds in the molecule (R.M. Silverstein, personal communication).

The structural similarity between the plant monoterpenes myrcene and α-pinene and the pheromone components ipsdienol and verbenol led early investigators to propose that bark beetles produced their pheromone components by the hydroxylation of plant-derived precursors (Hughes, 1973, 1974; Fig. 1). In addition, at that time monoterpene synthesis was associated with plant tissues and not described in Animalia. Studies in the 1970s and 1980s led to the conclusion that monoterpenoid pheromone components in bark beetles were produced from the carbon skeletons of host oleoresin precursors (Borden, 1985; Byers, 1981; Hendry et al., 1980; Hughes, 1973, 1974; Vanderwel et al., 1992). Hendry et al. (1980) appeared to have presented unequivocal evidence for the host tree origin of the monoterpene carbon skeleton when they showed that, when exposed to ^{2}H-myrcene, adult male *I. paraconfusus* converted this labelled host hydrocarbon to ^{2}H-ipsenol and ^{2}H-ipsdienol.

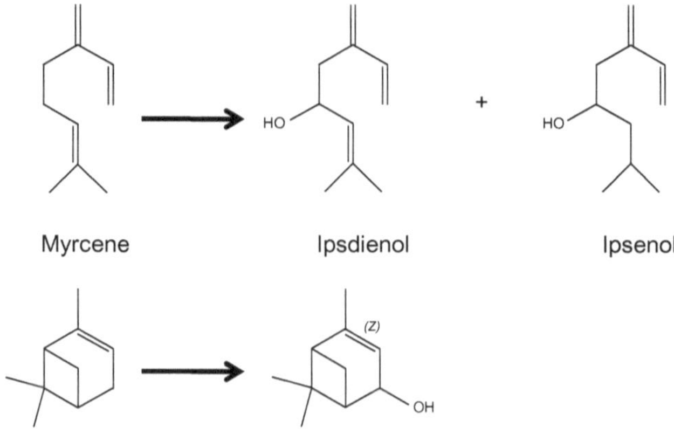

Fig. 1 The first bark beetle pheromone components identified were ipsdienol, ipsenol and verbenol (Silverstein et al., 1966). Because of their similarity to myrcene and α-pinene, it was suggested that the pheromone components arose from monoterpenes (Hughes, 1973, 1974).

Studies in the late 1980s and the 1990s began challenging the view that bark beetle terpenoid pheromone components were formed exclusively from host precursors. Labelled mevalonate injected into *I. typographus* was incorporated into volatile extracts, and radioactivity was associated with preparative GC fractions that co-eluted with the hemiterpenoid pheromone component, 2-methyl-3-buten-2-ol (Lanne et al., 1989). Byers and Birgersson (1990) presented evidence that there was not enough myrcene in the diet of some *Ips* species to account for the amount of pheromone produced. Theories were advanced that perhaps beetles sequestered myrcene or its hydroxylated products during larval stages, to provide sufficient monoterpenoids for pheromone production (Ivarsson et al., 1993; Vanderwel, 1994). Ivarsson et al. (1993) showed that compactin, a statin that inhibits 3-hydroxy-3-methylglutaryl-CoA reductase (HMGR) (Nakamura and Abeles, 1985), inhibited the production of ipsdienol and *E*-myrcenol in *I. duplicatus*, providing indirect evidence for de novo biosynthesis. Compactin did not inhibit the synthesis of *cis*-verbenol, consistent with *cis*-verbenol being derived from host tree α-pinene. A comparison of pheromone production in *I. pini* under various conditions showed that males feeding on freshly cut Jeffrey pine bolts produced abundant ipsdienol in a 98:2 (−)/(+) ratio (Fig. 2A). Unfed juvenile hormone III (JH III)-treated males produced abundant ipsdienol in a 90:10 (−)/(+) ratio (Fig. 2B). Unfed,

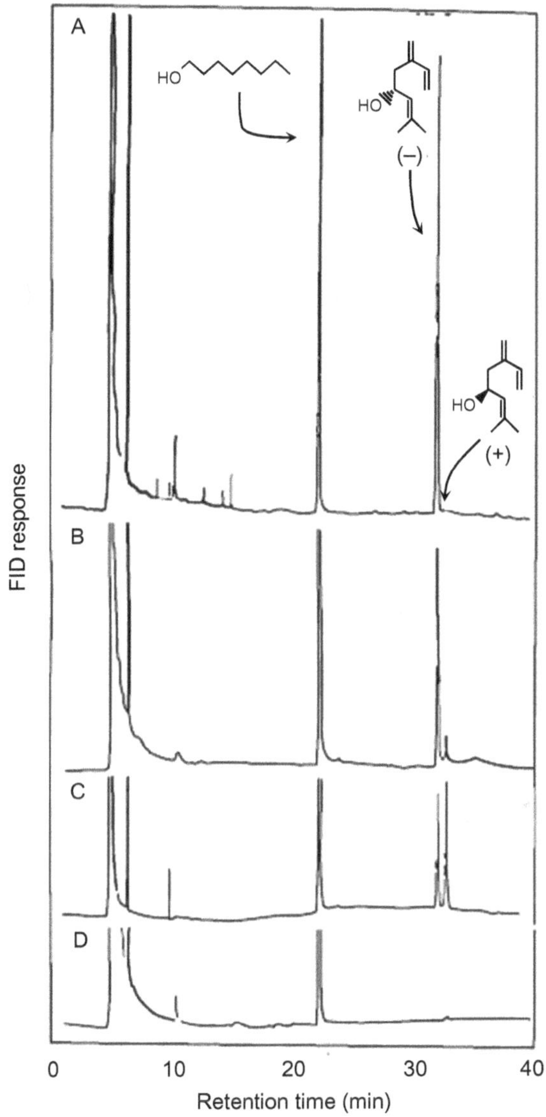

Fig. 2 Enantiomeric ratios of ipsdienol from male *Ips pini*. Figures show GC traces of extracts of males that have been allowed to feed in phloem (A), unfed males treated with JH III (B), unfed males exposed to a myrcene-saturated atmosphere(C) and unfed males that had not been treated with JH III or exposed to myrcene (D). The peak at 23 min is an *n*-octanol internal standard. *Adapted from Lu, F., 1999. Origin and endocrine regulation of pheromone biosynthesis in the pine bark beetles,* Ips pini *(Say) and* Ips paraconfusus *Lanier (Coleoptera: Scolytidae). Biochemistry. Reno, NV 89557, University of Nevada, Reno. Ph.D. 152 pp.*

non-JH III-treated males exposed to myrcene produced abundant ipsdienol in an approximate 50:50 (+)/(–) ratio (Lu, 1999) (Fig. 2C). Unfed, non-JH III-treated males produced minimal ipsdienol (Fig. 2D). The ratio of (–) to (+) enantiomers of pheromonal ipsdienol in western *I. pini* populations is approximately 95:5, and a 50:50 mixture does not function as a pheromone for western *I. pini*. These data are consistent with the de novo synthesis of pheromonal ipsdienol and also show how early workers who examined ipsdienol production without chiral columns (to separate (+)- and (–)-ipsdienol) came to the conclusion that it was derived from host tree-produced myrcene. It is difficult to determine exactly how much ipsdienol is derived from host tree-produced myrcene, but it appears to be a minor source.

Direct biochemical evidence that bark beetles synthesize the acyclic monoterpenoid pheromones ipsdienol and ipsenol de novo via the mevalonate pathway was obtained by demonstrating the incorporation of ^{14}C-acetate and ^{14}C-mevalonolactone into both components in *I. pini* and *I. paraconfusus* (Seybold et al., 1995a). Furthermore, the enantiomeric ratios of the radio-labelled components match the enantiomeric compositions found in nature (Seybold et al., 1995b). Work using radio-labelled precursors was extended to *D. jeffreyi* and other *Dendroctonus* species (Barkawi et al., 2003). Radioactivity from labelled acetate, mevalonate and isopentenol was incorporated into frontalin in *Dendroctonus* spp. It now appears that most bark beetle pheromone components are produced de novo, although a few, including verbenol, verbenone and verbenene (all derived from α-pinene), and 1- and 2-heptanol (derived from *n*-heptane) are still thought to be derived from host tree precursors.

2.1 Symbiotic Contributions

Earlier studies suggested that bark beetle pheromones could be produced by gut tract bacteria (Brand et al., 1975) or fungus (Brand et al., 1976). Endosymbiotic yeast resident in mountain pine beetle (*D. ponderosae*) alimentary canal can produce verbenone, an antiaggregation semiochemical, from *trans*-verbenol (Davis et al., 2013). The mutualistic fungus, *Grosmannia clavigera*, which is carried in the mycangia of *D. ponderosae* likely supports pheromone component biosynthesis indirectly by providing nutrients (Bentz and Six, 2006). For the spruce bark beetle (*Ips typographus*), the symbiotic blue stain fungi *G. penicillata* and *G. europhioides* can synthesize the aggregation pheromone component 2-methyl-3-buten-2-ol de novo

(Zhao et al., 2015). While these data suggest that bark beetles respond to symbiont-produced semiochemicals, it is also clear that major components are produced in insect tissues. All enzymes that have been characterized in pheromone biosynthetic pathways are eukaryotic, as indicated by the presence of poly(A) tails, alignments with genomic DNA sequences (Keeling et al., 2013) and expression studies, including qRT-PCR and in situ hybridizations, that localize mRNAs to bark beetle tissues (rev. in Blomquist et al., 2010). In addition, it is difficult to reconcile JH III-regulated pheromone production with a bacterial or fungal source of pheromone. With the exception of a recent study by Zhao et al. (2015), published evidence for symbiant contributions to pheromone biosynthesis is indirect at best; to date, there is little evidence that symbiants or associates contribute significantly to monoterpenoid pheromone components.

3. SITE OF PHEROMONE PRODUCTION

Many insects have specialized cells in the abdomen or use epidermal cells to produce pheromone (Ma and Ramaswami, 2003). Bark beetle pheromones are often associated with frass, suggesting a different site of synthesis. Early studies isolated pheromones from the hindguts of various bark beetles (Byers and Wood, 1981; Hughes, 1973; Pitman et al., 1969). The location of pheromone in the hindgut and the histology and anatomy of gut tissue, particularly the cellular diversity and features which show a secretory richness (Diaz et al., 2000), suggested the midgut as a possible location of pheromone production. Studies with *I. paraconfusus* examined incorporation of radiolabelled acetate into pheromone precursors throughout various body regions in an attempt to localize the site of pheromone synthesis (Ivarsson et al., 1998). Those studies implicated the metathorax, which would contain the midgut, as the site of pheromone biosynthesis of *I. paraconfusus*.

The site of pheromone synthesis in *I. pini* was determined through a combination of molecular and biochemical experiments. Inhibition of HMGR by the statin compactin in *I. duplicatus* showed reduction in ipsdienol synthesis (Ivarsson et al., 1993), and topical treatment of unfed male *I. pini* and *I. paraconfusus* with JH III resulted in a marked increase in HMGR mRNA (Tillman et al., 2004). Therefore, HMGR was used as a marker for mevalonate-derived pheromone biosynthesis in bark beetles.

Hall et al. (2002a) used in situ hybridization to show that HMGR mRNA was very abundant in midguts of JH III-treated male *I. pini*. *HMGR* expression increases in the anterior midgut when beetles are treated with JH

III, as compared with the acetone control (Fig. 3A and B; Hall et al., 2002a). To rule out other portions of the alimentary canal as pheromone-producing tissue, the authors isolated alimentary tracts from control (Fig. 3C) and JH III-treated beetles (Fig. 3D), which showed, similar to the sagittal view of the beetle, that JH III treatment raised HMGR transcript levels in the anterior midgut.

Biochemical assays then confirmed that the *I. pini* midgut makes the pheromone ipsdienol. Alimentary canals from fed and JH III-treated males were dissected into proventriculus, midgut and hindgut. The tissues were

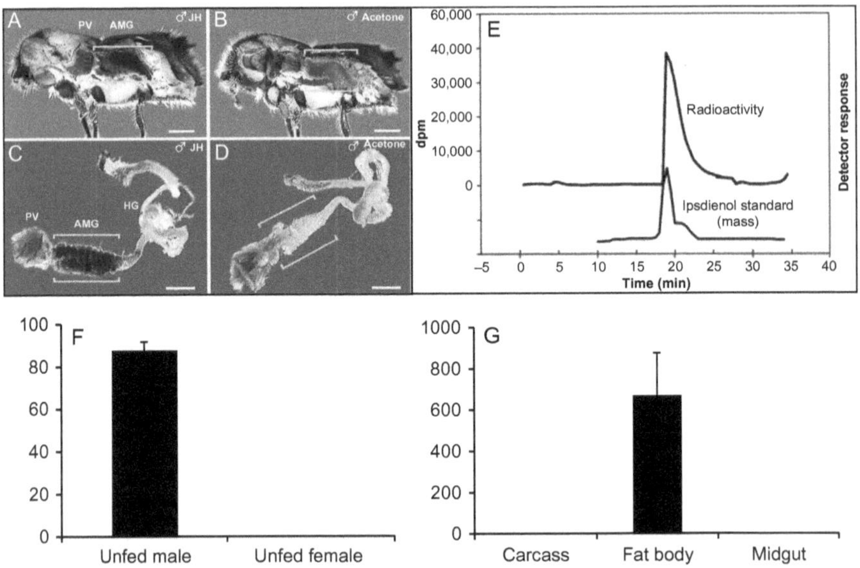

Fig. 3 Localization of pheromone production in *I. pini* and *D. ponderosae*. (A, B) In situ hybridization of JH III-induced HMGR expression in *I. pini* with the sagittal view exposed showing the proventriculus (PV) and anterior midgut (AMG) of a JH III-treated (A) and acetone treated (B) male. (C, D) Isolated alimentary canal isolated from a male *I. pini* that had been treated with JH III (C) or with acetone (D). Radio-HPLC (E) of pentane–diethyl ether extract of midgut tissue from five fed male *I. pini* incubated with [1-^{14}C] acetate confirms that ipsdienol is produced de novo in midgut tissues. (F, G) *exo*-Brevicomin is found in unfed male, but not female *D. ponderosae* (F), and localizes to the fat body (G). *Figures reproduced from Hall, G.M., Tittiger, C., Andrews, G.L., et al., 2002. Midgut tissue of male pine engraver,* Ips pini, *synthesizes monoterpenoid pheromone component ipsdienol de novo. Naturwissenschaften 89 (2), 79–83; Song, M., Delaplain, P., Nguyen, T.T., et al., 2014.* exo-*Brevicomin biosynthetic pathway enzymes from the Mountain Pine Beetle, Dendroctonus ponderosae. Insect Biochem. Mol. Biol. 53 (0), 73–80; and Song, M., Gorzalski, A., Nguyen, T., et al., 2014.* exo-*Brevicomin biosynthesis in the fat body of the mountain pine beetle,* Dendroctonus ponderosae. *J. Chem. Ecol. 40 (2), 181–189.* (See the colour plate.)

incubated with radiolabelled acetate, extracted and analysed by radio-HPLC. Most of the radioactivity in the pentane:ether extract from midgut tissue was associated with the peak corresponding to an ipsdienol standard (Fig. 3E), thus showing that pheromone biosynthesis occurs in the anterior midgut (Hall et al., 2002a). A similar approach showed that the monoterpenoid pheromone, frontalin, is also produced in midgut tissue in *D. jeffreyi* (Hall et al., 2002b; Nardi et al., 2002).

Nardi et al. (2002) performed an ultrastructural examination of the beetle midgut and the endodermal cells from both *D. jeffreyi* and *I. pini*. Using electron microscopy, the smooth endoplasmic reticulum (SER) of the JH III-treated male midgut cells was found to be large and arranged in microcrystalline arrays, structures not seen in the negative controls for both *I. pini* and *D. jeffreyi*. The formation of microcrystalline arrays of SER in the pheromone-producing cells is the main distinguishing subcellular characteristic inferring involvement of the SER in pheromone biosynthesis. Crystalline SER arrays are consistently observed in mammalian cells with high mevalonate pathway flux; so abundant SER in pheromone biosynthetic bark beetle midgut cells is consistent with elevated isoprenoid lipid biosynthesis. Cytochromes P450 involved in pheromone production are associated with the endoplasmic reticulum (Sandstrom et al., 2006, 2008), but at this point in time the subcellular localization of the other pheromone biosynthetic enzymes in bark beetles is unknown.

Unlike the precedent set for monoterpenoid pheromone production to the midgut, *exo*-brevicomin biosynthesis in *D. ponderosae* localizes to the fat body. Song et al. (2014a) used short-term tissue culture techniques to show that *exo*-brevicomin was detectable in the fat body of unfed males (ie, adults producing *exo*-brevicomin) only, but not in the fat bodies of females, or carcass or midgut of *exo*-brevicomin-producing males (Fig. 3F and G). Tissue incubations using deuterium-labelled precursors to some steps, including epoxidation and cyclization in the *exo*-brevicomin biosynthetic pathway, showed conversion to other pathway intermediates only in fat body tissues (Song et al., 2014a). Genes encoding enzymes implicated in *exo*-brevicomin biosynthesis show elevated mRNA levels in unfed male fat bodies compared to other tissues or females (Aw et al., 2010; Song et al., 2014b), a pattern similar to those involved in isoprenoid biosynthesis in *Ips* spp. These data strongly support that *exo*-brevicomin is synthesized and at least stored in the fat body prior to release through the digestive tract.

If these discoveries are reasonably extended to other bark beetle taxa, it appears that monoterpenoid pheromone components are produced in the alimentary canal, whereas fatty acid-derived components such as *exo*-brevicomin can be produced elsewhere. The fat body is a curious location because it is endodermally derived, whereas all other known insect pheromone biosynthetic glands or tissues, including the bark beetle midgut, are ectodermally derived (Ma and Ramaswami, 2003). The paradigm that pheromones are produced from ectodermally derived cells may still hold true for *exo*-brevicomin because the bark beetle fat body may be associated with other cell types, for example, oenocytes, which are ectodermally derived (Arrese and Soulages, 2010; Lockey, 1988). Further work is required to determine whether pheromone component biosynthesis is localized within fat body substructures. Pheromone component biosynthesis in this tissue raises the question of how the chemical is transported and deposited into the alimentary canal prior to release.

4. ENDOCRINE REGULATION

Endocrine regulation of pheromone biosynthesis in insects is mediated by pheromone biosynthesis-activating neuropeptide (PBAN) in Lepidoptera, by ecdysteroids in the house fly and presumably other Diptera, and by JH III in some Coleoptera, including some bark beetles (Blomquist et al., 2004). The endocrine regulation of pheromone production in bark beetles is complex and not fully understood. In male *I. pini*, feeding on phloem stimulates the *corpora allata* (CA) to synthesize and release JH III which in turn induces ipsdienol production (Tillman et al., 1998). Application of JH III to male *I. duplicatus* (Ivarrson and Birgersson, 1995) and *I. pini* (Tillman et al., 1998) results in increased pheromone production. However, in two other *Ips* species, *I. paraconfusus* and *I. confusus*, JH III does not induce pheromone production (Bearfield et al., 2009; Tillman et al., 2004).

Studies prior to the 1990s had the difficult task of reconciling the JH III-mediated increase in pheromone production with the expectation that the carbon skeleton of pheromone components was derived from the diet. It was known that bark beetles possess the mechanism to detoxify monoterpenes and it was thought that the products of these processes were used directly as pheromones. Since the recognition that ipsdienol and ipsenol are made de novo in bark beetles, work has concentrated on determining what steps in the process are controlled. Ivarsson and Birgersson (1995) showed that a JH analog increases ipsdienol and *E*-myrcenol production in *I. duplicatus*, whereas exposure to myrcene did not increase pheromone production. The same report also showed that PBAN from *Helicoverpa zea* has no effect on pheromone production in *I. duplicatus*. In *I. pini*, Tillman et al. (1998, 2004) showed that topically applied JH III increased pheromone production and markedly upregulated *HMGR*.

The complexity of hormonally regulated pheromone production becomes evident through comparisons of different species (Bearfield et al., 2009). HMGR transcript levels increased in both *I. pini* and *I. paraconfusus* when males of those species are treated with JH III, but HMGR enzyme activity remains low even though the transcript level increases in *I. paraconfusus*. This result suggests different posttranscriptional HMGR regulation by JH III in *I. paraconfusus* compared to *I. pini*. In addition to HMGR, other enzymes from the mevalonate pathway were also studied. The effect of JH III on 3-hydroxy-3-methylglutaryl-CoA synthase (HMGS) transcript levels in *I. pini* was studied using RT-PCR and Northern blotting (Bearfield et al., 2006). In both males and females, JH III stimulated mRNA levels in the first 8 h. While male mRNA levels continued to increase, female mRNA levels started to decrease after the first 8 h. Furthermore, male transcript levels were higher than those in females at every time point examined. These data suggested the coordinate regulation of mevalonate pathway genes by JH III in *I. pini*. Bearfield et al. (2009) further showed different regulatory mechanisms in comparisons between *I. pini* and *I. confusus*, another member of grandicollis group. By using radio-labelling studies, activities of key mevalonate pathway enzymes including HMGS, HMGR and geranyl diphosphate synthase (GPPS) were measured. Feeding upregulated these enzyme activities in both species, but JH III only increased their activities in *I. pini* and not in *I. confusus*. However, RT-PCR showed JH III and feeding upregulated the mRNA level in both species. These data suggest that at least an additional factor is functioning posttranscriptionally to regulate pheromone production.

Antennae also play a role in the regulation of pheromone production in *I. pini* (Ginzel et al., 2007). Male beetles without antenna had higher levels of pheromone, phenomena first demonstrated in the Colorado potato beetle (Dickens et al., 2002). Following JH III treatment, beetles with antennae removed showed higher enzyme activities for HMGS, HMGR and GPPS than beetles with intact antennae. mRNA levels are also higher in male beetles without antennae. This suggests negative feedback at the transcription level. It was proposed that this mechanism prevents the beetles from overcrowding.

5. *IPS* SPP. METABOLIC PATHWAYS

5.1 Ipsdienol and Ipsenol

In vivo radiochemical studies demonstrated that labelled acetate and mevalonolactone are incorporated into ipsdienol and ipsenol in several *Ips* species (Seybold et al., 1995a; Tillman et al., 1998). The most likely pathway to account for this was the mevalonate pathway (Fig. 4) to form isopentenyl pyrophosphate (IPP) and dimethylallyl pyrophosphate (DMAPP) which then condense to form geranyl diphosphate prior to being converted to myrcene and then to ipsdienol and ipsenol. Biochemical studies with isolated midgut tissue demonstrated the conversion of geranyl diphosphate to myrcene (Martin et al., 2003). Numerous earlier studies had demonstrated that myrcene was converted to both ipsdienol and ipsenol (rev. in Seybold and Tittiger, 2003). Fish et al. (1984) and D. Vanderwel (unpublished data) showed that ipsdienol could be converted to ipsdienone, reduced to ipsenone and then converted to ipsenol.

The enzymes that catalyse pheromone biosynthesis and their cognate genes are of interest for several reasons. From a management perspective, understanding their regulation and interactions is a first step to developing new strategies that specifically target chemical communication in these pests. These future strategies may include technologies like short interfering RNAs or chemical inhibitors that could reduce pheromone levels without impacting nontarget species. The identities and regulatory schemes of these enzymes in multiple taxa can be used to learn how pheromone systems have evolved and help predict responses to management pressures. Thus, identifying these genes was a high priority. Fortunately, biochemical and molecular data for mevalonate pathway genes in *I. pini* could be leveraged to identify late step enzymes through a combination of functional genomics, molecular biology and biochemical techniques. Clustering analyses of microarray data comparing gene expression profiles in fed (pheromone-producing) and unfed (control) males confirmed that mevalonate pathway genes are coordinately regulated by JH III (Fig. 5; Keeling et al., 2004, 2006). Other genes in the cluster were tagged as high priority candidates. Sequence analysis and qRT-PCR experiments to determine JH III regulation and mRNA localization in the midgut further narrowed the candidate pool to a few genes encoding enzymes with a high probability of catalysing late steps in ipsdienol biosynthesis. They are summarized later.

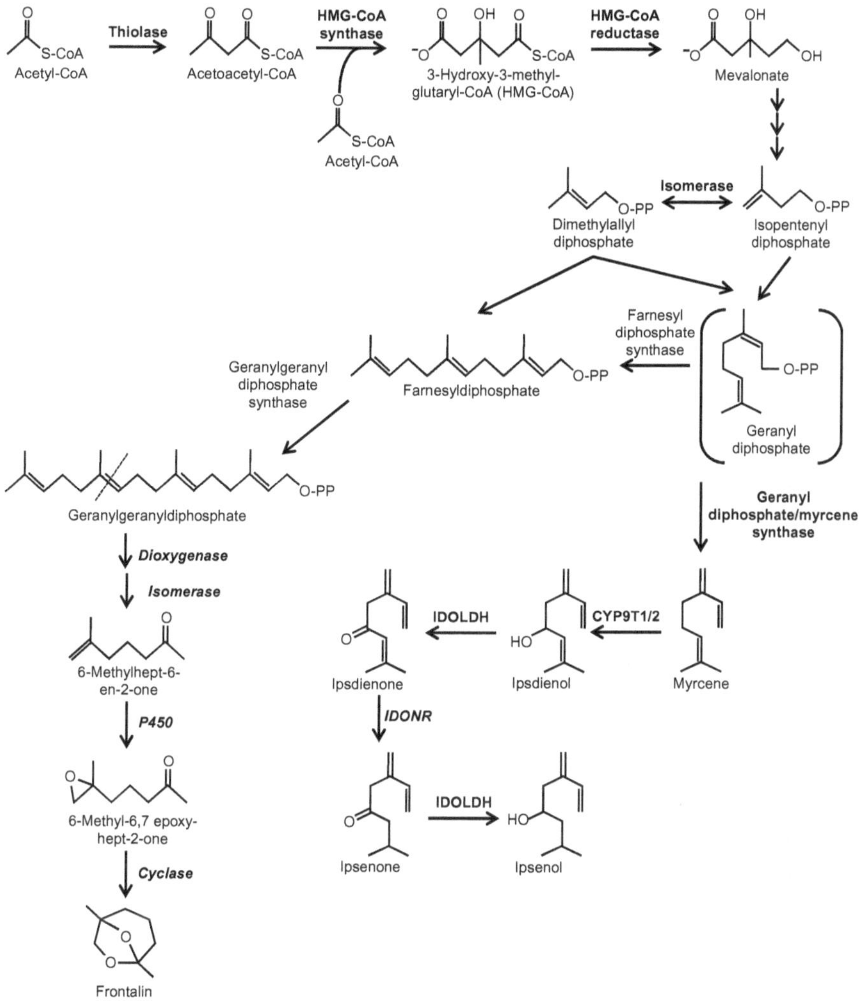

Fig. 4 De novo monoterpenoid pheromone biosynthesis. Intermediates and products are in *normal font*. Enzymes are in *bold*, and enzymes that have not been identified or characterized are in ***bold italics***.

In vertebrate systems, the two enzymes HMGR and HMGS are pivotal to the regulation of carbon flow to isoprenoids, and these were the first enzymes examined in bark beetle pheromone biosynthesis (Tittiger, 2003; Tittiger et al., 1999, 2003). Using PCR and primers designed from conserved regions for these genes from other metazoans, *HMGR* was isolated. Both feeding and JH III markedly upregulated HMGR transcript level

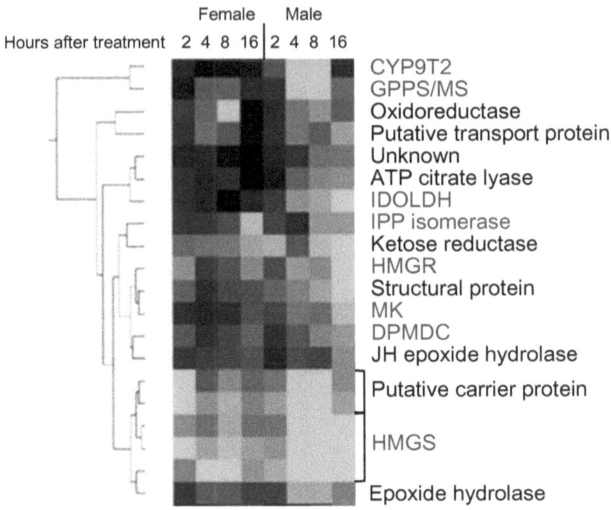

Fig. 5 Coordinated gene expression in *Ips pini* midguts. Heat map of genes induced upon topical treatment with JH III. Each column represents average relative expression ratios for each time point. Genes encoding mevalonate pathway and pheromone biosynthetic pathway enzymes are noted in *red. Adapted from Keeling, C.I., Bearfield, J.C., Young, S., Blomquist, G.J., Tittiger, C., 2006. Effects of juvenile hormone on gene expression in the pheromone-producing midgut of the pine engraver beetle,* Ips pini. *Insect Mol. Biol. 15 (2), 207–216.* (See the colour plate.)

and enzyme activity in *I. pini*, and based on the important regulatory role of *HMGR* in other organisms, it was assumed that *HMGR* was the key regulatory step in bark beetle pheromone production (Tillman et al., 2004). This was consistent with the finding by Ivarsson et al. (1993) who showed that the HMGR inhibitor, compactin, reduces ipsdienol production.

5.2 GPPS/Myrcene Synthase

GPPS catalyses the condensation of DMAPP and IPP to form geranyl diphosphate (GPP). GPP is the precursor of monoterpenes, a large family of naturally occurring C_{10} compounds predominately found in plants. Martin et al. (2003) showed myrcene synthase (MS) activity in *I. paraconfusus*. Conclusive evidence for de novo monoterpene biosynthesis in an animal was obtained by describing *I. pini* GPPS (Gilg et al., 2005). *GPPS* expression levels are regulated by JH III in a dose- and time-dependent manner, almost exclusively in the anterior midgut of male *I. pini* similar to other mevalonate pathway genes involved in pheromone biosynthesis. The functionally expressed recombinant enzyme produced GPP as its major product. A three-dimensional structural model of GPPS showed that the insect enzyme has the sequence and structural motifs common to *E*-isoprenyl diphosphate synthases, and moreover, interactions between key residues at or near the floor of the binding pocket limit the product size to C_{10} molecules (Fig. 6; Gilg et al., 2005; unpublished).

In pheromone-producing male *I. pini*, the acyclic monoterpene myrcene is required for the production of the major aggregation pheromone component, ipsdienol. Surprisingly, monoterpene synthase activity, first demonstrated in *I. confusus* (Martin et al., 2003), was shown to be associated with recombinant GPPS (Gilg et al., 2009). Enzyme assays were performed on recombinant GPPS to determine the presence of monoterpene synthase activity, and the reaction products were analysed by coupled GC–MS. The functionally expressed recombinant enzyme produced both GPP and myrcene from IPP and DMAPP, making it a bifunctional enzyme. This unique isoprenyl diphosphate synthase possesses the functional plasticity that is characteristic of terpene biosynthetic enzymes of plants, contributing toward the current understanding of product specificity of the isoprenoid pathway.

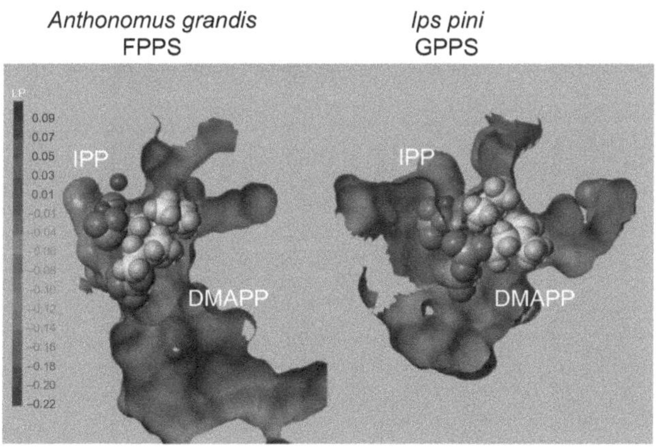

Fig. 6 Molecular models of prenyl transferase substrate binding sites. Cut-away views of the substrate binding domains for FPPS from the cotton boll weevil (*Anthonomus grandis*) and GPPS/MS from *Ips pini*, shaded to show relative hydrophobicity (*brown* to *blue* corresponding from most to least hydrophobic). IPP and DMAPP substrates are shown in space fill, hydrocarbon portions in *blue* and *white*, phosphate portions in *orange* and *red*. Mg^{2+} cofactors for each enzyme are in *purple*. The substrate binding domain for GPPS/MS is clearly smaller than that for FPPS. *Adapted from Gilg, A., 2004. The isolation and characterization of geranyl diphosphate synthase from the pine engraver, Ips pini (Coleoptera: Scolytidae). Dissertation. Biochemistry and Molecular Biology, University of Nevada, Reno Ph.D. 157 pp.* (See the colour plate.)

5.3 Cytochrome P450 (CYP9T2)

Cytochrome P450 monooxygenases (P450s) constitute a diverse superfamily of enzymes that play a crucial role in the metabolism of a wide range of both endogenous and foreign compounds. Insects generally have approximately 100 cytochrome P450 genes, so preliminary identification of *I. pini* myrcene hydroxylase relied on expression profiling. *CYP9T2* was the only P450 among at least four assayed that had an expression pattern consistent with pheromone biosynthesis. A full-length cDNA clone was isolated and expressed in Sf9 cells. A functional assay using microsomes of Sf9 cells infected with baculoviral constructs encoding CYP9T2 and housefly (*Musca domestica*) P450 reductase demonstrate that CYP9T2 is a myrcene hydroxylase that converts myrcene to ipsdienol (Sandstrom et al., 2006).

Male *I. pini* have two sources of myrcene, the host tree and de novo biosynthesis via the mevalonate pathway. The major monoterpenoid aggregation pheromone component released by *I. pini* is 95% (4*R*)-(−)-ipsdienol (Figs 2 and 7A), whereas recombinant CYP9T2 produced 81% *R*-(−)-ipsdienol (Fig. 7B). Sandstrom et al. (2008) demonstrated that the recombinant ortholog (CYP9T1) from *I. confusus* produces a similar ratio of *R*-(−)- and *S*-(+)-ipsdienol (85% *R*-(−)) as does CYP9T2, even though the pheromone blend from *I. confusus* is approximately a 10/90 *R*-(−)/*S*-(+) ratio (Fig. 7C and D). These data strongly suggest that enzymatic steps downstream from the hydroxylation step are required to produce the final enantiomeric blend.

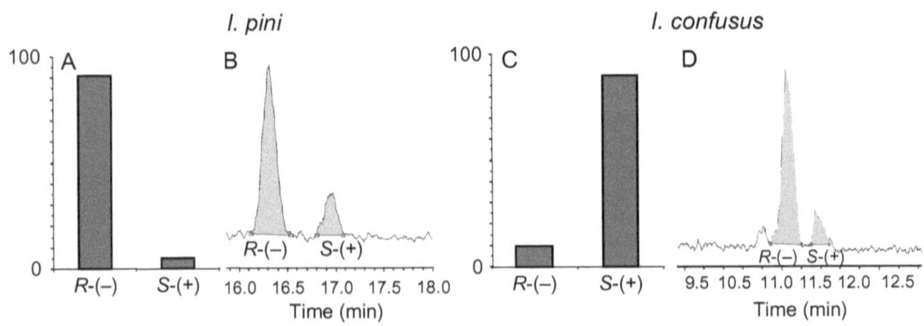

Fig. 7 Enantiomeric ratios of ipsdienol in two *Ips* spp. The enantiomeric composition of pheromonal ipsdienol recovered from feeding male *Ips pini* (A) is diametrically opposed to that of *I. confusus* (B), while GC traces of extracts of products from in vitro assays of myrcene hydroxylase from *I. pini* (CYP9T2, C) and *I. confusus* (CYP9T1, D) show that both enzymes produce preferentially (−)-ipsdienol. For both insects, the enantiomeric ratio resulting from myrcene hydroxylase activity is not the same as that of the pheromone. *Adapted from Sandstrom, P., Ginzel, M., Bearfield, J., Welch, W., Blomquist, G., Tittiger, C., 2008. Myrcene hydroxylases do not determine enantiomeric composition of pheromonal ipsdienol in* Ips *spp. J. Chem. Ecol. 34 (12), 1584–1592.*

5.4 Postmyrcene Hydroxylation Modifications (Evidence)

Three lines of evidence argue for a role of ipsdienone (Fig. 4) in achieving the final 95:5 (–/+) enantiomeric composition in the western *I. pini* pheromone. First, in examining pheromone-producing tissue, Ivarsson et al. (1997) isolated much of the labelled C_{10} product as ipsdienone. Second, D. Vanderwel et al. (unpublished) demonstrated that deuterium-labelled *S*-(+)-ipsdienol was converted to ipsdienone, and that ipsdienone was selectively converted to the *R*-(–)-ipsdienol. Finally, the demonstration that CYP9T2 produced an 81:19 (–/+)-ipsdienol composition (Fig. 7B) argues for other genes to control the final enantiomeric composition. Domingue et al. (2006) selected lines of *I. pini* with either primarily the *R*-(–) enantiomer (western *I. pini*) or the *S*-(+) enantiomer (eastern *I. pini*) and created F1 and F2 progeny. In the analysis of the results from this work, they conclude that dominance at one autosomal locus explains much of the variation in ipsdienol blend between the divergently selected lines, although later work (Domingue and Teale, 2008) suggested the situation might be more complex.

5.5 Ipsdienol Dehydrogenase

A male *I. pini* midgut tissue microarray hybridized to cDNA from JH III-treated beetles identified three oxidoreductases whose expression patterns were similar to known pheromone biosynthetic genes (Fig. 5). The gene represented by EST IPG012D04 had the third highest basal transcript level in males compared to females and was specific to the fed male anterior midgut. Full-length cDNAs of this oxidoreductase, now called ipsdienol dehydrogenase (IDOLDH) and homologues from eastern *I. pini* and *I. confusus*, were cloned and sequenced. Homology searches of IDOLDH identified it as a short-chain dehydrogenase/reductase (SDR). SDRs constitute a large family of NAD(P)(H)-dependent oxidoreductases and have important roles in the metabolism of lipids, amino acids, carbohydrates, cofactors, hormones and xenobiotics as well as in redox sensor mechanisms (Kavanagh et al., 2008). An NADP(H) binding assay and coupled GC–MS analysis demonstrated that recombinant IDOLDH could readily oxidize *R*-(–)-ipsdienol to ipsdienone and to stereospecifically catalyse the reverse reaction. The recombinant enzyme also accepts the pheromone precursor ipsenone to *R*-(–)-ipsenol, but other terpenoid alcohols and substrate analogs are not substrates, suggesting very high specificity for pheromone components (R. Figueroa Teran, H. Pak, G.J. Blomquist, and C. Tittiger, unpublished data). Further studies are being conducted to measure the kinetic properties of this protein.

While the production and activity of IDOLDH in *I. pini* suggest that the enzyme has a key role in determining the final stereochemistry of ipsdienol and ipsenol, other elements are very likely involved. Kinetic analysis shows that the oxidation reaction catalysed by IDOLDH is highly favoured compared to the reduction reaction. Given the relatively high levels of ipsdienone observed in vivo (Ivarsson et al., 1997), the implication is that ipsdienone is produced from (–)-ipsdienol by IDOLDH. This would effectively increase the relative abundance of (+)-ipsdienol—which is in contrast to the observed extreme enantiomeric excess of (–)-ipsdienol released as pheromone by western populations. Thus, other proteins or enzymes likely interact with ipsdienol and/or ipsdienone to affect the enantiomeric blend, possibly by acting directly on products or substrates and/or modifying IDOLDH activity. To this end, other uncharacterized genes that cluster with *GPPS/MS*, *IDOLDH* and *Cyp9T2* based on microarray (Fig. 5) and qRT-PCR profiling may have important roles.

The genes encoding all of the enzymes listed earlier share common regulatory schemes: they all show higher basal expression levels in males compared to females, mRNA levels increase following treatment with JH III or feeding, and JH III-mediated increases in mRNA are localized to the midgut, where ipsdienol is synthesized. Significantly, *GPPS/MS* regulation shows one unique difference compared to the rest: it alone does not show elevated mRNA levels in females following JH III treatment (Keeling et al., 2004). Given that GPPS/MS diverts carbon from the mevalonate pathway to ipsdienol, the fact that this enzyme is not elevated in female midguts may explain why females do not produce ipsdienol despite the fact that the rest of the biosynthetic pathway is apparently present. Female *I. pini* may well be competent to produce ipsdienol, but do not because they do not shunt carbon into that branch of the pathway.

5.6 Molecular Evolution Considerations

Tight coordinate regulation of genes encoding pheromone biosynthetic enzymes with mevalonate pathway genes may indicate an evolutionary history of how pheromone production is regulated. *Ips pini* GPPS/MS has significant sequence homology with GGPPS, with amino acid differences near the substrate binding domain probably contributing to differences in product profiles between the two enzymes (Gilg et al., 2005). GGPPS may reflect an ancestral enzyme, with modification(s) leading to a new function: the ability to produce myrcene. This would have resulted in endogenous myrcene (toxin) production, which an ancestral beetle presumably would have been metabolically equipped to handle due to the monoterpene hydroxylase enzymes such as CYP9T2 already dealing with ingested tree myrcene. This would exert strong selective pressure to coordinately express *GPPS/MS* and *CYP9T2*, just to keep the beetle healthy. Thus, if the myrcene hydroxylase gene were induced by feeding, MS gene regulation may have been tuned accordingly, providing a convenient boost to the hydroxylated monoterpenoid signal (generated by resin detoxification) that a beetle was actively engaging a new host. Further selective pressure may have linked this activity to the mevalonate pathway, particularly *HMGR* and *HMGS*, which regulate flux through the pathway and thus indirectly control myrcene/ipsdienol production. Better control of the signal may have subsequently arisen through endocrine regulation (JH III) and post-transcriptional mechanisms (Blomquist et al., 2010).

6. *DENDROCTONUS* SPP.: METABOLIC PATHWAYS

The bicyclic acetals frontalin and *exo*- and *endo*-brevicomin along with hydroxylation products of α-pinene are often part of *Dendroctonus* spp. pheromones. Frontalin is an antiaggregation pheromone is some species and an aggregation pheromone in others (Borden 1985). In *D. ponderosae* the pheromone is still not fully characterized, although three major components are known: frontalin is an antiaggregation pheromone, while *exo*-brevicomin and *trans*-verbenol are aggregation pheromone components (Conn et al., 1984; Greis et al., 1990; Hunt et al., 1986; Pureswaran et al., 2000). The roles of other components, eg, myrcenol and *endo*-brevicomin, are not as clearly understood (Hunt et al., 1986). *exo*-Brevicomin, frontalin and verbenol are apparently synthesized via different metabolic pathways.

6.1 Frontalin

Frontalin [(1*S*,5*R*)-1,5-dimethyl-6,8-dioxabicyclo[3.2.1]octane], a bicyclic acetal, is produced by *D. ponderosae* males after they arrive in the host tree and may function as a spacing factor signalling 'the tree is full' to other beetles (Pureswaran et al., 2000). It is made using the mevalonate pathway in the anterior midgut of male beetles (Hall et al., 2002b). Barkawi et al. (2003) showed that both ^{14}C-labelled mevalonolactone and isopentenol injected into the abdomen of pheromone-producing male *D. jeffreyi* result in production of ^{14}C-labelled frontalin. Furthermore, in situ hybridization showed that *HMGR*, which encodes a key regulatory enzyme in the mevalonate pathway, has high transcript levels mostly in midgut tissue of *D. jeffreyi*, showing that the midgut is the site of pheromone biosynthesis (Hall et al., 2002b). Prior to these studies, the precursor to frontalin was shown to be 6-methyl-hept-6-en-2-one (6MHO) (Perez et al., 1996; Vanderwel et al., 1992).

Most of the enzymes downstream from the mevalonate pathway that contribute to frontalin biosynthesis have not been characterized. Indeed, the only characterized enzyme is a geranylgeranyl diphosphate synthase (GGPP synthase) at the terminus of the mevalonate pathway. The *D. ponderosae* genome encodes two isoprenyl transferases, a GPP/FPP which produces either C10 or C15 products depending upon the IPP and DMAPP concentrations, and a GGPP synthase, which produces a C20 product (Keeling et al., 2013). Transcript levels in various tissues and life stages suggested that GGPP rather than GPP or FPP is used as the precursor to frontalin. Reduction of transcript level by RNA interference of the isoprenyl diphosphate synthases identified GGPP synthase as having the largest effect on frontalin production, suggesting that frontalin is derived from the 20-carbon isoprenoid precursor rather than from the 10- or 15-carbon precursors (Fig. 8) (Keeling et al., 2013).

The mechanism converting GGPP to frontalin remain uncharacterized. Sulcatone is a likely intermediate and may arise from dioxygenase activity on GGPP (Fig. 4; Blomquist et al., 2010; Keeling et al., 2013). Thereafter, sulcatone is likely isomerized to 6MHO, which can be epoxidized (likely by a cytochrome P450) and cyclized to produce frontalin. The identities of enzymes catalysing these reactions are being actively pursued.

6.2 *exo*-Brevicomin

exo-Brevicomin (*exo*-7-ethyl-5-methyl-6,8-dioxabicyclo[3.2.1]octane) is a bicyclic acetal that synergizes the aggregation effect of female-produced *trans*-verbenol. It is produced in the fat body of newly emerged males (but not females), and production drops when the males arrive at a new tree and mate (Pureswaran et al., 2000).

Early studies showed that 6-(*Z*)-nonen-2-one is a precursor (Francke et al., 1996; Vanderwel et al., 1992) and predicted that it was derived from fatty acids either via desaturation and limited β-oxidation of long-chain fatty acyl precursors, as precedented in pheromone biosynthesis in Lepidoptera (Rafaeli and Jurenka, 2003), or else directly via a thioesterase II-mediated termination of fatty acid biosynthesis at a 10-carbon intermediate, followed by desaturation to (*Z*)-7-decenoyl-CoA. Chain shortening of a longer

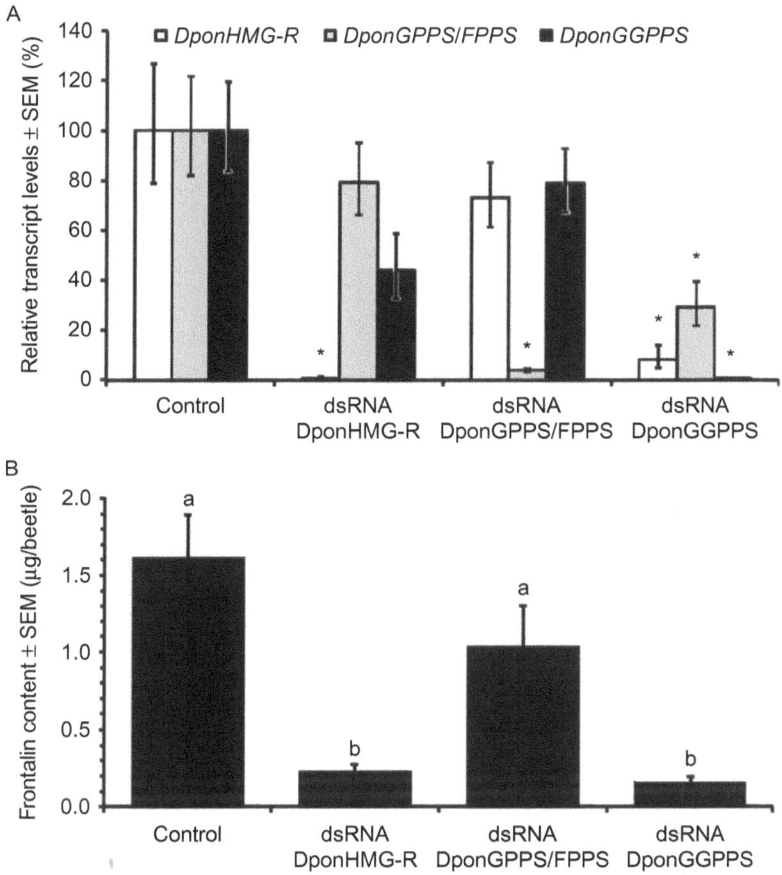

Fig. 8 RNAi of frontalin biosynthesis in *D. ponderosae.* (A) Transcript levels for DponHMG-R, DponGPPS/FPPS and DponGGPPS in juvenile hormone-treated male adults injected with dsRNA for DponHMG-R, DponGPPS/FPPS and DponGGPPS. *Asterisks* indicate treatments that were significantly different from the control ($P < 0.05$, by two-tailed *t*-test with Holm–Bonferroni correction on the $\Delta\Delta CT$ values, after significant one-way ANOVA tests; $n = 10$). (B) Frontalin levels in juvenile hormone-treated male adults injected with dsRNA for DponHMG-R, DponGPPS/FPPS and DponGGPPS (one-way ANOVA: $F3{,}44 = 12.1383$, $P < 0.00001$, $n = 12$). Treatments with the same letter are not significantly different ($P < 0.05$, by two-tailed *t*-test with Holm–Bonferroni correction). *Reproduced from Keeling, C., Yuen, M., Liao, N., et al., 2013. Draft genome of the mountain pine beetle, Dendroctonus ponderosae Hopkins, a major forest pest. Genome Biol. 14 (3), R27.*

precursor appears to be the more likely scenario because unfed male fat bodies incubated with decanoic acid do not show evidence of a short-chain desaturase activity (Song et al., 2014a), which would be necessary if chain elongation from acetate were stopped at C10. The entrance into the pathway is thus ω-3-decenoyl-CoA. Here there are two proposed branches (Fig. 9). One

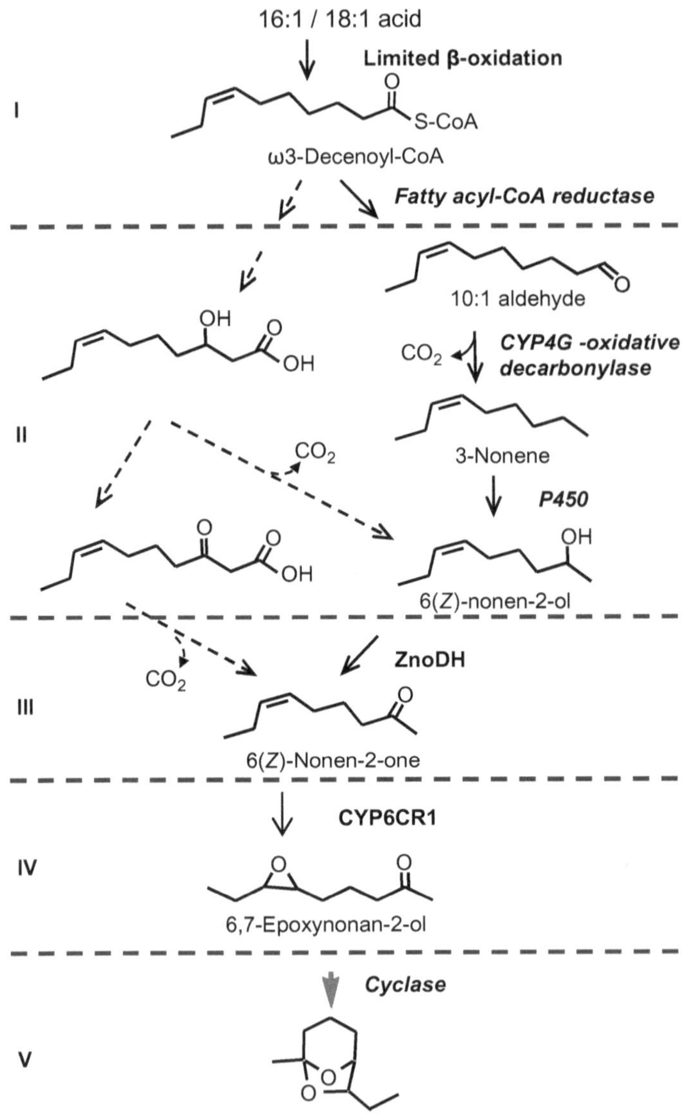

Fig. 9 *exo*-Brevicomin biosynthetic pathway. The pathway is divided into different steps identified by Roman numerals for ease of reference. For Step II, the 3-nonen-2-one intermediate may arise from decarboxylation of β-oxidation intermediates by α-oxidation (*left, dashed arrows*) or via hydrocarbon biosynthesis followed by hydroxylation to 3-nonen-2-ol and oxidation by ZnoDH (*right, solid arrows*). Intermediates and products are in *normal font*. Enzymes are in *bold*, and enzymes that have not been identified or characterized are in *bold italics*. *Adapted from Song, M., Gorzalski, A., Nguyen, T., et al., 2014. exo-Brevicomin biosynthesis in the fat body of the mountain pine beetle,* Dendroctonus ponderosae. *J. Chem. Ecol. 40 (2), 181–189.*

follows further β-oxidation reactions to 3-hydroxy-(*Z*)-7-decenoyl-CoA or 3-keto-(*Z*)-7-decenoyl-CoA which are thought to be converted to the free acid and then decarboxylated by α-oxidation (Vanderwel et al., 1992), as precedented in the production of (*Z*)-10-heptadecen-2-one in *D. mulleri* (Skiba and Jackson 1994). Alternatively, Song et al. (2014a) proposed that a fatty acyl-CoA reductase and an oxidative decarbonylase (Cyp4G family cytochrome P450) (Qiu et al., 2012) convert ω-3-decenoyl-CoA to (*Z*)-3-nonene. This hydrocarbon could then be hydroxylated by a P450 to (*Z*)-6-nonen-2-ol and then converted to (*Z*)-6-nonen-2-one through the activity of an oxidoreductase.

As with efforts to recover pheromone biosynthetic enzymes in *Ips* spp., identification and characterization of *exo*-brevicomin biosynthetic genes in *D. ponderosae* relied on functional genomic data to first generate a candidate list, followed by expression studies and enzyme assays of recombinant proteins to confirm functions. Clustering analysis of microarray data surveying gene expression profiles of multiple biological states, ranging from larvae through mated adults, revealed apparent coordination of three genes with an expression pattern consistent with *exo*-brevicomin production (Aw et al., 2010). mRNA levels were low in immature stages and in females and males that had been allowed to mate, but high in mature males that had exited the brood tree but not yet entered a host tree ('unfed' males). Two of the three genes had open reading frames, while the third did not and thus may produce a noncoding RNA (Aw et al., 2010). mRNA levels for the two protein-coding genes were highest in fat bodies compared to other tissues (Song 2014b).

One of the open reading frames encoded a short-chain oxidoreductase and the other encoded a cytochrome P450. Each was cloned separately into a baculoviral expression vector and assayed in vitro for potential roles in *exo*-brevicomin biosynthesis (Song et al., 2014b). The oxidoreductase converted deuterium-labelled (*Z*)-6-nonen-2-ol to the corresponding secondary ketone, using both NAD^+ and $NADP^+$ as an electron acceptor, and was thus named 'ZnoDH' for (*Z*)-6-*n*onen-2-ol *deh*ydrogenase. While substrate profiling for this enzyme is only preliminary, incubations of C10 fatty acids with recombinant enzyme under various conditions suggest that β-oxidation intermediates are not substrates (Song et al., 2014b). Thus, ZnoDH likely produces (*Z*)-non-6-en-2-one from a secondary alcohol, supporting the suggestion that *exo*-brevicomin is produced from a C9 hydrocarbon intermediate. The cytochrome P450, CYP6CR1, converts (*Z*)-non-6-en-2-one to 6,7-epoxynonan-2-ol—the predicted substrate for the final cyclization reaction. ZnoDH and CYP6CR1 are thus thought to act in the antepenultimate and penultimate steps of the pathway in *D. ponderosae*.

ZnoDH and CYP6CR1 seem well adapted to *exo*-brevicomin biosynthesis. In addition to mRNA levels that correlate closely with *exo*-brevicomin production, recombinant CYP6CR1 has an apparently very high substrate specificity: no products were observed after incubations with (*E*)-non-6-en-2-one or with the frontalin precursor, 6-methyl-hept-6-en-2-one (Song et al., 2014b). Thus, despite a superficial similarity in structures and intermediates, frontalin and *exo*-brevicomin are likely synthesized by different enzymes with similar activities.

The final step, cyclization of the epoxy-ketone, is likely enzyme-catalysed (Song et al., 2014a), though the enzyme responsible remains uncharacterized.

6.3 *trans*-Verbenol

trans-Verbenol (4,7,7-trimethylbicyclo[3.1.1]hept-3-en-2-ol) is a bicyclic monoterpenoid alcohol. It is the major female-produced aggregation pheromone component. Production likely begins upon feeding on a new host tree and ceases upon mating (Pureswaran et al., 2000). It is most likely produced via cytochrome P450-mediated hydroxylation of the host tree resin component, α-pinene. Female *D. ponderosae* produce a range of hydroxylated α-pinene products, including *cis*-verbenol and myrtenol (Greis et al., 1990; Pierce et al., 1987), indicating multiple P450 enzymes, one of which produces the pheromone component, (–)-*trans*-verbenol. There is evidence supporting de novo biosynthesis in *D. frontalis* (Renwick et al., 1973), so de novo biosynthesis, likely via the mevalonate pathway, is a formal possibility in *D. ponderosae* as well (C. Keeling, personal communication).

6.4 Other Components

In addition to the three major chemicals described earlier, other volatiles are present in *D. ponderosae* with more or less well-defined roles. *endo*-Brevicomin is produced by males in a pattern parallel to that of *exo*-brevicomin, but at much lower levels (Pureswaran et al., 2000). Given that its precursor, 6-(*E*)-nonen-2-one, is a stereoisomer of the *exo*-brevicomin precursor (Vanderwel et al., 1992), precursors to *endo*- and *exo*-brevicomin may be produced by the same enzymes, though the epoxidation steps may by catalysed by different P450s because CYP6CR1 does not appear to accept 6-(*E*)-nonen-2-one as a substrate (Song et al., 2014b) The role of *endo*-brevicomin as a semiochemical is unclear, though it is a significant aggregation pheromone component in *D. frontalis* (Sullivan et al., 2007). Verbenone is thought to be an auto-oxidation product of verbenol and may work as an antiaggregation signal along with frontalin (Lindgren and Miller, 2002). Other components, including *cis*-verbenol, ipsdienol and myrcenol (Hunt et al., 1986; Pierce et al., 1987), have less well-defined roles. Host tree volatiles also synergize beetle-produced pheromone components (Pureswaran et al., 2000).

7. SUMMARY

The combination of biochemical, genomic and postgenomic work has allowed tremendous advances in our understanding of pheromone production in pine bark beetles in the last two decades. The obvious next steps are to complete our understanding of the genes involved in *Ips* and *Dendroctonus* pheromone production and undertake a more complete gene expression profile using RNA-seq techniques, and these studies are underway in our laboratories. Comparative studies of orthologous genes such as CYP6CR1 and an expected ortholog accepting 6-(*E*)-nonen-2-one in *D. frontalis*, or variants of IDOLDH in *Ips* spp. could inform how genomes evolve and thus be useful to learn how bark beetles adapt over evolutionary time. Very little is known about the upregulation of enzyme activity during pheromone production and the different of effect of juvenile hormone on pheromone production in different *Ips* species likely involves addition regulation at the translational level. A huge challenge will be to use the information on pheromone production to develop techniques to control bark beetle infestations. Possible approaches will likely involve RNA interference, specific inhibitors or techniques not yet developed to modulate pheromone production in pine bark beetles.

REFERENCES

Arrese, E.L., Soulages, J.L., 2010. Insect fat body: energy, metabolism, and regulation. Annu. Rev. Entomol. 55, 207–225.

Aw, T., Schlauch, K., Keeling, C.I., et al., 2010. Functional genomics of mountain pine beetle (Dendroctonus ponderosae) midguts and fat bodies. BMC Genomics 11, 215.

Barkawi, L.S., Francke, W., Blomquist, G.J., Seybold, S.J., 2003. Frontalin: de novo biosynthesis of an aggregation pheromone component by *Dendroctonus* spp. bark beetles (Coleoptera: Scolytidae). Insect Biochem. Mol. Biol. 33, 773–788.

Bearfield, J.C., Keeling, C.I., Young, S., Blomquist, G.J., Tittiger, C., 2006. Isolation, endocrine regulation, and mRNA distribution of the 3-hydroxy-3-methylglutaryl coenzyme

A synthase (HMG-S) gene from the pine engraver, *Ips pini*. Insect Mol. Biol. 15 (2), 187–195.

Bearfield, J.C., Henry, A.G., Tittiger, C., Blomquist, G.J., Ginzel, M.D., 2009. Two regulatory mechanisms of monoterpenoid pheromone production in Ips spp. of bark beetles. J. Chem. Ecol. 35 (6), 689–697.

Bentz, B.J., Six, D.L., 2006. Ergosterol content of fungi associated with Dendroctonus ponderosae and Dendroctonus rufipennis (Coleoptera: Curculionidae, Scolytinae). Ann. Entomol. Soc. Am. 99 (2), 189–194.

Blomquist, G.J., Jurenka, R.A., Schal, C., Tittiger, C., 2004. Biochemistry and molecular biology of pheromone production. In: Gilbert, L.I., Iatrou, K., Gill, S.S. (Eds.), Comprehensive Insect Physiology, Biochemistry, Pharmacology, and Molecular Biology. Elsevier, Amsterdam.

Blomquist, G.J., Figueroa-Teran, R., Aw, M., et al., 2010. Pheromone production in bark beetles. Insect Biochem. Mol. Biol. 40 (10), 699–712.

Borden, J.H., 1985. Aggregation pheromones. In: Kerkut, G.A., Gilbert, L.I. (Eds.), Comprehensive Insect Physiology, Biochemistry and Pharmacology. 9, Pergamon Press, Oxford, pp. 257–285.

Brand, J.M., Bracke, J.W., Markovetz, A.J., Wood, D.L., Browne, L.E., 1975. Production of verbenol pheromone by a bacterium isolated from bark beetles. Nature 254 (5496), 136–137.

Brand, J.M., Bracke, J.W., Britton, L.N., Markovetz, A.J., Barras, S.J., 1976. Bark beetle pheromones: production of verbenone by a mycangial fungus of Dendroctonus frontalis. J. Chem. Ecol. 2 (2), 195–199.

Byers, J., 1981. Pheromone biosynthesis in the pine bark beetle, *Ips paraconfusus*, during feeding or exposure to vapours of host plant precursors. Insect Biochem. 11, 563–569.

Byers, J., Birgersson, G., 1990. Pheromone production in a bark beetle independent of myrcene precursor in host pine species. Naturwissenschaften 77, 385–387.

Byers, J., Wood, D., 1981. Antibiotic-induced inhibition of pheromone biosynthesis in a bark beetle. Science 213, 763–764.

Conn, J.E., Borden, J.H., Hunt, D.W.A., Holman, J., Whitney, H.S., 1984. Pheromone production by axenically reared *Dendroctonus ponderosae* and *Ips paraconfusus* (Coleoptera: Scolytidae). J. Chem. Ecol. 10 (2), 281–290.

Davis, T., Crippen, T., Hofstetter, R., Tomberlin, J., 2013. Microbial volatile emissions as insect semiochemicals. J. Chem. Ecol. 39 (7), 840–859.

Diaz, E., Cisneros, R., Zuniga, G., 2000. Comparative anatomical and histological study of the alimentary canal of the *Dendroctonus frontalis* (Coleoptera: Scolytidae) complex. Ann. Entomol. Soc. Am. 93 (2), 303–311.

Dickens, J.C., Oliver, J.E., Hollister, B., Davis, J.C., Klun, J.A., 2002. Breaking a paradigm: male-produced aggregation pheromone for the Colorado potato beetle. J. Exp. Biol. 205 (Pt 13), 1925–1933.

Domingue, M., Teale, S., 2008. The genetic architecture of pheromone production between populations distant from the hybrid zone of the pine engraver, *Ips pini*. Chemoecology 17, 255–262.

Domingue, M., Starmer, W., Teale, S., 2006. Genetic control of the enantiomeric composition of ipsdienol in the pine engraver, *Ips pini*. J. Chem. Ecol. 32 (5), 1005–1026.

Fish, R., Browne, L.E., Bergot, B.J., 1984. Pheromone biosynthetic pathways: conversion of ipsdienone to (-)-ipsdienol, a mechanism for enantioselective reduction in the male bark beetle, *Ips paraconfusus*. J. Chem. Ecol. 10, 1057–1064.

Francke, W., Schroder, F., Philipp, P., Meyer, H., Sinnwell, V., Gries, G., 1996. Identification and synthesis of new bicyclic acetals from the mountain pine beetle, Dendroctonus ponderosae Hopkins (Col.: Scol.). Bioorg. Med. Chem. 4 (3), 363–374.

Gilg, A.B., Bearfield, J.C., Tittiger, C., Welch, W.H., Blomquist, G.J., 2005. Isolation and functional expression of the first animal geranyl diphosphate synthase and its role in bark beetle pheromone biosynthesis. Proc. Natl. Acad. Sci. U.S.A. 102 (28), 9760–9765.

Gilg, A., Tittiger, C., Blomquist, G., 2009. Unique animal prenyltransferase with monoterpene synthase activity. Naturwissenschaften 96 (6), 731–735.

Ginzel, M.D., Bearfield, J.C., Keeling, C.I., McCormack, C.C., Blomquist, G.J., Tittiger, C., 2007. Antennally-mediated negative-feedback regulation of pheromone production in the pine engraver beetle, *Ips pini*. Naturwissenschaften 94, 61–64.

Greis, G., Leufven, A., LaFontaine, J.P., et al., 1990. New metabolites of a-pinene produced by the mountain pine beetle, *Dendroctonus ponderosae* (Coleoptera: Scolytidae). Insect Biochem. 20, 365–367.

Hall, G.M., Tittiger, C., Andrews, G.L., et al., 2002a. Midgut tissue of male pine engraver, *Ips pini*, synthesizes monoterpenoid pheromone component ipsdienol de novo. Naturwissenschaften 89 (2), 79–83.

Hall, G.M., Tittiger, C., Blomquist, G.J., et al., 2002b. Male jeffrey pine beetle, *Dendroctonus jeffreyi*, synthesizes the pheromone component frontalin in anterior midgut tissue. Insect Biochem. Mol. Biol. 32 (11), 1525–1532.

Hendry, L.B., Piatek, B., Browne, L.E., et al., 1980. *In vivo* conversion of a labelled host plant chemical to pheromones of the bark beetle, *Ips paraconfusus*. Nature 284, 485.

Hughes, P.R., 1973. *Dendroctonus*: production of pheromones and related compounds in response to host monoterpenes. Z. Angew. Entomol. 73, 294–312.

Hughes, P.R., 1974. Myrcene: a precursor of pheromones in Ips beetles. J. Insect Physiol. 20 (7), 1274–1275.

Hunt, D.W.A., Borden, J.H., Pierce Jr., H.D., Slessor, K.N., King, G.G.S., Czyzewska, E.K., 1986. Sex-specific production of ipsdienol and myrcenol by *Dendroctonus ponderosae* (Coleoptera: Scolytidae) exposed to myrcene vapors. J. Chem. Ecol. 12 (7), 1579–1586.

Ivarsson, P., Birgersson, G., 1995. Regulation and biosynthesis of pheromone components in the double spined bark beetle, *Ips duplicatus* (Coleoptera: Scolytidae). J. Insect Physiol. 41 (10), 843–849.

Ivarsson, P., Schlyter, F., Birgersson, G., 1993. Demonstration of de novo pheromone biosynthesis in *Ips duplicatus* (Coleoptera: Scolytidae): inhibition of ipsdienol and *E*-myrcenol production by compactin. Insect Biochem. Mol. Biol. 23, 655–662.

Ivarsson, P., Blomquist, G.J., Seybold, S.J., 1997. In vitro production of the pheromone intermediates ipsdienone and ipsenone by the bark beetles *Ips pini* (Say) and *I. paraconfusus* Lanier (Coleoptera: Scolytidae). Naturwissenschaften 84, 454–457.

Ivarsson, P., Tittiger, C., Blomquist, C., Borgeson, C.E., Seybold, S.J., Blomquist, G.J., 1998. Pheromone precursor synthesis is localized in the metathorax of *Ips paraconfusus* Lanier (Coleoptera: Scolytidae). Naturwissenschaften 85, 507–511.

Kavanagh, K., Jörnvall, H., Persson, B., Oppermann, U., 2008. Medium- and short-chain dehydrogenase/reductase gene and protein families. Cell. Mol. Life Sci. 65 (24), 3895–3906.

Keeling, C.I., Blomquist, G.J., Tittiger, C., 2004. Coordinated gene expression for pheromone biosynthesis in the pine engraver beetle, *Ips pini* (Coleoptera: Scolytidae). Naturwissenschaften 91, 324–328.

Keeling, C.I., Bearfield, J.C., Young, S., Blomquist, G.J., Tittiger, C., 2006. Effects of juvenile hormone on gene expression in the pheromone-producing midgut of the pine engraver beetle, *Ips pini*. Insect Mol. Biol. 15 (2), 207–216.

Keeling, C., Yuen, M., Liao, N., et al., 2013. Draft genome of the mountain pine beetle, Dendroctonus ponderosae Hopkins, a major forest pest. Genome Biol. 14 (3), R27.

Kurz, W.A., Dymond, C.C., Stinson, G., et al., 2008. Mountain pine beetle and forest carbon feedback to climate change. Nature 452 (7190), 987–990.

Lanne, B.S., Ivarsson, P., Johnsson, P., Bergstrom, G., Wassgren, A.B., 1989. Biosynthesis of 2-methyl-3-buten-2-ol, a pheromone component of Ips typographus (Coleoptera: Scolytidae). Insect Biochem. 19, 163–167.

Lindgren, B.S., Miller, D.R., 2002. Effect of verbenone on five species of bark beetles (Coleoptera: Scolytidae) in lodgepole pine forests. Environ. Entomol. 31 (5), 759–765.

Lockey, K.H., 1988. Lipids of the insect cuticle: origin, composition and function. Comp. Biochem. Physiol. B Biochem. Mol. Biol. 89 (4), 595–645.

Lu, F., 1999. Origin and endocrine regulation of pheromone biosynthesis in the pine bark beetles, Ips pini (Say) and Ips paraconfusus Lanier (Coleoptera: Scolytidae). Biochemistry, Ph.D. University of Nevada, Reno, NV. 152pp.

Ma, P., Ramaswami, S., 2003. Biology and ultrastructure of sex pheromone-producing tissue. In: Blomquist, G.J., Vogt, R.J. (Eds.), Insect Pheromone Biochemistry and Molecular Biology. Elsevier Academic Press, Amsterdam, pp. 19–51.

Martin, D., Bohlmann, J., Gershenzon, J., Francke, W., Seybold, S.J., 2003. A novel sex-specific and inducible monoterpene synthase activity associated with a pine bark beetle, the pine engraver, *Ips pini*. Naturwissenschaften 90 (4), 173–179.

Nakamura, C.E., Abeles, R.H., 1985. Mode of interaction of beta-hydroxy-beta-methylglutaryl coenzyme A reductase with strong binding inhibitors: compactin and related compounds. Biochemistry 24 (6), 1364–1376.

Nardi, J., Young, A., Ujhelyi, E., Tittiger, C., Lehane, M., Blomquist, G.J., 2002. Specialization of midgut cells for synthesis of male isoprenoid pheromone components in two scolytid beetles, *Dendroctonus jeffreyi* and *Ips pini*. Tissue Cell 34 (4), 221–231.

Perez, A.L., Gries, R., Gries, G., Oehlschlager, A.C., 1996. Transformation of presumptive precursors to frontalin and exo-brevicomin by bark beetles and the West Indian sugarcane weevil (Coleoptera). Bioorg. Med. Chem. 4 (3), 445–450.

Phillips, M.A., Croteau, R.B., 1999. Resin-based defenses in conifers. Trends Plant Sci. 4 (5), 184–190.

Pierce, H.D., Conn, J.E., Oehlschlager, A.C., Borden, J.H., 1987. Monoterpene metabolism in female mountain pine beetles, *Dendroctonus ponderosae* Hopkins, attacking ponderosa pine. J. Chem. Ecol. 13 (6), 1455–1480.

Pitman, G.B., Vite, J.P., Kinzer, G.W., Fentiman, A.F., 1969. Specificity of population-aggregating pheromones in *Dendroctonus*. J. Insect Physiol. 15, 363–366.

Pureswaran, D.S., Gries, R., Borden, J.H., Pierce Jr., H.D., 2000. Dynamics of pheromone production and communication in the mountain pine beetle, Dendroctonus ponderosae Hopkins, and the pine engraver, Ips pini (Say) (Coleoptera: Scolytidae). Chemoecology 10, 153–168.

Qiu, Y., Tittiger, C., Wicker-Thomas, C., et al., 2012. An insect-specific P450 oxidative decarbonylase for cuticular hydrocarbon biosynthesis. Proc. Natl. Acad. Sci. U.S.A. 109 (37), 14858–14863.

Rafaeli, A., Jurenka, R.A., 2003. PBAN regulation of pheromone biosynthesis in female moths. In: Blomquist, G.J., Vogt, R.J. (Eds.), Insect Pheromone Biochemistry and Molecular Biology. Elsevier Academic Press, Amsterdam, pp. 107–134.

Renwick, J.A.A., Hughes, P.R., Ty, T.D., 1973. Oxidation products of pinene in bark beetle, *Dendroctonus frontalis*. J. Insect Physiol 19 (9), 1735–1750.

Rudinsky, J., 1962. Ecology of Scolytidae. Annu. Rev. Entomol. 27, 411–446.

Sandstrom, P., Welch, W.H., Blomquist, G.J., Tittiger, C., 2006. Functional expression of a bark beetle cytochrome P450 that hydroxylates myrcene to ipsdienol. Insect Biochem. Mol. Biol. 36 (11), 835–845.

Sandstrom, P., Ginzel, M., Bearfield, J., Welch, W., Blomquist, G., Tittiger, C., 2008. Myrcene hydroxylases do not determine enantiomeric composition of pheromonal ipsdienol in Ips spp. J. Chem. Ecol. 34 (12), 1584–1592.

Seybold, S.J., Tittiger, C., 2003. Biochemistry and molecular biology of *de novo* isoprenoid pheromone production in the Scolytidae. Annu. Rev. Entomol. 48, 425–453.

Seybold, S.J., Ohtsuka, T., Wood, D.L., Kubo, I., 1995a. The enantiomeric composition of ipsdienol: a chemotaxonomic character for North American populations of *Ips* spp. in the *pini* subgeneric group (Coleoptera: Scolytidae). J. Chem. Ecol. 21, 995–1016.

Seybold, S.J., Quilici, D.R., Tillman, J.A., Vanderwel, D., Wood, D.L., Blomquist, G.J., 1995b. De novo biosynthesis of the aggregation pheromone components ipsenol and ipsdienol by the pine bark beetles *Ips paraconfusus* Lanier and *Ips pini* (Say) (Coleoptera: Scolytidae). Proc. Natl. Acad. Sci. U.S.A. 92 (18), 8393–8397.

Seybold, S.J., Bohlmann, J., Raffa, K.F., 2000. Biosynthesis of coniferophagous bark beetle pheromones and conifer isoprenoids: evolutionary perspective and synthesis. Can. Entomol. 132, 697–753.

Silverstein, R.M., Rodin, J.O., Wood, D.L., 1966. Sex attractants in frass produced by male Ips confusus in Ponderosa Pine. Science 154 (3748), 509–510.

Skiba, P.J., Jackson, L.L., 1994. Fatty acid elongation in the biosynthesis of (Z)-10-heptadecen-2-one and 2-tridecanone in ejaculatory bulb microsomes of Drosophila buzzatii. Insect Biochem. Mol. Biol. 24 (8), 847–853.

Song, M., Delaplain, P., Nguyen, T.T., et al., 2014a. exo-Brevicomin biosynthetic pathway enzymes from the mountain pine beetle, Dendroctonus ponderosae. Insect Biochem. Mol. Biol. 53, 73–80.

Song, M., Gorzalski, A., Nguyen, T., et al., 2014b. exo-Brevicomin biosynthesis in the fat body of the mountain pine beetle, Dendroctonus ponderosae. J. Chem. Ecol. 40 (2), 181–189.

Steele, C.L., Lewinsohn, E., Croteau, R., 1995. Induced oleoresin biosynthesis in grand fir as a defense against bark beetles. Proc. Natl. Acad. Sci. U.S.A. 92 (10), 4164–4168.

Sullivan, B., Shepherd, W., Pureswaran, D., Tashiro, T., Mori, K., 2007. Evidence that (+)-endo-Brevicomin is a male-produced component of the southern pine beetle aggregation pheromone. J. Chem. Ecol. 33 (8), 1510–1527.

Tillman, J.A., Holbrook, G.L., Dallara, P.L., et al., 1998. Endocrine regulation of de novo aggregation pheromone biosynthesis in the pine engraver, *Ips pini* (Say) (Coleoptera: Scolytidae). Insect Biochem. Mol. Biol. 28 (9), 705–715.

Tillman, J.A., Lu, F., Staehle, L., et al., 2004. Juvenile hormone regulates *de novo* isoprenoid aggregation pheromone biosynthesis in pine bark beetles, *Ips* spp. (Coleoptera: Scolytidae), through transcriptional control of HMG-CoA reductase. J. Chem. Ecol. 30 (12), 2459–2494.

Tittiger, C., 2003. Molecular biology of bark beetle pheromone production and endocrine regulation. In: Blomquist, G.J., Vogt, R. (Eds.), Insect Pheromone Biochemistry and Molecular Biology. Elsevier, San Diego, pp. 201–230.

Tittiger, C., Blomquist, G.J., Ivarsson, P., Borgeson, C.E., Seybold, S.J., 1999. Juvenile hormone regulation of HMG-R gene expression in the bark beetle *Ips paraconfusus* (Coleoptera: Scolytidae): implications for male aggregation pheromone biosynthesis. Cell. Mol. Life Sci. 55 (1), 121–127.

Tittiger, C., Barkawi, L.S., Bengoa, C.S., Blomquist, G.J., Seybold, S.J., 2003. Structure and juvenile hormone-mediated regulation of the HMG-CoA reductase gene from the Jeffrey pine beetle, *Dendroctonus jeffreyi*. Mol. Cell. Endocrinol. 199 (1–2), 11–21.

Vanderwel, D., 1994. Factors affecting pheromone production in beetles. Arch. Insect Biochem. Physiol. 25 (4), 347–362.

Vanderwel, D., Gries, G., Singh, S.M., Borden, J.H., Oehlschlager, A.C., 1992. (*E*)- and (*Z*)-6-nonen-2-one: biosynthetic precursors of *endo*;- and *exo*-brevicomin in two bark beetles (Coleoptera: Scolytidae). J. Chem. Ecol. 18 (8), 1389–1404.

Zhao, T., Axelsson, K., Krokene, P., Borg-Karlson, A.-K., 2015. Fungal symbionts of the spruce bark beetle synthesize the beetle aggregation pheromone 2-methyl-3-buten-2-ol. J. Chem. Ecol. 41 (9), 848–852.

CHAPTER SEVEN

Bark Beetle Research in the Postgenomic Era

C.I. Keeling
Simon Fraser University, Burnaby, BC, Canada

Contents

Abstract

Bark beetles (Coleoptera: Scolytinae) are pests of many forests around the world. The rapid development of molecular tools and high-throughput sequencing has permitted the examination of the biological processes of pheromone biosynthesis and host colonization at the genomic and molecular levels. I briefly review the available information on host defence toxicity and then review the currently available transcriptome and genome data available in bark beetles. Using the cytochrome P450 gene family as an example, I show how new sequence information and functional genomics can bring insights into how bark beetles have adapted to their highly defended host tissue where they spend most of their life. Some of the remaining questions to be answered in bark beetle research are also discussed.

ABBREVIATIONS

% e.e. percent enantiomeric excess
GSTs glutathione *S*-transferases
P450s cytochromes P450
PCWDE plant cell wall-degrading enzymes

Advances in Insect Physiology, Volume 50
ISSN 0065-2806
http://dx.doi.org/10.1016/bs.aiip.2015.12.004

1. INTRODUCTION

Bark beetles are species in the subfamily Scolytinae in the family Curculionidae (weevils) of the insect order Coleoptera. Coleoptera contains over 400,000 described species (Hunt et al., 2007), with an estimated 1.5 million in total (Stork et al., 2015), including 220 genera and 6000 species in Scolytinae. Many bark beetles are major forest pests, including the mountain pine beetle (*Dendroctonus ponderosae*) in western North America, the southern pine beetle (*Dendroctonus frontalis*) in south eastern North America and Central America, the European spruce bark beetle (*Ips typographus*), the European elm bark beetle (*Scolytus multistriatus*), and the coffee berry borer beetle (*Hypothenemus hampei*). As the name suggests, most bark beetles inhabit the bark and intracortical layer of trees where adults feed and reproduce, and where larvae develop (although the coffee berry borer beetle is one of the exceptions). This location includes the phloem tissue, in which sugars and amino acids are being transported, and the actively growing tissue of the cambium zone. The bark and intracortical tissues are protected by host chemical defences (Lindgren and Raffa, 2013), especially terpenoids in conifer trees (Franceschi et al., 2005; Keeling and Bohlmann, 2006; Zulak and Bohlmann, 2010). However, bark beetles have mechanisms to tolerate this harsh environment and successfully colonize their host. Pheromone-mediated aggregation and tolerance of host defences are the focus of much bark beetle research, particularly those beetles that infest conifer hosts. As we enter the postgenomic era of chemical ecology (Berenbaum, 2002), we can use the new tools of molecular biology and functional genomics to help answer our questions about bark beetle biology.

2. HOST DEFENCE TERPENE TOXICITY

The adaptation of bark beetles to intracortical life must come with a certain level of tolerance to host tree defences that is not likely to be found in insects that do not inhabit this niche. Conifer host trees display many different forms of physical and chemical defences against insects (Franceschi et al., 2005), of which terpenoids in the form of resin or volatiles have been most thoroughly studied with respect to bark beetles (Raffa, 2014). Terpenoids can act as chemical defences as well as physical defences when the flow of resin 'pitches out' the beetles or entombs them. Early work showed that *Dendroctonus brevicomis*, *D. monticolae*, and *D. jeffreyi* have higher tolerance to host- vs nonhost volatiles (Smith, 1963). Many host terpenoids also are semiochemicals for bark beetle behaviour (Seybold et al., 2006).

Much of the bark beetle literature assumes that host terpenoids are toxic to bark beetles. However, surprisingly little published data support this claim. Acute toxicity is usually tested by vapour exposure, topical application, or dietary supplementation in a controlled laboratory setting. Testing for nonacute affects is even more limited (eg, Manning and Reid, 2013). The data are unfortunately far from comprehensive in any bark beetle species. Also, only a small subset of compounds in the complex blend of monoterpenes, sesquiterpenes, and diterpenes conifers produce has been tested (Hall et al., 2013; Keeling and Bohlmann, 2006; Keeling et al., 2011). Coniferophagus bark beetles appear to have much higher tolerance to conifer host monoterpenes when compared to other insect species that do not encounter these compounds at such high levels in their natural environment (eg, Everaerts et al., 1988). The toxicity of monoterpenes varies between bark beetle species (Table 1). For instance, *Dendroctonus micans* and *Ips sexdentatus* appear to be more tolerant to host monoterpenes than other species (Delorme and Lieutier, 1990; Everaerts et al., 1988). However, such comparisons are often confounded by the different testing methods used for each species in different publications (eg, volatile exposure, topical application, dietary supplementation, measuring lethal time at a specific dose vs lethal dose at a specific time, etc.).

The absolute and relative toxicity of specific host terpenoids differ for each beetle species. Limonene, myrcene, α-pinene, and 3-carene often appear to be the most toxic of the compounds tested, but this is not true for all beetle species. Many of these terpenoid compounds are chiral but the enantiomers are not always tested separately nor is the % enantiomeric excess (% e.e.) specified. In addition to monoterpenes, conifer oleoresin also contains similar quantities of diterpene resin acids and lesser quantities of sesquiterpenes (Keeling and Bohlmann, 2006). A more thorough examination of the toxicity of terpenoids to bark beetles should include these other components, as the current literature on these compounds is quite limited (Kopper et al., 2005).

The mechanism of acute toxicity has not been established for these terpenoids in bark beetles nor have the mechanisms of any possible nonacute effects. However, in other beetle species, the mechanism of acute toxicity for monoterpenes is the inhibition of acetylcholinesterase (López and Pascual-Villalobos, 2010, 2015). Terpenes are found at such high concentrations in conifers that they may also act as lipophilic solvents that disrupt

Table 1 Summary of Experiments Examining the Toxicity of Host Defence Terpenoids in Bark Beetles

Compound	*Dendroctonus armandi*	*D. brevicomis*	*D. frontalis*	*D. micans*	*D. ponderosae*	*D. rufipennis*	*D. simplex*	*Ips pini*	*I. sexdentatus*	*I. typographus*	*Scolytus ventralis*	*Tomicus piniperda*
α-Pinene	a	b	c	[a]–d	e, f	g	**g**	**h**	[a]–j	**d**	**k**	j
β-Pinene	a	b	c	[a]–d		g	g	**h**	[a]–j	d	k	j
Limonene	a	**b**	**c**	[a]–d	e, f	**g**	**g**	h	[a]–j	**d**	**k**	**j**
Limonene oxide			c									
Limonene dioxide			**c**									
Myrcene		**b**	c	[a]–d	e	**g**	**g**	h		**d**	**k**	
Terpinolene					e			**h**				
3-Carene	a	**b**	**c**	[a]–d		g	g		[a]–j	d	k	**j**
Camphene			c			g	g		[a]–j		k	**j**
α-Phellandrene			c									
β-Phellandrene			c			**g**	g				k	
α-Terpinene				[a]–d						**d**		
Abietic acid								[a]–i				
Dehydroabietic acid								[a]–i				
Isopimaric acid								[a]–i				

[a]Found to be nontoxic under conditions and concentrations tested.
When study provided relative toxicity data between tested compounds, the top three most toxic compounds are highlighted with bold letters.
The stereochemistry of the chiral compounds assayed was indicated in a few, but not all studies, and differences in toxicity between enantiomers were sometimes observed. Enantiomeric forms are grouped together here.
a, Dai et al. (2015b); b, Smith (1965); c, Coyne and Lott (1976); d, Everaerts et al. (1988); e, Reid and Purcell (2011); f, Manning and Reid (2013); g, Werner (1995); h, Raffa and Smalley (1995); i, Kopper et al. (2005); j, Delorme and Lieutier (1990); k, Raffa et al. (1985).

membranes (Mendanha and Alonso, 2015). Terpenoid compounds in several conifer species have juvenile hormone activity against non-Scolytinae insects such as *Pyrrhocoris apterus* bugs (Sláma and Williams, 1965). In addition, two diterpenes identified from the angiosperm species *Lindera erythrocarpa* and *Solidago serotina* disrupt the juvenile hormone–receptor complex in mosquitos, causing both acute and reproductive effects (Lee et al., 2015). Juvenile hormone itself is a terpenoid, and it is not known whether other host terpenoids bark beetles are exposed to have juvenile hormone activity. What is known in bark beetles is that ultrastructural changes in midgut tissue occur when *Dendroctonus valens* adults are exposed to α-pinene or myrcene (López et al., 2011).

Tolerance implies that the beetles have one or more biochemical adaptations to allow them to survive in a host defence-rich environment. The biochemical tolerance mechanisms may include detoxification of compounds via oxidation (eg, with cytochrome P450 enzymes (P450s)) to more readily excreted derivatives, inactivation by conjugation to transport proteins (eg, with glutathione *S*-transferases (GSTs)), or the evolution of proteins or processes to prevent the cellular uptake of host defence compounds in the first place. Oxidation by P450 enzymes is an important detoxification mechanism and bark beetle P450 enzymatic activity appears to vary with host (Sturgeon and Robertson, 1985). Studies examining the volatile metabolites of bark beetles when exposed to host monoterpenes or phloem typically see an array of oxidized products (eg, Hughes, 1975; Libbey et al., 1985; Pierce et al., 1987; Renwick et al., 1976), implying that multiple P450s with some substrate selectivity are involved (Fig. 1).

The distinction between host defence detoxification and pheromone biosynthesis can sometimes be blurred. In several instances, a host defence compound has been adopted by a bark beetle species as the precursor for the biosynthesis of an aggregation pheromone or other semiochemical (eg, *cis*- and *trans*-verbenol, ipsdienol, ipsenol, verbenone, and myrcenol; Seybold et al., 2000). The use of a modified host defence compound as an aggregation pheromone is more than just ironic, it tells conspecific beetles that a host tree is present and its defences are being overcome. In some cases, the beetles are able to make these terpenoid precursors themselves *de novo*, such as myrcene in ipsdienol biosynthesis (Gilg et al., 2009). This ability may impart an advantage to those beetles arriving first on the tree. The pioneer beetle's behaviour is risky, with beetle energy status, arrival time, and arrival sequence (early pioneer beetles vs those that arrive later) all playing a role in success (Latty and Reid, 2009, 2010). An adaptation to produce an

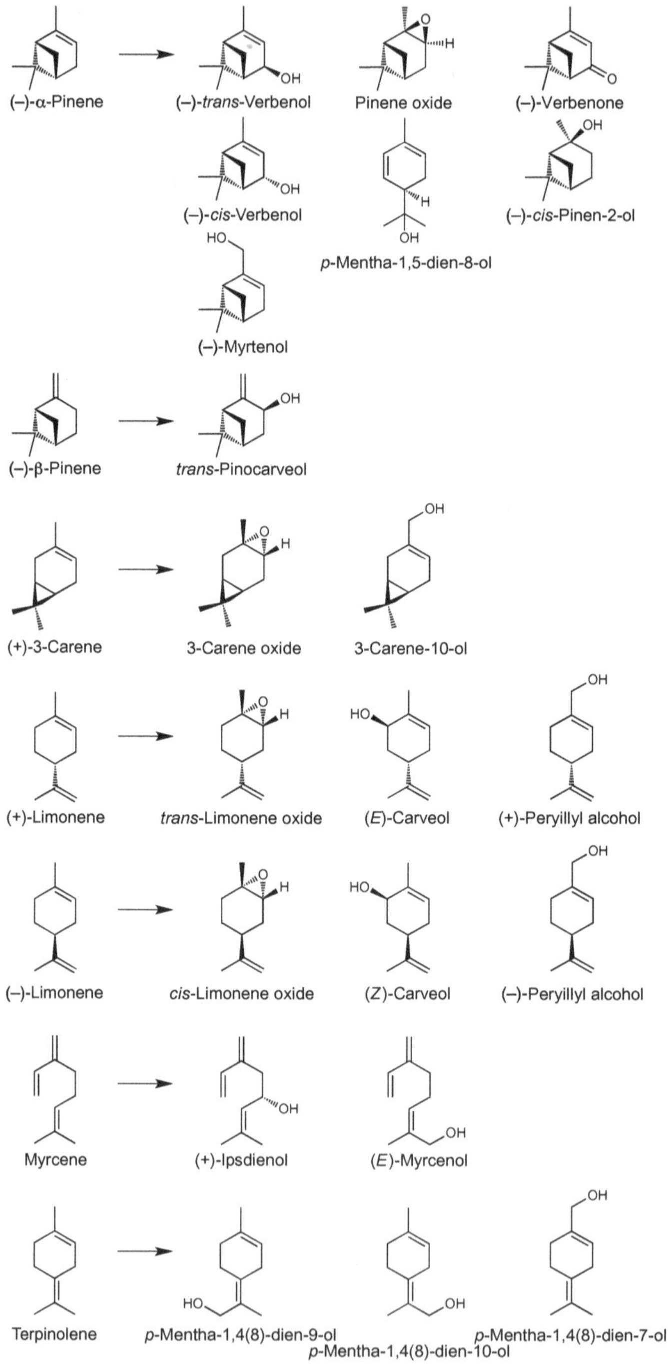

Fig. 1 Examples of monoterpene metabolites from bark beetles. Chemical structures of representative host monoterpenes and their metabolites from bark beetles. Information taken predominantly from Pierce et al. (1987) and Keeling et al. (2013b). The metabolites show evidence for the hydroxylation, epoxidation, and oxidation of host monoterpenes in regio- and stereospecific ways. These are all modifications that P450 enzymes can catalyse.

aggregation pheromone *de novo* without having to consume and digest the host first could confer an advantage in host colonization, especially to the pioneer beetles. How this developed is an interesting evolutionary question.

Bark beetles live with a milieu of microorganisms under the bark including bacteria (Adams et al., 2013; Morales-Jiménez et al., 2009, 2012), yeast (Zúñiga, 2016), fungi (Khadempour et al., 2012; Lee et al., 2006), and nematodes (Grucmanová and Holuša, 2013). At least a few of these have symbiotic relationships with bark beetles and they may reduce bark beetles' exposure to tree defence compounds (Six, 2012). Therefore, understanding the effects of host compounds on bark beetle colonization should include the effects these compounds have on these other players in the bark beetle system. Some microorganisms are sensitive to terpenoids (Kopper et al., 2005), while others are able to metabolize them (Adams et al., 2013; Boone et al., 2013). Some of the genes involved in fungal tolerance to terpenoids are being characterized (Lah et al., 2013; Wang et al., 2013, 2014).

3. GENOMICS IN BARK BEETLES

3.1 Genome Sequences in Bark Beetles and Other Beetles

The rapid advancement of sequencing and other high-throughput technologies in genomics, proteomics, and metabolomics have begun to provide new information and new resources in Coleoptera. To date, only nine genomes have been sequenced in Coleoptera (Table 2). We are fortunate that two of the coleopteran genomes sequenced to date are from Scolytinae: the mountain pine beetle (*D. ponderosae*) (Keeling et al., 2013c) and the coffee borer beetle (*H. hampei*) (Vega et al., 2015). The genome sequence for the red flour beetle (*Tribolium castaneum*) has been available for some time (Richards et al., 2008). The other coleopteran genome sequences are only recently publically available on NCBI and are just beginning to be published. These include the Asian long-horned beetle (*Anoplophora glabripennis*), the bull-headed dung beetle (*Onthophagus taurus*), a burying beetle (*Nicrophorus vespilloides*) (Cunningham et al., 2015), the Colorado potato beetle (*Leptinotarsa decemlineata*), the emerald ash borer (*Agrilus planipennis*), and a scarab beetle (*Oryctes borbonicus*). We expect to see many more beetle genome sequences in the near future as the i5K (i5K Consortium, 2013) and other insect genome sequencing initiatives begin to bear fruit. We are thus on the cusp of a deluge of additional sequence data in Coleoptera that will allow us to further functional genomics studies in beetles. These genome

Table 2 Summary of Available Genome Assemblies in Coleoptera

Species	Common Name	NCBI Accession	References
Agrilus planipennis	Emerald ash borer	PRJNA230921	Unpublished
Anoplophora glabripennis	Asian long-horned beetle	PRJNA167479	Unpublished
Dendroctonus ponderosae	**Mountain pine beetle**	**PRJNA162621 (male)** and **PRJNA179493 (female)**	**Keeling et al. (2013c)**
Hypothenemus hampei	**Coffee borer beetle**	**PRJNA279497**	**Vega et al. (2015)**
Leptinotarsa decemlineata	Colorado potato beetle	PRJNA171749	Unpublished
Nicrophorus vespilloides	Burying beetle	PRJNA284849	Cunningham et al. (2015)
Onthophagus taurus	Bull-headed dung beetle	PRJNA167478	Unpublished
Oryctes borbonicus	Scarab beetle	PRJNA293509	Unpublished
Tribolium castaneum	Red flour beetle	PRJNA12540	Richards et al. (2008)

The bark beetles are indicated in bold.

assemblies allow genome-wide analyses of sequence variation between species for evolution studies and within species for population studies (eg, Janes et al., 2014).

3.2 Transcriptome Data in Bark Beetles

More abundant than genome sequences in bark beetle research are transcriptome data, predominantly from conifer-feeding beetles. This data usually originate from tissues of pheromone biosynthesis, detoxification, and olfaction and provide a comprehensive description of the expressed genes in the tissues of interest. The first such study was for the midguts of *Ips pini* (Eigenheer et al., 2003). This was followed by transcript sequencing in *D. ponderosae*, first for the midgut and fat bodies (Aw et al., 2010), and then for a more comprehensive panel from multiple tissue sources, life stages, and treatments (Keeling et al., 2012), and then several other species (Table 3).

Table 3 Summary of Available Transcriptome Assemblies and Other Transcriptome Data in Scolytinae

Species	Common Name	NCBI Accession	References	Tissues
Dendroctonus armandi	Chinese white pine beetle	Not available	Dai et al. (2015b)	Larvae, pupae, and adults
D. frontalis	Southern pine beetle	PRJNA187208	Unpublished	Panel of life stages and treatments
D. ponderosae	Mountain pine beetle	PRJNA37293	Aw et al. (2010)	Adult midgut and fat bodies
		PRJNA189792 and PRJNA189795	Keeling et al. (2012)	Panel of life stages and treatments
		PRJNA269763	Keeling et al. (2016)	Adult midgut and fat bodies
		PRJNA203305 (SRA data only)	Robert et al. (2013)	Adults
		PRJNA28530 (SRA data only)	Unpublished	Larvae
D. valens	Red turpentine beetle	Not available	Gu et al. (2015)	Antennae
Euwallacea fornicatus	Polyphagous shot hole borer	PRJNA260703	Unpublished	Unspecified
Hypothenemus hampei	Coffee borer beetle	PRJNA279497	Vega et al. (2015)	Unspecified
Ips pini	Pine engraver beetle	CB407466–CB409136 (ESTs only)	Eigenheer et al. (2003)	Adult midgut
I. typographus	European spruce beetle	PRJNA178930	Andersson et al. (2013)	Adult antennae
Tomicus yunnanensis	Yunnan shoot borer	PRJDB2098 (SRA data only)	Zhu et al. (2012a)	Adult head
		PRJNA175397 (SRA data only)	Zhu et al. (2012b)	Adults

Unless otherwise specified, NCBI accession number refers to BioProject accession for the transcript assembly.

Targeted identification of GSTs in *Dendroctonus armandi* (Dai et al., 2015a), and P450s in *Ips paraconfusus* (Huber et al., 2007), *D. armandi* (Dai et al., 2014, 2015b), *D. valens* (López et al., 2013), and *Dendroctonus rhizophagus* (Cano-Ramírez et al., 2013) have also been completed.

The behaviour of bark beetles is mediated in large part by semiochemicals including kairomones from host and nonhost trees, pheromones from conspecifics or allospecifics, and volatiles produced by symbiotic fungi. Bark beetle antennae are sensitive to compounds from all of these various sources (Andersson et al., 2009; Jaramillo et al., 2013; McLeod et al., 2005; Mustaparta et al., 1985; Njihia et al., 2014; Pureswaran and Borden, 2004; Pureswaran et al., 2000, 2004; Tommeras et al., 1984; Whitehead et al., 1989; Wilson et al., 1996). Thus, identifying and characterizing the genes of the olfactory system of bark beetles are important components in the study of host colonization. Transcriptomes from olfactory tissues have been sequenced in *D. ponderosae* (Keeling et al., 2012), *D. valens* (Gu et al., 2015), *I. typographus* (Andersson et al., 2013), and *Tomicus yunnanensis* (whole head with antennae; Zhu et al., 2012a). These studies have described many of the genes and gene families in odorant perception, such as pheromone/odorant receptors, gustatory receptors, ionotropic receptors, pheromone/odorant-binding proteins, chemosensory proteins, and sensory neuron membrane proteins (de Bruyne and Baker, 2008; Renou, 2014). In addition, a few studies have identified genes that may be involved in odorant degradation (Cano-Ramírez et al., 2013; Gu et al., 2015; Keeling et al., 2013b). These sequence resources provide the important raw information to inform functional genomic studies and comparative studies of many bark beetle processes.

4. FUNCTIONAL GENOMICS IN BARK BEETLES

Functional genomics studies have predominantly focussed on the *Ips* and *Dendroctonus* genera in bark beetles. Following from the prior biochemical studies in pheromone biosynthesis in *Ips* species (reviewed by Blomquist et al., 2010), the functional genomics era began in bark beetle research with the cloning of the first gene in pheromone biosynthesis in *I. paraconfusus*, 3-hydroxy-3-methylglutaryl-CoA reductase gene (*HMG-R*), which encodes a highly regulated enzyme in the mevalonate pathway (Tittiger et al., 1999). Subsequent to this study, many studies examining the genes of pheromone biosynthesis were undertaken in *D. jeffreyi* and several *Ips* species (Hall et al., 2002a,b; Keeling et al., 2004; Tittiger et al., 1999, 2000, 2003).

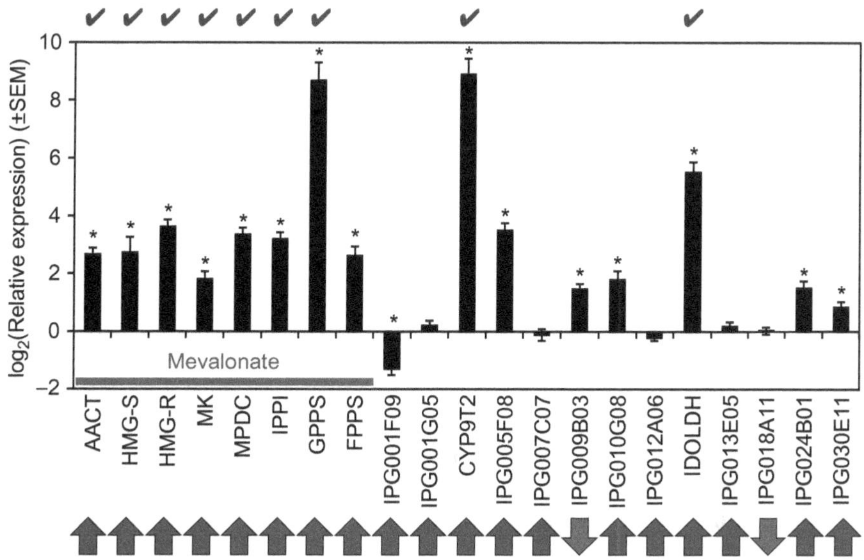

Fig. 2 Candidate genes in *Ips pini* pheromone biosynthesis. qRT–PCR analysis of relative basal mRNA levels in midguts between sexes (males/females) of acetone and juvenile hormone III-treated beetles. Expression values are $\log_2$ for the ratio of the transcript levels of acetone-treated males over acetone-treated female beetles. *Asterisks* indicate significant difference between sexes. Checkmarks indicate genes with confirmed role in ipsdienol biosynthesis. *Blue line* indicates genes of the mevalonate pathway involved in the early steps of ipsdienol biosynthesis. *Arrows* indicate the direction of the significant change in transcript levels when males are treated with juvenile hormone III, compared to acetone-treated males. *Adapted from Keeling, C.I., Bearfield, J.C., Young, S., Blomquist, G.J., Tittiger, C., 2006. Effects of juvenile hormone on gene expression in the pheromone-producing midgut of the pine engraver beetle,* Ips pini. *Insect Mol. Biol., 15, 207–216.* (See the colour plate.)

The first 'omics' study, a small scale expressed sequence tag (EST) sequencing project in *I. pini*, was completed in 2003 (Eigenheer et al., 2003). This sequencing project permitted the production of cDNA microarrays used to examine changes in gene expression with phloem feeding (Tittiger et al., 2005) and juvenile hormone III treatment (Keeling et al., 2006). The latter study identified candidate genes in ipsdienol pheromone biosynthesis based upon sex-specificity and induction with juvenile hormone III treatment (Fig. 2). Some of these candidates were later confirmed to play a role in pheromone biosynthesis through heterologous expression and functional characterization (Figueroa-Teran et al., 2012; Gilg et al., 2005, 2009; Sandstrom et al., 2006).

Building on the successes with *I. pini*, a similar microarray approach in *D. ponderosae* identified candidate genes in pheromone biosynthesis (Aw et al., 2010). Two candidate genes (*CYP6CR1* and *ZnoDH*) were subsequently confirmed to be involved in *exo*-brevicomin pheromone biosynthesis (Song et al., 2014). With the arrival of RNA-seq for gene expression studies (Weber, 2015), additional studies in pheromone biosynthesis and host colonization were completed (Keeling et al., 2016; Robert et al., 2013), identifying more candidate genes in pheromone biosynthesis, juvenile hormone signalling, digestion, and detoxification. Concurrent with these later studies, the development of quantitative proteomics techniques such as iTRAQ (Ross et al., 2004) permitted the analysis of protein levels in different tissues and life stages for pheromone biosynthesis, host colonization, and larval cold tolerance in *D. ponderosae* (Bonnett et al., 2012; Keeling et al., 2016; Pitt et al., 2014).

5. CYTOCHROMES P450 IN BARK BEETLES

The accumulating transcriptome and genome data in Scolytinae are allowing researchers to begin to understand how bark beetles have adapted to their environment and how they may differ from other insects. The heme-containing cytochrome P450 gene family with predominantly monooxygenase activity is one large gene family central to this process. Although some P450 enzymes in insects are responsible for hormone biosynthesis and other conserved biochemical processes, many are associated with plant–insect interactions (Feyereisen, 2011; Schuler, 2011) and are thought to play an integral role in host defence detoxification and pheromone biosynthesis in bark beetles. Consequently, the P450 gene family has been studied the most in bark beetles, followed by GSTs and plant cell wall-degrading enzymes (Dai et al., 2015a; Keeling et al., 2013c; Pauchet et al., 2010).

Over half of the P450s described to date in Coleoptera originate from Scolytinae. The description of P450 gene families in the different species have not been completed to the same level; some are based on genome annotation, whereas others are from transcriptome analysis, typically from specific rather than comprehensive tissue sources. From Scolytinae genome sequences, 86 P450s have been identified in *D. ponderosae* (Keeling et al., 2013c) and 54 in the coffee borer beetle (Vega et al., 2015). These are less than the 134 described in *T. castaneum* (Feyereisen, 2006, 2011) but within the range found in other insects (Nelson, 2011). Even with these sampling limitations and biases, we can begin to hypothesize about P450 functions based upon their presence/absence in different species, as well as their expression patterns within a species. A phylogeny of beetle P450s, along with two out-group species *Apis mellifera* (European honey bee) and *Bombyx mori* (silk moth), is shown in Fig. 3.

Insects contain four clades of P450s: the mitochondrial, and the CYP2, CYP3, and CYP4 clades (Feyereisen, 2006). The mitochondrial and CYP2 clades show 1:1 orthology between the different species, reflecting the conserved roles that most of these P450s have in hormone biosynthesis and metabolism (Feyereisen, 2011). The CYP3 and CYP4 clades, however, show apparently lineage-specific diversification (Fig. 3). In the CYP3 clade, several bark beetle and/or Curculionidae-specific (based upon information available at this time) families are apparent, particularly in the CYP6, CYP9, CYP345, and CYP347 families. Because the only non-Scolytinae Curculionidae with comprehensive P450 information, the white pine weevil *Pissodes strobi* (NCBI PRJNA186387 and unpublished data), is also a conifer bark-feeding insect, it is hard to distinguish between phylogenetic vs functional conservation. For example, in several cases, a clade of P450s has the coffee borer beetle P450 more basal than the *P. strobi* P450. This does not make sense based purely on beetle phylogeny, but may indicate that this clade of P450s has evolved differently for purposes unique to the habitat of phloem feeding. A larger dataset of P450s from more species within Curculionidae with different habitats is needed to provide strength to these types of hypotheses. The CYP4 clade also shows diversification in beetles and some lineage-specific clades. Although no CYP4s have been functionally characterized in bark beetles, CYP4s in insects are implicated in pheromone metabolism, insecticide resistance, and detoxification (Feyereisen, 2006), and CYP4Gs are involved in cuticular hydrocarbon biosynthesis (Qiu et al., 2012). The CYP4 clade is present in organisms other than insects and is often involved in fatty acid metabolism such as the ω-hydroxylation of fatty acids (Hardwick, 2008).

Several studies have shown that juvenile hormone III regulates P450 genes and other genes in pheromone biosynthesis (Keeling et al., 2006, 2016; Seybold and Vanderwel, 2003; Tillman et al., 2004; Yu et al., 2015), but how are the expression levels of genes encoding P450s involved in detoxification regulated? Phloem feeding and/or topical juvenile hormone III treatment induce the expression of P450s in *I. paraconfusus* (Huber et al., 2007), *I. pini* (Keeling et al., 2006; Sandstrom et al., 2008; Tittiger et al., 2005), *D. ponderosae* (Aw et al., 2010; Keeling et al., 2016;

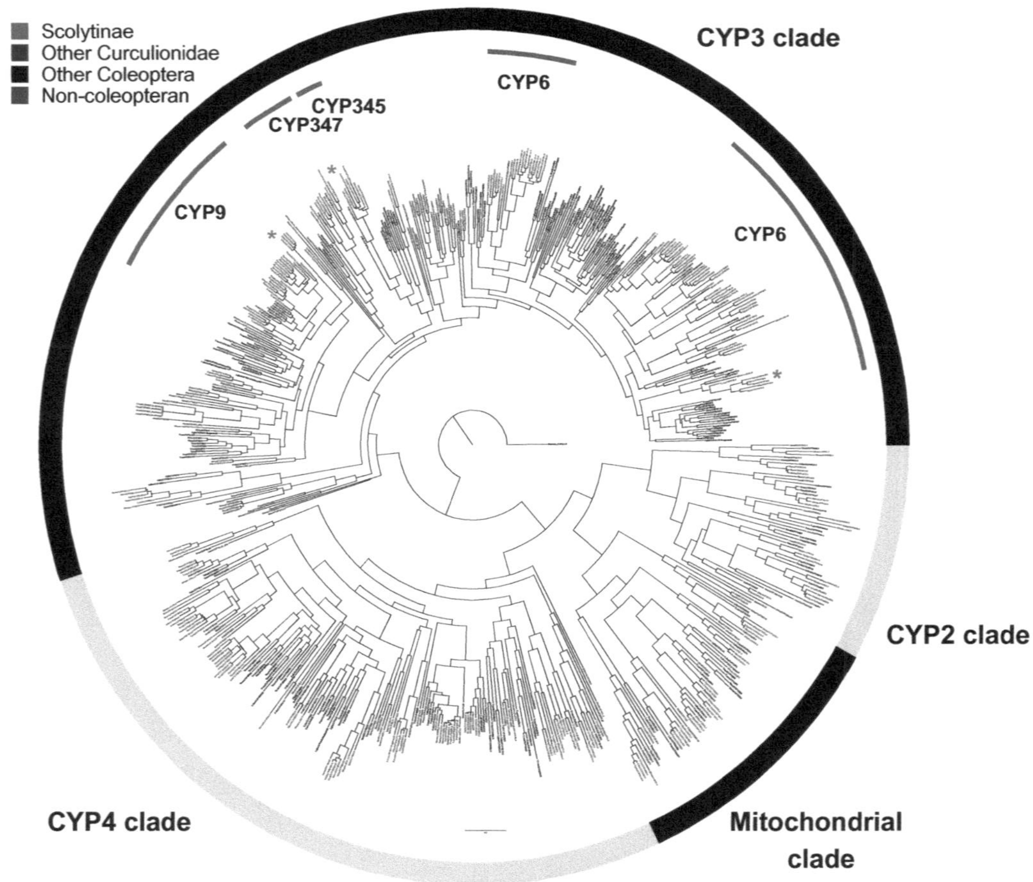

Fig. 3 See legend on opposite page.

Robert et al., 2013), and *D. rhizophagus* and *D. valens* (Obregón-Molina et al., 2015). Exposing beetles to monoterpenes induces the expression of specific P450s in *D. armandi* (Dai et al., 2015b), and in the antenna and gut of *D. rhizophagus* (Cano-Ramírez et al., 2013) and *D. valens* (López et al., 2013). These studies show that certain host compounds can induce expression of specific P450s in tissue- and sex-specific ways. However, no biochemical data are available yet to show that these host compounds are substrates for the P450s they induce. The mechanism for substrate-induced P450 gene expression is not known in bark beetles, nor has it been explored to date. A similar observation is seen with the induction of human P450s with various drugs and xenobiotics, and the mechanism is still far from understood (Zhu, 2010). It is likely to involve a suite of nuclear and intracellular receptors with specificity to a range of substrates that orchestrate the regulation of the specific P450 transcripts. Because a bark beetle is exposed to a much greater diversity of host defence compounds than the number of possible receptors, we would predict that the receptors may bind a range of substrates, and some of these may not be substrates for the P450s these receptors ultimately induce. *Tribolium castaneum* has 21 identified nuclear receptors and many of these are involved in metamorphosis and reproduction (Tan and Palli, 2008). *Dendroctonus ponderosae* has approximately 30 putative nuclear receptors, (27 and 33 gene models in the male and female genomes, respectively) (Keeling et al., 2013c). Many of these *D. ponderosae* receptors are orthologous to *T. castaneum* and other insect receptors, but some orphan receptors may be involved in the induction of gene expression due to host defence compounds.

Fig. 3 Cytochrome P450 proteins in bark beetles and other insects. Phylogeny of P450s from bark beetles, other beetles, the honey bee, and the silk moth. An alignment was prepared with MAFFT version 7.123 (Katoh and Standley, 2013) and the phylogeny subsequently prepared with FastTree 2 version 2.1.6 (Price et al., 2010) and displayed with FigTree version 1.4.2 (http://tree.bio.ed.ac.uk/software/figtree/). Proteins less than 200 amino acids were excluded from the alignment and phylogeny. The phylogeny was rooted with human CYP3A4. Abbreviations: Agla, *Anoplophora glabripennis*; Amel, *Apis mellifera*; Baen, *Brassicogethes aeneus*; Blon, *Brontispa longissima*; Bmor, *Bombyx mori*; Cmon, *Cryptolaemus montrouzieri*; Darm, *Dendroctonus armandi*; Dfro, *Dendroctonus frontalis*; Dhel, *Dastarcus helophoroides*; Dpon, *Dendroctonus ponderosae*; Drhi, *Dendroctonus rhizophagus*; Dval, *Dendroctonus valens*; Dvir, *Diabrotica virgifera virgifera*; Hham, *Hypothenemus hampei*; Icon, *Ips confusus*; Ipar, *Ips paraconfusus*; Ipin, *Ips pini*; Ldec, *Leptinotarsa decemlineata*; Pdiv, *Phyllopertha diversa*; Pjap, *Propylea japonica*; Pstr, *Pissodes strobi*; Tcas, *Tribolium castaneum*; Tmol, *Tenebrio molitor*; Tyun, *Tomicus yunnanensis*. The functionally characterized bark beetle P450s discussed in the text are highlighted with *green asterisks*. (See the colour plate.)

6. FUNCTIONAL CHARACTERIZATION OF BARK BEETLE GENES

The nucleotide sequences and transcript levels are only the beginning in describing bark beetle genes. Information on the actual function of the gene products through heterologous expression and *in vitro* assays, or targeted knock-out or knock-down *in vivo*, are necessary to fully describe gene function. Relatively few enzymes have been functionally characterized to date in bark beetles. Most of these are involved in pheromone biosynthesis, including an alcohol dehydrogenase (IDOLDH) (Figueroa-Teran et al., 2012) and a unique geranyl diphosphate synthase/myrcene synthase (Gilg et al., 2005, 2009) involved in *I. pini* ipsdienol pheromone biosynthesis; a geranylgeranyl diphosphate synthase and a bifunctional geranyl diphosphate synthase/farnesyl diphosphate synthase involved in *D. ponderosae* frontalin biosynthesis (Keeling et al., 2013a); and a (*Z*)-6-nonen-2-ol dehydrogenase (ZnoDH) involved in *D. ponderosae exo*-brevicomin biosynthesis (Song et al., 2014). In addition, three P450s have been functionally characterized in bark beetles, CYP9T2 (CYP9T1 and CYP9T3 orthologues), CYP345E2, and CYP6CR1.

Like *IDOLDH* (Figueroa-Teran et al., 2012), geranyl diphosphate synthase/myrcene synthase (Gilg et al., 2005), and other members of the mevalonate pathway (Fig. 2), the *CYP9T2* transcript stood out in the *I. pini* microarray studies as being male-specific and juvenile hormone III inducible (Keeling et al., 2006), consistent with ipsdienol pheromone biosynthesis. This P450, from the western North American population of *I. pini*, hydroxylates myrcene to produce the chiral male aggregation pheromone component, ipsdienol (Fig. 1), with 62% e.e. (−) (Sandstrom et al., 2006). CYP9T3, from the eastern North American population of *I. pini*, produces 77.4% e.e. (−) ipsdienol (Song et al., 2013). This study also found that CYP9T2 and CYP9T3 had *in vitro* activity with (+)- and (−)-α-pinene, (+)-3-carene, and (+)-limonene, but not α-phellandrene, (−)-β-pinene, γ-terpinene, or terpinolene. The orthologous CYP9T1 in *Ips confusus* also produces ipsdienol when assayed with myrcene, but with 70% e.e. (−) (Sandstrom et al., 2008). These significant differences in product stereochemistry result from just a few sequence differences between enzymes. The P450 phylogeny (Fig. 3) shows that to date, only *Ips* species have CYP9Ts, suggesting that this may be a specific P450 for pheromone biosynthesis in *Ips*. However, *D. ponderosae* and *D. brevicomis* produce ipsdienol with >94% e.e. (+) (Byers, 1982; Hunt et al., 1986), suggesting that this product must originate from other P450s in *Dendroctonus*.

Interestingly, the % e.e. of ipsdienol in *Ips* beetles does not reflect the stereospecificity of the P450s, but rather there is a distribution of different % e.e. across each population. Western populations of *I. pini* produce an average of 90% e.e. (−) ipsdienol, but beetles in south western British Columbia average of 32% e.e. (+), and eastern populations averaged 14% e.e. (+) (Miller et al., 1989). *I. paraconfusus* releases 87.6% e.e. (+) ipsdienol (Fish et al., 1984). The population differences in ipsdienol stereochemistry in *I. pini* originate from a complicated autosomal dominance and an X-linked effect (Domingue and Teale, 2007; Domingue et al., 2006). A mechanism for these differences was originally hypothesized by Fish et al. (1984) to involve the achiral ketone ipsdienone. Microarray gene expression analysis in *I. pini* (Keeling et al., 2006) identified a candidate male-specific alcohol dehydrogenase (IDOLDH) that was subsequently shown to specifically oxidize (−)-ipsdienol to the achiral ipsdienone (Figueroa-Teran et al., 2012).

Using a similar functional genomics approach but with *D. ponderosae* microarrays, Aw et al. (2010) identified a candidate P450 in *exo*-brevicomin pheromone biosynthesis. This CYP6CR1 was later established to catalyse the epoxidation of the penultimate precursor in *exo*-brevicomin biosynthesis (Song et al., 2014). This protein clusters with other members of the CYP6CR subfamily from other *D. ponderosae*, *D. frontalis*, other bark beetles, the white pine weevil, and the Asian long-horned beetle (Fig. 3). Although not all of these beetles are known to produce *exo*-brevicomin, it is possible that this subfamily of P450s have activities with other substrates. Interestingly, the non-Scolytinae members in this P450 subfamily all are also bark-feeding beetles.

The *CYP345E2* gene was identified through tissue-specific EST sequencing to be a highly expressed antenna-specific P450 in *D. ponderosae* (Keeling et al., 2012). It can epoxidize or hydroxylate host terpene odorants such as (+)-3-carene, (−)-camphene, and terpinolene, and both enantiomers of α-pinene, β-pinene, and limonene (Keeling et al., 2013b). *D. ponderosae* also has a CYP345E1 and a CYP345E3, but these are not antenna specific. Members of the CYP345E subfamily can be found in other *Dendroctonus* species as well as the weevil *P. strobi* (Fig. 3) and are expected to have similar functions in terpene modification, in the antenna and other tissues.

From the P450 phylogeny (Fig. 3), it is obvious that more functional information is needed for bark beetle P450s to fully appreciate their role in pheromone biosynthesis and host defence detoxification. It also remains to be established whether specific P450s have evolved to modify host compounds into pheromones (eg, *trans*-verbenol in *Dendroctonus* species; Seybold et al., 2000), or whether they are merely part of the repertoire of detoxifying P450s.

7. FUTURE DIRECTIONS

The influx of sequence information in bark beetles and the increased availability of high-throughput technologies in genome and transcriptome sequencing, and quantitative transcriptomics and proteomics, have initiated a renewed vigour in research aimed at understanding the molecular mechanisms underlying bark beetle behaviour and physiology. As more bark beetle and other beetle genomes are sequenced, we will be able to explore which genes are unique to specific clades of bark beetles and which occur more diversely. As briefly described earlier, research on the processes of olfaction, host defence detoxification, host colonization, and pheromone biosynthesis are already using this new information. However, many questions are still to be answered about these processes. These include:

- What are the substrates for the pheromone/odorant receptors and pheromone/odorant-binding proteins already identified?
- How are odorants degraded or otherwise cleared in olfactory tissues?
- What are the acute and longer-term effects of host defences?
- What are the mechanisms of acute toxicity and longer-term effects of host defences?
- What are the mechanisms of host defence tolerance and detoxification?
- How are the processes of detoxification regulated (by hormonal or xenobiotic inducers)?
- What mechanisms do beetles have for cold tolerance (see also Bonnett et al., 2012; Huber and Wiebe, 2016)?
- What are the mechanisms of pheromone biosynthesis (see also Blomquist, 2016)?
- How is pheromone biosynthesis regulated (by juvenile hormone and other factors)?

Other areas for further research include the enzymes of lignocellulose digestion. The wood-boring Asian long-horned beetle, *A. glabripennis*, has both enzymes and microbial help in digesting lignocellulose (Geib et al., 2008; Scully et al., 2013). Unique plant cell wall-digesting enzymes have been identified in bark beetles but so far have not been functionally characterized

(Keeling et al., 2013c; Pauchet et al., 2010). In addition, some bark beetles have adopted microbial carbohydrate metabolizing genes via horizontal gene transfer (Acuña et al., 2012; Keeling et al., 2013c; Kirsch et al., 2014). Because of the unique habitat where bark beetles live, there is potential in the bark beetle sequence resources to discover unique enzymes that could have commercial application in the biosynthesis or degradation of biomaterials (Oppert et al., 2010; Watanabe and Tokuda, 2010).

With greater sequence resources, genomics-enhanced population studies can better delineate differences in populations across the large ranges that many bark beetle species inhabit (Agata et al., 2011; Bracewell et al., 2011; Davis et al., 2009; Janes et al., 2014; Sperling et al., 2016) and can also compare different bark beetle species across the same geographic or host range (Bertheau et al., 2013; Mayer et al., 2015). These sequences resources also increase our ability to examine genome evolution, chromosome structure, and karyology (Zúñiga et al., 2002a,b), such as in the genus *Dendroctonus*, where karyotype varies greatly (Zúñiga et al., 2002a). All these studies help to understand the evolution and diversification of bark beetles (eg, Reeve et al., 2012).

The rapidly advancing tools to study gene function *in vivo* are beginning to be used in bark beetles. The beetles examined so far typically have a robust systemic RNA interference (RNAi) response (eg, Bodemann et al., 2012; Ulrich et al., 2015), including the bark beetle, *D. ponderosae* (Keeling et al., 2013a). Successfully rearing bark beetles in the laboratory can be a challenge when they spend most of their lives under the bark and have symbiotic relationships with various fungi and bacteria. However, this ability will be needed to use RNAi and the CRISPR/Cas gene editing system for studying the function of genes *in vivo* (Sander and Joung, 2014; Gilles et al., 2015). Practical studies to perfect bark beetle husbandry so that eggs and early life stages can be injected or otherwise treated and then allowed to develop to adulthood and reproduce in the laboratory is an important component to these *in vivo* techniques. Future bark beetle research will continue to profit from advances in sequencing and other 'omics' technologies and *in vivo* gene editing and knock-down tools. The application of these new techniques will help us to answer some of the long-standing questions in bark beetle biology.

ACKNOWLEDGEMENTS

I thank Christine Chiu (University of British Columbia) for helpful discussions and David R. Nelson (University of Tennessee Health Science Center) for his expertise in naming P450s.

REFERENCES

Acuña, R., Padilla, B.E., Flórez-Ramos, C.P., Rubio, J.D., Herrera, J.C., Benavides, P., Lee, S.J., Yeats, T.H., Egan, A.N., Doyle, J.J., Rose, J.K., 2012. Adaptive horizontal transfer of a bacterial gene to an invasive insect pest of coffee. Proc. Natl. Acad. Sci. U.S.A. 109, 4197–4202.

Adams, A.S., Aylward, F.O., Adams, S.M., Erbilgin, N., Aukema, B.H., Currie, C.R., Suen, G., Raffa, K.F., 2013. Mountain pine beetles colonizing historical and naive host trees are associated with a bacterial community highly enriched in genes contributing to terpene metabolism. Appl. Environ. Microbiol. 79, 3468–3475.

Agata, K., Alasaad, S., Almeida-Val, V.M., Alvarez-Dios, J.A., Barbisan, F., Beadell, J.S., Beltran, J.F., Benitez, M., Bino, G., Bleay, C., Bloor, P., Bohlmann, J., Booth, W., Boscari, E., Caccone, A., Campos, T., Carvalho, B.M., Climaco, G.T., Clobert, J., Congiu, L., Cowger, C., Dias, G., Doadrio, I., Farias, I.P., Ferrand, N., Freitas, P.D., Fusco, G., Galetti, P.M., Gallardo-Escarate, C., Gaunt, M.W., Gomez Ocampo, Z., Goncalves, H., Gonzalez, E.G., Haye, P., Honnay, O., Hyseni, C., Jacquemyn, H., Jowers, M.J., Kakezawa, A., Kawaguchi, E., Keeling, C.I., Kwan, Y.S., La Spina, M., Lee, W.O., Lesniewska, M., Li, Y., Liu, H., Liu, X., Lopes, S., Martinez, P., Meeus, S., Murray, B.W., Nunes, A.G., Okedi, L.M., Ouma, J.O., Pardo, B.G., Parks, R., Paula-Silva, M.N., Pedraza-Lara, C., Perera, O.P., Pino-Querido, A., Richard, M., Rossini, B.C., Samarasekera, N.G., Sanchez, A., Sanchez, J.A., Santos, C.H., Shinohara, W., Soriguer, R.C., Sousa, A.C., Sousa, C.F., Stevens, V.M., Tejedo, M., Valenzuela-Bustamante, M., Van De Vliet, M.S., Vandepitte, K., Vera, M., Wandeler, P., Wang, W., Won, Y.J., Yamashiro, A., Yamashiro, T., Zhu, C., 2011. Permanent genetic resources added to Molecular Ecology Resources Database 1 December 2010–31 January 2011. Mol. Ecol. Resour. 11, 586–589.

Andersson, M.N., Larsson, M.C., Schlyter, F., 2009. Specificity and redundancy in the olfactory system of the bark beetle *Ips typographus*: single-cell responses to ecologically relevant odors. J. Insect Physiol. 55, 556–567.

Andersson, M.N., Grosse-Wilde, E., Keeling, C.I., Bengtsson, J.M., Yuen, M.M.S., Li, M., Hillbur, Y., Bohlmann, J., Hansson, B.S., Schlyter, F., 2013. Antennal transcriptome analysis of the chemosensory gene families in the tree killing bark beetles, *Ips typographus* and *Dendroctonus ponderosae* (Coleoptera: Curculionidae: Scolytinae). BMC Genomics 14, 198.

Aw, T., Schlauch, K., Keeling, C.I., Young, S., Bearfield, J.C., Blomquist, G.J., Tittiger, C., 2010. Functional genomics of mountain pine beetle (*Dendroctonus ponderosae*) midguts and fat bodies. BMC Genomics 11, 215.

Berenbaum, M.R., 2002. Postgenomic chemical ecology: from genetic code to ecological interactions. J. Chem. Ecol. 28, 873–896.

Bertheau, C., Schuler, H., Arthofer, W., Avtzis, D.N., Mayer, F., Krumbock, S., Moodley, Y., Stauffer, C., 2013. Divergent evolutionary histories of two sympatric spruce bark beetle species. Mol. Ecol. 22, 3318–3332.

Blomquist, G.J., 2016. Biochemistry and molecular biology of pheromone production. In: Tittiger, C., Blomquist, G.J. (Eds.), Advances in Insect Physiology. Elsevier, Philadelphia, PA, USA.

Blomquist, G.J., Figueroa-Teran, R., Aw, M., Song, M., Gorzalski, A., Abbott, N.L., Chang, E., Tittiger, C., 2010. Pheromone production in bark beetles. Insect Biochem. Mol. Biol. 40, 699–712.

Bodemann, R.R., Rahfeld, P., Stock, M., Kunert, M., Wielsch, N., Groth, M., Frick, S., Boland, W., Burse, A., 2012. Precise RNAi-mediated silencing of metabolically active proteins in the defence secretions of juvenile leaf beetles. Proc. R. Soc. Lond. B 279, 4126–4134.

Bonnett, T.R., Robert, J.A., Pitt, C., Fraser, J.D., Keeling, C.I., Bohlmann, J., Huber, D.P., 2012. Global and comparative proteomic profiling of overwintering and developing mountain pine beetle, *Dendroctonus ponderosae* (Coleoptera: Curculionidae), larvae. Insect Biochem. Mol. Biol. 42, 890–901.

Boone, C.K., Keefover-Ring, K., Mapes, A.C., Adams, A.S., Bohlmann, J., Raffa, K.F., 2013. Bacteria associated with a tree-killing insect reduce concentrations of plant defense compounds. J. Chem. Ecol. 39, 1003–1006.

Bracewell, R.R., Pfrender, M.E., Mock, K.E., Bentz, B.J., 2011. Cryptic postzygotic isolation in an eruptive species of bark beetle (*Dendroctonus ponderosae*). Evolution 65, 961–975.

Byers, J.A., 1982. Male-specific conversion of the host plant compound, myrcene, to the pheromone, (+)-ipsdienol, in the bark beetle, *Dendroctonus brevicomis*. J. Chem. Ecol. 8, 363–371.

Cano-Ramírez, C., López, M.F., Cesar-Ayala, A.K., Pineda-Martínez, V., Sullivan, B.T., Zúñiga, G., 2013. Isolation and expression of cytochrome P450 genes in the antennae and gut of pine beetle *Dendroctonus rhizophagus* (Curculionidae: Scolytinae) following exposure to host monoterpenes. Gene 520, 47–63.

i5K Consortium, 2013. The i5K initiative: advancing arthropod genomics for knowledge, human health, agriculture, and the environment. J. Hered. 104, 595–600.

Coyne, J.F., Lott, L.H., 1976. Toxicity of substances in pine oleoresin to southern pine beetles. J. Ga. Entomol. Soc. 11, 301–305.

Cunningham, C.B., Ji, L., Wiberg, R.A., Shelton, J., Mckinney, E.C., Parker, D.J., Meagher, R.B., Benowitz, K.M., Roy-Zokan, E.M., Ritchie, M.G., Brown, S.J., Schmitz, R.J., Moore, A.J., 2015. The genome and methylome of a beetle with complex social behavior, *Nicrophorus vespilloides* (Coleoptera: Silphidae). Genome Biol. Evol. 7 (12), 3383–3396. http://dx.doi.org/10.1093/gbe/evv194.

Dai, L., Wang, C., Zhang, X., Yu, J., Zhang, R., Chen, H., 2014. Two CYP4 genes of the Chinese white pine beetle, *Dendroctonus armandi* (Curculionidae: Scolytinae), and their transcript levels under different development stages and treatments. Insect Mol. Biol. 23, 598–610.

Dai, L., Ma, J., Ma, M., Zhang, H., Shi, Q., Zhang, R., Chen, H., 2015a. Characterisation of GST genes from the Chinese white pine beetle *Dendroctonus armandi* (Curculionidae: Scolytinae) and their response to host chemical defence. Pest Manag. Sci. http://dx.doi.org/10.1002/ps.4059.

Dai, L., Ma, M., Wang, C., Shi, Q., Zhang, R., Chen, H., 2015b. Cytochrome P450s from the Chinese white pine beetle, *Dendroctonus armandi* (Curculionidae: Scolytinae): expression profiles of different stages and responses to host allelochemicals. Insect Biochem. Mol. Biol. 65, 35–46.

Davis, C.S., Mock, K.E., Bentz, B.J., Bromilow, S.M., Bartell, N.V., Murray, B.W., Roe, A.D., Cooke, J.E.K., 2009. Isolation and characterization of 16 microsatellite loci in the mountain pine beetle, *Dendroctonus ponderosae* Hopkins (Coleoptera: Curculionidae: Scolytinae). Mol. Ecol. Resour. 9, 1071–1073.

De Bruyne, M., Baker, T.C., 2008. Odor detection in insects: volatile codes. J. Chem. Ecol. 34, 882–897.

Delorme, L., Lieutier, F., 1990. Monoterpene composition of the preformed and induced resins of Scots pine, and their effect on bark beetles and associated fungi. Eur. J. For. Pathol. 20, 304–316.

Domingue, M.J., Teale, S.A., 2007. The genetic architecture of pheromone production between populations distant from the hybrid zone of the pine engraver, *Ips pini*. Chemoecology 17, 255–262.

Domingue, M.J., Starmer, W.T., Teale, S.A., 2006. Genetic control of the enantiomeric composition of ipsdienol in the pine engraver, *Ips pini*. J. Chem. Ecol. 32, 1005–1026.

Eigenheer, A.L., Keeling, C.I., Young, S., Tittiger, C., 2003. Comparison of gene representation in midguts from two phytophagous insects, *Bombyx mori* and *Ips pini*, using expressed sequence tags. Gene 316, 127–136.

Everaerts, C., Grégoire, J.-C., Merlin, J., 1988. The toxicity of Norway spruce monoterpenes to two bark beetle species and their associates. In: Mattson, W.J., Levieux, J., Bernard-Dagan, C. (Eds.), Mechanisms of Woody Plant Defenses Against Insects. Springer, New York.

Feyereisen, R., 2006. Evolution of insect P450. Biochem. Soc. Trans. 34, 1252–1255.

Feyereisen, R., 2011. Arthropod CYPomes illustrate the tempo and mode in P450 evolution. Biochim. Biophys. Acta 1814, 19–28.

Figueroa-Teran, R., Welch, W.H., Blomquist, G.J., Tittiger, C., 2012. Ipsdienol dehydrogenase (IDOLDH): a novel oxidoreductase important for *Ips pini* pheromone production. Insect Biochem. Mol. Biol. 42, 81–90.

Fish, R.H., Browne, L.E., Bergot, B.J., 1984. Pheromone biosynthetic pathways: conversion of ipsdienone to (–)-ipsdienol, a mechanism for enantioselective reduction in the male bark beetle, *Ips paraconfusus*. J. Chem. Ecol. 10, 1057–1064.

Franceschi, V.R., Krokene, P., Christiansen, E., Krekling, T., 2005. Anatomical and chemical defenses of conifer bark against bark beetles and other pests. New Phytol. 167, 353–376.

Geib, S.M., Filley, T.R., Hatcher, P.G., Hoover, K., Carlson, J.E., Jimenez-Gasco Mdel, M., Nakagawa-Izumi, A., Sleighter, R.L., Tien, M., 2008. Lignin degradation in wood-feeding insects. Proc. Natl. Acad. Sci. U.S.A. 105, 12932–12937.

Gilg, A.B., Bearfield, J.C., Tittiger, C.R., Welch, W.H., Blomquist, G.J., 2005. Isolation and functional expression of an animal geranyl diphosphate synthase and its role in bark beetle pheromone biosynthesis. Proc. Natl. Acad. Sci. U.S.A. 102, 9760–9765.

Gilg, A.B., Tittiger, C., Blomquist, G.J., 2009. Unique animal prenyltransferase with monoterpene synthase activity. Naturwissenschaften 96, 731–735.

Gilles, A.F., Schinko, J.B., Averof, M., 2015. Efficient CRISPR-mediated gene targeting and transgene replacement in the beetle *Tribolium castaneum*. Development 142, 2832–2839.

Grucmanová, S.A.R., Holuša, J., 2013. Nematodes associated with bark beetles, with focus on the genus *Ips* (Coleoptera: Scolytinae) in Central Europe. Acta Zool. Bulgar. 65, 547–556.

Gu, X.C., Zhang, Y.N., Kang, K., Dong, S.L., Zhang, L.W., 2015. Antennal transcriptome analysis of odorant reception genes in the red turpentine beetle (RTB), *Dendroctonus valens*. PLoS One 10, e0125159.

Hall, G., Tittiger, C., Andrews, G., Mastick, G., Kuenzli, M., Luo, X., Seybold, S., Blomquist, G., 2002a. Midgut tissue of male pine engraver, *Ips pini*, synthesizes monoterpenoid pheromone component ipsdienol *de novo*. Naturwissenschaften 89, 79–83.

Hall, G.M., Tittiger, C., Blomquist, G.J., Andrews, G.L., Mastick, G.S., Barkawi, L.S., Bengoa, C., Seybold, S.J., 2002b. Male jeffrey pine beetle, *Dendroctonus jeffreyi*, synthesizes the pheromone component frontalin in anterior midgut tissue. Insect Biochem. Mol. Biol. 32, 1525–1532.

Hall, D.E., Yuen, M.M., Jancsik, S., Quesada, A.L., Dullat, H.K., Li, M., Henderson, H., Arango-Velez, A., Liao, N.Y., Docking, R.T., Chan, S.K., Cooke, J.E., Breuil, C., Jones, S.J., Keeling, C.I., Bohlmann, J., 2013. Transcriptome resources and functional characterization of monoterpene synthases for two host species of the mountain pine beetle, lodgepole pine (*Pinus contorta*) and jack pine (*Pinus banksiana*). BMC Plant Biol. 13, 80.

Hardwick, J.P., 2008. Cytochrome P450 omega hydroxylase (CYP4) function in fatty acid metabolism and metabolic diseases. Biochem. Pharmacol. 75, 2263–2275.

Huber, D.P.W., Wiebe, P., 2016. Proteomic and transcriptomic investigations of host choice and overwintering physiology. In: Tittiger, C., Blomquist, G.J. (Eds.), Advances in Insect Physiology. Elsevier, Philadelphia, PA, USA.

Huber, D.P., Erickson, M.L., Leutenegger, C.M., Bohlmann, J., Seybold, S.J., 2007. Isolation and extreme sex-specific expression of cytochrome P450 genes in the bark beetle, *Ips paraconfusus*, following feeding on the phloem of host ponderosa pine, *Pinus ponderosa*. Insect Mol. Biol. 16, 335–349.

Hughes, P.R., 1975. Pheromones of *Dendroctonus*: origin of alpha-pinene oxidation products present in emergent adults. J. Insect Physiol. 21, 687–691.

Hunt, D.W., Borden, J.H., Pierce Jr., H.D., Slessor, K.N., King, G.G., Czyzewska, E.K., 1986. Sex-specific production of ipsdienol and myrcenol by *Dendroctonus ponderosae* (Coleoptera: Scolytidae) exposed to myrcene vapors. J. Chem. Ecol. 12, 1579–1586.

Hunt, T., Bergsten, J., Levkanicova, Z., Papadopoulou, A., John, O.S., Wild, R., Hammond, P.M., Ahrens, D., Balke, M., Caterino, M.S., Gómez-Zurita, J., Ribera, I., Barraclough, T.G., Bocakova, M., Bocak, L., Vogler, A.P., 2007. A comprehensive phylogeny of beetles reveals the evolutionary origins of a superradiation. Science 318, 1913–1916.

Janes, J.K., Li, Y., Keeling, C.I., Yuen, M.M., Boone, C.K., Cooke, J.E., Bohlmann, J., Huber, D.P., Murray, B.W., Coltman, D.W., Sperling, F.A., 2014. How the mountain pine beetle (*Dendroctonus ponderosae*) breached the Canadian Rocky Mountains. Mol. Biol. Evol. 31, 1803–1815.

Jaramillo, J., Torto, B., Mwenda, D., Troeger, A., Borgemeister, C., Poehling, H.M., Francke, W., 2013. Coffee berry borer joins bark beetles in coffee klatch. PLoS One 8, e74277.

Katoh, K., Standley, D.M., 2013. MAFFT multiple sequence alignment software version 7: improvements in performance and usability. Mol. Biol. Evol. 30, 772–780.

Keeling, C.I., Bohlmann, J., 2006. Genes, enzymes and chemicals of terpenoid diversity in the constitutive and induced defence of conifers against insects and pathogens. New Phytol. 170, 657–675.

Keeling, C.I., Blomquist, G.J., Tittiger, C., 2004. Coordinated gene expression for pheromone biosynthesis in the pine engraver beetle, *Ips pini* (Coleoptera: Scolytidae). Naturwissenschaften 91, 324–328.

Keeling, C.I., Bearfield, J.C., Young, S., Blomquist, G.J., Tittiger, C., 2006. Effects of juvenile hormone on gene expression in the pheromone-producing midgut of the pine engraver beetle, *Ips pini*. Insect Mol. Biol. 15, 207–216.

Keeling, C.I., Weisshaar, S., Ralph, S.G., Jancsik, S., Hamberger, B., Dullat, H.K., Bohlmann, J., 2011. Transcriptome mining, functional characterization, and phylogeny of a large terpene synthase gene family in spruce (*Picea* spp.). BMC Plant Biol. 11, 43.

Keeling, C.I., Henderson, H., Li, M., Yuen, M., Clark, E.L., Fraser, J.D., Huber, D.P.W., Liao, N.Y., Docking, T.R., Birol, I., Chan, S.K., Taylor, G.A., Palmquist, D., Jones, S.J.M., Bohlmann, J., 2012. Transcriptome and full-length cDNA resources for the mountain pine beetle, *Dendroctonus ponderosae* Hopkins, a major insect pest of pine forests. Insect Biochem. Mol. Biol. 42, 525–536.

Keeling, C.I., Chiu, C.C., Aw, T., Li, M., Henderson, H., Tittiger, C., Weng, H.B., Blomquist, G.J., Bohlmann, J., 2013a. Frontalin pheromone biosynthesis in the mountain pine beetle, *Dendroctonus ponderosae*, and the role of isoprenyl diphosphate synthases. Proc. Natl. Acad. Sci. U.S.A. 110, 18838–18843.

Keeling, C.I., Henderson, H., Li, M., Dullat, H.K., Ohnishi, T., Bohlmann, J., 2013b. CYP345E2, an antenna-specific cytochrome P450 from the mountain pine beetle, *Dendroctonus ponderosae* Hopkins, catalyses the oxidation of pine host monoterpene volatiles. Insect Biochem. Mol. Biol. 43, 1142–1151.

Keeling, C.I., Yuen, M.M.S., Liao, N.Y., Docking, T.R., Chan, S.K., Taylor, G.A., Palmquist, D.L., Jackman, S.D., Nguyen, A., Li, M., Henderson, H., Janes, J.K., Zhao, Y., Pandoh, P., Moore, R., Sperling, F.A.H., Huber, D.P.W., Birol, I., Jones, S.J.M., Bohlmann, J., 2013c. Draft genome of the mountain pine beetle, *Dendroctonus ponderosae* Hopkins, a major forest pest. Genome Biol. 14, R27.

Keeling, C.I., Li, M., Dullat, H.K., Henderson, H., Yuen, M.M.S., Bohlmann, J., 2016. Quantitative metabolome, proteome and transcriptome analysis of midgut and fat body tissues in the mountain pine beetle, *Dendroctonus ponderoase* Hopkins, and insights into pheromone biosynthesis. Insect Biochem. Mol. Biol. 70, 170–183.

Khadempour, L., Lemay, V., Jack, D., Bohlmann, J., Breuil, C., 2012. The relative abundance of mountain pine beetle fungal associates through the beetle life cycle in pine trees. Microb. Ecol. 64, 909–917.

Kirsch, R., Gramzow, L., Theissen, G., Siegfried, B.D., Ffrench-Constant, R.H., Heckel, D.G., Pauchet, Y., 2014. Horizontal gene transfer and functional diversification of plant cell wall degrading polygalacturonases: key events in the evolution of herbivory in beetles. Insect Biochem. Mol. Biol. 52, 33–50.

Kopper, B.J., Illman, B.L., Kersten, P.J., Klepzig, K.D., Raffa, K.F., 2005. Effects of diterpene acids on components of a conifer bark beetle-fungal interaction: tolerance by *Ips pini* and sensitivity by its associate *Ophiostoma ips*. Environ. Entomol. 34, 486–493.

Lah, L., Haridas, S., Bohlmann, J., Breuil, C., 2013. The cytochromes P450 of *Grosmannia clavigera*: genome organization, phylogeny, and expression in response to pine host chemicals. Fungal Genet. Biol. 50, 72–81.

Latty, T.M., Reid, M.L., 2009. First in line or first in time? Effects of settlement order and arrival date on reproduction in a group-living beetle *Dendroctonus ponderosae*. J. Anim. Ecol. 78, 549–555.

Latty, T.M., Reid, M.L., 2010. Who goes first? Condition and danger dependent pioneering in a group-living bark beetle (*Dendroctonus ponderosae*). Behav. Ecol. Sociobiol. 64, 639–646.

Lee, S., Kim, J.-J., Breuil, C., 2006. Diversity of fungi associated with the mountain pine beetle *Dendroctonus ponderosae* and infested lodgepole pines in British Columbia. Fungal Divers. 22, 91–105.

Lee, S.H., Oh, H.W., Fang, Y., An, S.B., Park, D.S., Song, H.H., Oh, S.R., Kim, S.Y., Kim, S., Kim, N., Raikhel, A.S., Je, Y.H., Shin, S.W., 2015. Identification of plant compounds that disrupt the insect juvenile hormone receptor complex. Proc. Natl. Acad. Sci. U.S.A. 112, 1733–1738.

Libbey, L.M., Ryker, L.C., Yandell, K.L., 1985. Laboratory and field studies of volatiles released by *Dendroctonus ponderosae* Hopkins (Coleoptera, Scolytidae). Z. Angew. Entomol. 100, 381–392.

Lindgren, B.S., Raffa, K.F., 2013. Evolution of tree killing in bark beetles (Coleoptera: Curculionidae): trade-offs between the maddening crowds and a sticky situation. Can. Entomol. 145, 471–495.

López, M.D., Pascual-Villalobos, M.J., 2010. Mode of inhibition of acetylcholinesterase by monoterpenoids and implications for pest control. Ind. Crop. Prod. 31, 284–288.

López, M.D., Pascual-Villalobos, M.J., 2015. Are monoterpenoids and phenylpropanoids efficient inhibitors of acetylcholinesterase from stored product insect strains? Flavour Frag. J. 30, 108–112.

López, M.F., Cano-Ramírez, C., Shibayama, M., Zúñiga, G., 2011. alpha-Pinene and myrcene induce ultrastructural changes in the midgut of *Dendroctonus valens* (Coleoptera: Curculionidae: Scolytinae). Ann. Entomol. Soc. Am. 104, 553–561.

López, M.F., Cano-Ramírez, C., Cesar-Ayala, A.K., Ruiz, E.A., Zúñiga, G., 2013. Diversity and expression of P450 genes from *Dendroctonus valens* LeConte (Curculionidae: Scolytinae) in response to different kairomones. Insect Biochem. Mol. Biol. 43, 417–432.

Manning, C.G., Reid, M.L., 2013. Sub-lethal effects of monoterpenes on reproduction by mountain pine beetles. Agric. For. Entomol. 15, 262–271.

Mayer, F., Piel, F.B., Cassel-Lundhagen, A., Kirichenko, N., Grumiau, L., Okland, B., Bertheau, C., Grégoire, J.C., Mardulyn, P., 2015. Comparative multilocus phylogeography of two Palaearctic spruce bark beetles: influence of contrasting ecological strategies on genetic variation. Mol. Ecol. 24, 1292–1310.

McLeod, G., Gries, R., Von Reuss, S.H., Rahe, J.E., Mcintosh, R., Konig, W.A., Gries, G., 2005. The pathogen causing Dutch elm disease makes host trees attract insect vectors. Proc. R. Soc. Lond. B 272, 2499–2503.

Mendanha, S.A., Alonso, A., 2015. Effects of terpenes on fluidity and lipid extraction in phospholipid membranes. Biophys. Chem. 198, 45–54.

Miller, D.R., Borden, J.H., Slessor, K.N., 1989. Inter- and intrapopulation variation of the pheromone, ipsdienol produced by male pine engravers, *Ips pini* (Say) (Coleoptera: Scolytidae). J. Chem. Ecol. 15, 233–247.

Morales-Jiménez, J., Zúñiga, G., Villa-Tanaca, L., Hernández-Rodríguez, C., 2009. Bacterial community and nitrogen fixation in the red turpentine beetle, *Dendroctonus valens* LeConte (Coleoptera: Curculionidae: Scolytinae). Microb. Ecol. 58, 879–891.

Morales-Jiménez, J., Zúñiga, G., Ramírez-Saad, H.C., Hernández-Rodríguez, C., 2012. Gut-associated bacteria throughout the life cycle of the bark beetle *Dendroctonus rhizophagus* Thomas and Bright (Curculionidae: Scolytinae) and their cellulolytic activities. Microb. Ecol. 64, 268–278.

Mustaparta, H., Tommeras, B.A., Lanier, G.N., 1985. Pheromone receptor cell specificity in interpopulational hybrids of *Ips pini* (Coleoptera: Scolytidae). J. Chem. Ecol. 11, 999–1007.

Nelson, D.R., 2011. Progress in tracing the evolutionary paths of cytochrome P450. Biochim. Biophys. Acta 1814, 14–18.

Njihia, T.N., Jaramillo, J., Murungi, L., Mwenda, D., Orindi, B., Poehling, H.M., Torto, B., 2014. Spiroacetals in the colonization behaviour of the coffee berry borer: a 'push-pull' system. PLoS One 9, e111316.

Obregón-Molina, G., Cesar-Ayala, A.K., López, M.F., Cano-Ramírez, C., Zúñiga, G., 2015. Comparison of orthologous cytochrome P450 genes relative expression patterns in the bark beetles *Dendroctonus rhizophagus* and *Dendroctonus valens* (Curculionidae: Scolytinae) during host colonization. Insect Mol. Biol. 24, 649–661.

Oppert, C., Klingeman, W.E., Willis, J.D., Oppert, B., Jurat-Fuentes, J.L., 2010. Prospecting for cellulolytic activity in insect digestive fluids. Comp. Biochem. Physiol. B Biochem. Mol. Biol. 155, 145–154.

Pauchet, Y., Wilkinson, P., Chauhan, R., Ffrench-Constant, R.H., 2010. Diversity of beetle genes encoding novel plant cell wall degrading enzymes. PLoS One 5, e15635.

Pierce, H.D., Conn, J.E., Oehlschlager, A.C., Borden, J.H., 1987. Monoterpene metabolism in female mountain pine beetles, *Dendroctonus ponderosae* Hopkins, attacking ponderosa pine. J. Chem. Ecol. 13, 1455–1480.

Pitt, C., Robert, J.A., Bonnett, T.R., Keeling, C.I., Bohlmann, J., Huber, D.P., 2014. Proteomics indicators of the rapidly shifting physiology from whole mountain pine beetle, *Dendroctonus ponderosae* (Coleoptera: Curculionidae), adults during early host colonization. PLoS One 9, e110673.

Price, M.N., Dehal, P.S., Arkin, A.P., 2010. FastTree 2—approximately maximum-likelihood trees for large alignments. PLoS One 5, e9490.

Pureswaran, D.S., Borden, J.H., 2004. New repellent semiochemicals for three species of *Dendroctonus* (Coleoptera: Scolytidae). Chemoecology 14, 67–75.

Pureswaran, D.S., Gries, R., Borden, J.H., Pierce Jr., H.D., 2000. Dynamics of pheromone production and communication in the mountain pine beetle, *Dendroctonus ponderosae* Hopkins, and the pine engraver, *Ips pini* (Say) (Coleoptera: Scolytidae). Chemoecology 10, 153–168.

Pureswaran, D.S., Gries, R., Borden, J.H., 2004. Antennal responses of four species of tree-killing bark beetles (Coleoptera: Scolytidae) to volatiles collected from beetles, and their host and nonhost conifers. Chemoecology 14, 59–66.

Qiu, Y., Tittiger, C., Wicker-Thomas, C., Le Goff, G., Young, S., Wajnberg, E., Fricaux, T., Taquet, N., Blomquist, G.J., Feyereisen, R., 2012. An insect-specific P450 oxidative decarbonylase for cuticular hydrocarbon biosynthesis. Proc. Natl. Acad. Sci. U.S.A. 109, 14858–14863.

Raffa, K.F., 2014. Terpenes tell different tales at different scales: glimpses into the chemical ecology of conifer—bark beetle—microbial interactions. J. Chem. Ecol. 40, 1–20.

Raffa, K.F., Smalley, E.B., 1995. Interaction of pre-attack and induced monoterpene concentrations in host conifer defense against bark beetle fungal complexes. Oecologia 102, 285–295.

Raffa, K.F., Berryman, A.A., Simasko, J., Teal, W., Wong, B.L., 1985. Effects of grand fir monoterpenes on the fir engraver, *Scolytus ventralis* (Coleoptera: Scolytidae), and its symbiotic fungus. Environ. Entomol. 14, 522–556.

Reeve, J.D., Anderson, F.E., Kelley, S.T., 2012. Ancestral state reconstruction for *Dendroctonus* bark beetles: evolution of a tree killer. Environ. Entomol. 41, 723–730.

Reid, M.L., Purcell, J.R.C., 2011. Condition-dependent tolerance of monoterpenes in an insect herbivore. Arthropod Plant Interact. 5, 331–337.

Renou, M., 2014. Pheromones and general odor perception in insects. In: Mucignat-Caretta, C. (Ed.), Neurobiology of Chemical Communication. CRC Press/Taylor & Francis, Boca Raton, FL.

Renwick, J.A., Hughes, P.R., Krull, I.S., 1976. Selective production of *cis*- and *trans*-verbenol from (-)-and (+)-alpha by a bark beetle. Science 191, 199–201.

Richards, S., Gibbs, R.A., Weinstock, G.M., Brown, S.J., Denell, R., Beeman, R.W., Gibbs, R., Beeman, R.W., Brown, S.J., Bucher, G., Friedrich, M., Grimmelikhuijzen, C.J., Klingler, M., Lorenzen, M., Richards, S., Roth, S., Schroder, R., Tautz, D., Zdobnov, E.M., Muzny, D., Gibbs, R.A., Weinstock, G.M., Attaway, T., Bell, S., Buhay, C.J., Chandrabose, M.N., Chavez, D., Clerk-Blankenburg, K.P., Cree, A., Dao, M., Davis, C., Chacko, J., Dinh, H., Dugan-Rocha, S., Fowler, G., Garner, T.T., Garnes, J., Gnirke, A., Hawes, A., Hernandez, J., Hines, S., Holder, M., Hume, J., Jhangiani, S.N., Joshi, V., Khan, Z.M., Jackson, L., Kovar, C., Kowis, A., Lee, S., Lewis, L.R., Margolis, J., Morgan, M., Nazareth, L.V., Nguyen, N., Okwuonu, G., Parker, D., Richards, S., Ruiz, S.J., Santibanez, J., Savard, J., Scherer, S.E., Schneider, B., Sodergren, E., Tautz, D., Vattahil, S., Villasana, D., White, C.S., Wright, R., Park, Y., Beeman, R.W., Lord, J., Oppert, B., Lorenzen, M., Brown, S., Wang, L., Savard, J., Tautz, D., Richards, S., Weinstock, G., Gibbs, R.A., Liu, Y., Worley, K., Weinstock, G., Elsik, C.G., Reese, J.T., Elhaik, E., Landan, G., Graur, D., Arensburger, P., Atkinson, P., Beeman, R.W., Beidler, J., Brown, S.J., Demuth, J.P., Drury, D.W., Du, Y.Z., Fujiwara, H., Lorenzen, M., Maselli, V., et al., 2008. The genome of the model beetle and pest *Tribolium castaneum*. Nature 452, 949–955.

Robert, J.A., Pitt, C., Bonnett, T.R., Yuen, M.M., Keeling, C.I., Bohlmann, J., Huber, D.P., 2013. Disentangling detoxification: gene expression analysis of feeding mountain pine beetle illuminates molecular-level host chemical defense detoxification mechanisms. PLoS One 8, e77777.

Ross, P.L., Huang, Y.N., Marchese, J.N., Williamson, B., Parker, K., Hattan, S., Khainovski, N., Pillai, S., Dey, S., Daniels, S., Purkayastha, S., Juhasz, P., Martin, S., Bartlet-Jones, M., He, F., Jacobson, A., Pappin, D.J., 2004. Multiplexed protein quantitation in *Saccharomyces cerevisiae* using amine-reactive isobaric tagging reagents. Mol. Cell. Proteomics 3, 1154–1169.

Sander, J.D., Joung, J.K., 2014. CRISPR-Cas systems for editing, regulating and targeting genomes. Nat. Biotechnol. 32, 347–355.
Sandstrom, P., Welch, W.H., Blomquist, G.J., Tittiger, C., 2006. Functional expression of a bark beetle cytochrome P450 that hydroxylates myrcene to ipsdienol. Insect Biochem. Mol. Biol. 36, 835–845.
Sandstrom, P., Ginzel, M.D., Bearfield, J.C., Welch, W.H., Blomquist, G.J., Tittiger, C., 2008. Myrcene hydroxylases do not determine enantiomeric composition of pheromonal ipsdienol in *Ips* spp. J. Chem. Ecol. 34, 1584–1592.
Schuler, M.A., 2011. P450s in plant-insect interactions. Biochim. Biophys. Acta 1814, 36–45.
Scully, E.D., Hoover, K., Carlson, J.E., Tien, M., Geib, S.M., 2013. Midgut transcriptome profiling of *Anoplophora glabripennis*, a lignocellulose degrading cerambycid beetle. BMC Genomics 14, 850.
Seybold, S.J., Vanderwel, D., 2003. Biosynthesis and endocrine regulation of pheromone production in the Coleoptera. In: Blomquist, G.J., Vogt, R.G. (Eds.), Insect Pheromone Biochemistry and Molecular Biology. Elsevier Academic Press, London.
Seybold, S.J., Bohlmann, J., Raffa, K.F., 2000. Biosynthesis of coniferophagous bark beetle pheromones and conifer isoprenoids: evolutionary perspective and synthesis. Can. Entomol. 132, 697–753.
Seybold, S.J., Huber, D.P.W., Lee, J.C., Graves, A.D., Bohlmann, J., 2006. Pine monoterpenes and pine bark beetles: a marriage of convenience for defense and chemical communication. Phytochemistry 5, 143–178.
Six, D.L., 2012. Ecological and evolutionary determinants of bark beetle—fungus symbioses. Insects 3, 339–366.
Sláma, K., Williams, C.M., 1965. Juvenile hormone activity for the bug *Pyrrhocoris apterus*. Proc. Natl. Acad. Sci. U.S.A. 54, 411–414.
Smith, R.H., 1963. Toxicity of pine resin vapors to 3 species of *Dendroctonus* bark beetles. J. Econ. Entomol. 56, 827–831.
Smith, R.H., 1965. Effect of monoterpene vapors on western pine beetle. J. Econ. Entomol. 58, 509–510.
Song, M., Kim, A.C., Gorzalski, A.J., Maclean, M., Young, S., Ginzel, M.D., Blomquist, G.J., Tittiger, C., 2013. Functional characterization of myrcene hydroxylases from two geographically distinct *Ips pini* populations. Insect Biochem. Mol. Biol. 43, 336–343.
Song, M., Delaplain, P., Nguyen, T.T., Liu, X., Wickenberg, L., Jeffrey, C., Blomquist, G.J., Tittiger, C., 2014. *exo*-Brevicomin biosynthetic pathway enzymes from the mountain pine beetle, *Dendroctonus ponderosae*. Insect Biochem. Mol. Biol. 53, 73–80.
Sperling, F.A.H., Janes, J.K., Batista, P.D., 2016. Mountain pine beetle population structure. In: Tittiger, C., Blomquist, G.J. (Eds.), Advances in Insect Physiology. Elsevier, Philadelphia, PA, USA.
Stork, N.E., McBroom, J., Gely, C., Hamilton, A.J., 2015. New approaches narrow global species estimates for beetles, insects, and terrestrial arthropods. Proc. Natl. Acad. Sci. U.S.A. 112, 7519–7523.
Sturgeon, K.B., Robertson, J.L., 1985. Microsomal polysubstrate monooxygenase activity in western and mountain pine beetles (Coleoptera, Scolytidae). Ann. Entomol. Soc. Am. 78, 1–4.
Tan, A., Palli, S.R., 2008. Identification and characterization of nuclear receptors from the red flour beetle, *Tribolium castaneum*. Insect Biochem. Mol. Biol. 38, 430–439.
Tillman, J.A., Lu, F., Goddard, L.M., Donaldson, Z.R., Dwinell, S.C., Tittiger, C., Hall, G.M., Storer, A.J., Blomquist, G.J., Seybold, S.J., 2004. Juvenile hormone regulates *de novo* isoprenoid aggregation pheromone biosynthesis in pine bark beetles, *Ips* spp., through transcriptional control of HMG-CoA reductase. J. Chem. Ecol. 30, 2459–2494.

Tittiger, C., Blomquist, G.J., Ivarsson, P., Borgeson, C.E., Seybold, S.J., 1999. Juvenile hormone regulation of HMG-R gene expression in the bark beetle *Ips paraconfusus* (Coleoptera: Scolytidae): implications for male aggregation pheromone biosynthesis. Cell. Mol. Life Sci. 55, 121–127.

Tittiger, C., O'Keeffe, C., Bengoa, C.S., Barkawi, L.S., Seybold, S.J., Blomquist, G.J., 2000. Isolation and endocrine regulation of an HMG-CoA synthase cDNA from the male Jeffrey pine beetle, *Dendroctonus jeffreyi* (Coleoptera: Scolytidae). Insect Biochem. Mol. Biol. 30, 1203–1211.

Tittiger, C., Barkawi, L.S., Bengoa, C.S., Blomquist, G.J., Seybold, S.J., 2003. Structure and juvenile hormone-mediated regulation of the HMG-CoA reductase gene from the Jeffrey pine beetle, *Dendroctonus jeffreyi*. Mol. Cell. Endocrinol. 199, 11–21.

Tittiger, C., Keeling, C.I., Blomquist, G.J., 2005. Some insights into the remarkable metabolism of the bark beetle midgut. In: Romeo, J.T. (Ed.), Chemical Ecology and Phytochemistry of Forest Ecosystems. Elsevier, Toronto.

Tommeras, B.A., Mustaparta, H., Grégoire, J.C., 1984. Receptor cells in *Ips typographus* and *Dendroctonus micans* specific to pheromones of the reciprocal genus. J. Chem. Ecol. 10, 759–769.

Ulrich, J., Dao, V.A., Majumdar, U., Schmitt-Engel, C., Schwirz, J., Schultheis, D., Ströhlein, N., Troelenberg, N., Grossmann, D., Richter, T., Dönitz, J., Gerischer, L., Leboulle, G., Vilcinskas, A., Stanke, M., Bucher, G., 2015. Large scale RNAi screen in *Tribolium* reveals novel target genes for pest control and the proteasome as prime target. BMC Genomics 16, 674.

Vega, F.E., Brown, S.M., Chen, H., Shen, E., Nair, M.B., Ceja-Navarro, J.A., Brodie, E.L., Infante, F., Dowd, P.F., Pain, A., 2015. Draft genome of the most devastating insect pest of coffee worldwide: the coffee berry borer, *Hypothenemus hampei*. Sci. Rep. 5, 12525.

Wang, Y., Lim, L., Diguistini, S., Robertson, G., Bohlmann, J., Breuil, C., 2013. A specialized ABC efflux transporter GcABC-G1 confers monoterpene resistance to *Grosmannia clavigera*, a bark beetle-associated fungal pathogen of pine trees. New Phytol. 197, 886–898.

Wang, Y., Lim, L., Madilao, L., Lah, L., Bohlmann, J., Breuil, C., 2014. Gene discovery for enzymes involved in limonene modification or utilization by the mountain pine beetle-associated pathogen *Grosmannia clavigera*. Appl. Environ. Microbiol. 80, 4566–4576.

Watanabe, H., Tokuda, G., 2010. Cellulolytic systems in insects. Annu. Rev. Entomol. 55, 609–632.

Weber, A.P., 2015. Discovering new biology through sequencing of RNA. Plant Physiol. 169, 1524–1531.

Werner, R.A., 1995. Toxicity and repellency of 4-allylanisole and monoterpenes from white spruce and tamarack to the spruce beetle and eastern larch beetle (Coleoptera: Scolytidae). Environ. Entomol. 24, 372–379.

Whitehead, A.T., Scott, D.T., Schmitz, R.F., Mori, K., 1989. Electroantennograms by mountain pine beetles, *Dendroctonus ponderosae* Hopkins, exposed to selected chiral semiochemicals. J. Chem. Ecol. 15, 2089–2099.

Wilson, I.M., Borden, J.H., Gries, R., Gries, G., 1996. Green leaf volatiles as antiaggregants for the mountain pine beetle, *Dendroctonus ponderosae* Hopkins (Coleoptera: Scolytidae). J. Chem. Ecol. 22, 1861–1875.

Yu, J., Dai, L., Zhang, R., Li, Z., Pham, T., Chen, H., 2015. Isolation and expression of HMG-CoA synthase and HMG-CoA reductase genes in different development stages, tissues and treatments of the Chinese white pine beetle, *Dendroctonus armandi* (Curculionidae: Scolytinae). Comp. Biochem. Physiol. B Biochem. Mol. Biol. 187, 62–70.

Zhu, B.T., 2010. On the general mechanism of selective induction of cytochrome P450 enzymes by chemicals: some theoretical considerations. Expert Opin. Drug Metab. Toxicol. 6, 483–494.

Zhu, J.-Y., Zhao, N., Yang, B., 2012a. Global transcriptional analysis of olfactory genes in the head of pine shoot beetle, *Tomicus yunnanensis*. Comp. Funct. Genomics 2012, 491748. http://dx.doi.org/10.1155/2012/491748.

Zhu, J.-Y., Zhao, N., Yang, B., 2012b. Global transcriptome profiling of the pine shoot beetle, *Tomicus yunnanensis* (Coleoptera: Scolytinae). PLoS One 7, e32291.

Zulak, K.G., Bohlmann, J., 2010. Terpenoid biosynthesis and specialized vascular cells of conifer defense. J. Integr. Plant Biol. 52, 86–97.

Zúñiga, G., 2016. Bark beetles and associated bacteria and yeast. In: Tittiger, C., Blomquist, G.J. (Eds.), Advances in Insect Physiology. Elsevier, Philadelphia, PA, USA.

Zúñiga, G., Cisneros, R., Hayes, J.L., Macias-Samano, J., 2002a. Karyology, geographic distribution, and origin of the genus *Dendroctonus* Erichson (Coleoptera: Scolytidae). Ann. Entomol. Soc. Am. 95, 267–275.

Zúñiga, G., Salinas-Moreno, Y., Hayes, J.L., Grégoire, J.C., Cisneros, R., 2002b. Chromosome number in *Dendroctonus micans* and karyological divergence within the genus *Dendroctonus* (Coleoptera: Scolytidae). Can. Entomol. 134, 503–510.

INDEX

Note: Page numbers followed by "*f*" indicate figures, and "*t*" indicate tables.

K.F. Raffa *et al.*, Fig. 1 Life cycle of bark beetles: (A) Adult North American spruce beetle, *Dendroctonus rufipennis*, on outer bark during host selection. (B) Brood adult *Ips typographus* preparing to emerge from *Picea abies* to seek new hosts. Note larval tunnels that begin at egg niches along ovipositional gallery in phloem, radiate outward, and terminate in pupal chambers. Note also staining by microorganisms, some of which facilitate host colonization and utilization, and others that inhibit brood development. *Panel (A): Photo by Y. Cardoza, UW-Madison. Panel (B): Photo by SLU, Alnarp.*

K.F. Raffa *et al.*, Fig. 2 Adult bark beetles killed by trees during failed colonization attempts. (A) Red turpentine beetle, *Dendroctonus valens* in *Pinus resinosa*. (B) Mountain pine beetle, *Dendroctonus ponderosae* in *Pinus contorta*. (C) European spruce beetles, *Ips typographus* in *Picea abies*. *Panel (A): Photo by K. Raffa. Panel (B): Photo by A. Lerch, UW-Madison. Panel (C): Photo by F. Schlyter.*

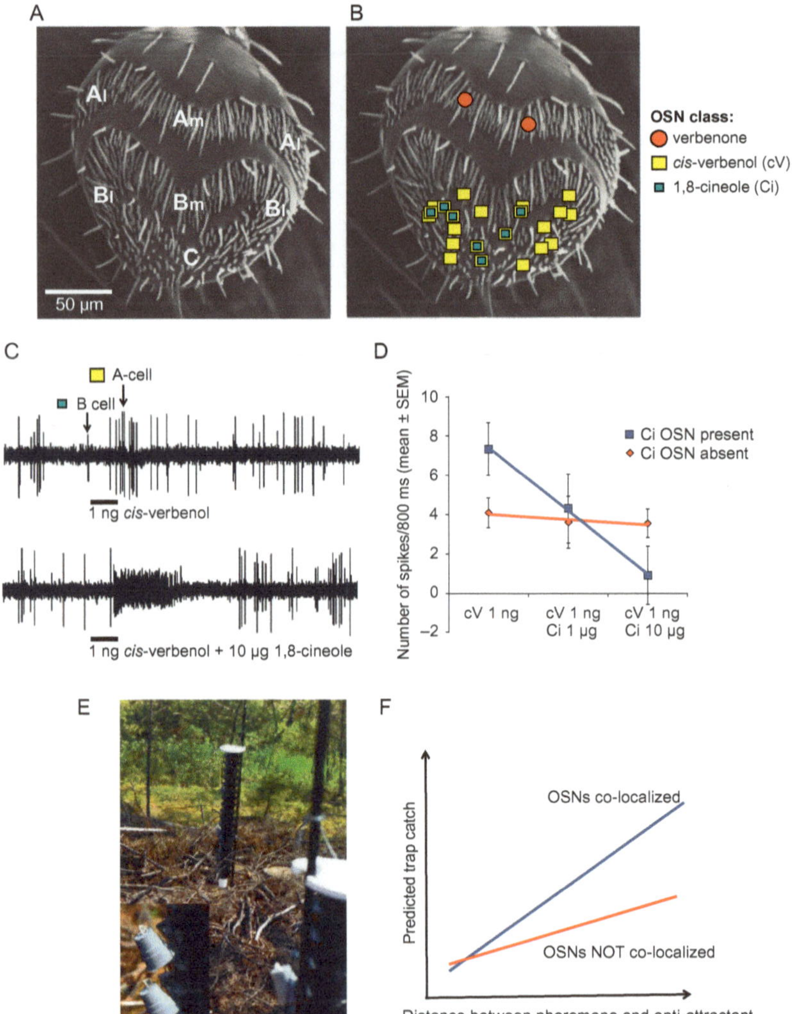

K.F. Raffa *et al.*, Fig. 3 (A) Ventral surface of the antennal club of *Ips typographus* showing the arrangement of olfactory sensilla in three main areas (A, B, and C). Based on the presence of certain functional classes of olfactory sensory neurons (OSNs) exclusively on the lateral sides of the A and B areas, these areas are subdivided into lateral (A_l and B_l) and medial (A_m and B_m) regions. (B) Approximate locations of sensilla containing OSNs for verbenone (filled *red circles*), *cis*-verbenol (cV, *large yellow squares*), and 1,8-cineole (Ci, *small blue squares*), the latter OSN class always co-localized with cV-OSNs. Ca. half of the cV-OSNs are co-localized with an uncharacterized OSN. (C) Raw spike trains showing the large amplitude cV A-neuron and the small amplitude Ci B-neuron in the same sensillum. *Upper trace* shows the excitation of the cV-neuron in response to

K.F. Raffa *et al.*, Fig. 3—Cont'd stimulation with 1 ng cV. *Lower trace* shows the inhibition of the cV response in the A-neuron and the simultaneous excitation of the Ci B-neuron due to stimulation with a binary mixture of cV and Ci. *Black horizontal bars* indicate the 0.5 s odour pulse. (D) Mean response of the cV A-neuron to 1 ng cV presented alone or in binary combinations with Ci ($N=10$–12). The cV response is inhibited by Ci only in sensilla in which the cV OSN is paired with the Ci OSN (*blue line*). (E) Lindgren multiple-funnel traps with vertically spaced pheromone and antiattractant dispensers (grey cups) used in the odour source spacing experiment (Binyameen et al., 2014). (F) Predicted trap catch increase in response to odour source spacing of pheromone and antiattractants detected by co-localized (*blue line*) vs non-co-localized OSNs (*red line*), respectively. *Panels (A) and (B): Modified from Andersson, M.N., Larsson, M.C., Schlyter, F., 2009. Specificity and redundancy in the olfactory system of the bark beetle* Ips typographus*: single-cell responses to ecologically relevant odors. J. Insect Physiol. 55, 556–567. Panel (B): Micrograph provided by E. Hallberg. Panels (C) and (D): Redrawn from Andersson, M.N., Larsson, M.C., Blaženec, M., Jakuš, R., Zhang, Q.-H., Schlyter, F., 2010. Peripheral modulation of pheromone response by inhibitory host compound in a beetle. J. Exp. Biol. 213, 3332–3339. Panel (E): Photo by F. Schlyter.*

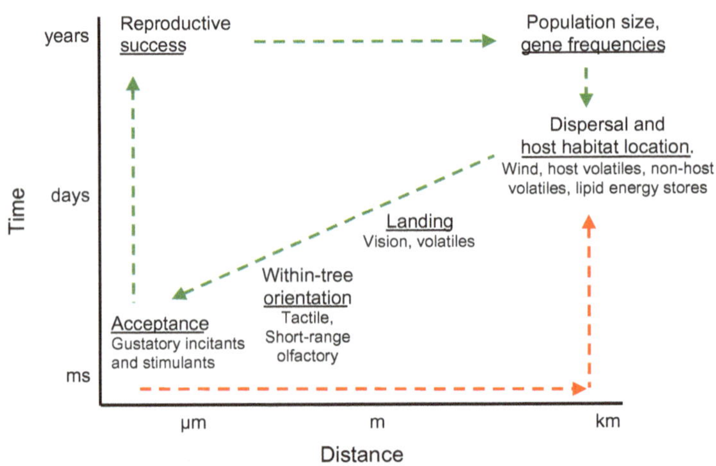

K.F. Raffa *et al.*, Fig. 5 Conceptual overview of host selection sequence by bark beetles. Beetles disperse upon emergence, and orient using anemotactic and chemical cues to locate host habitats. They land in response to vertical silhouettes and general host compounds such as monoterpenes and ethanol. Upon landing, they orient under bark crevices and in cracks based on thigmotactic stimuli that also elicit responsiveness to short-distance olfactory stimuli and biting. Tunnelling is elicited by chemosensory stimuli perceived in sensilla, with different chemical classes, compounds, and concentrations interacting to process host recognition and host acceptance reactions. The sequence may be interrupted (red—negative feedback) at any stage prior to oviposition. Reproductive success is determined by the interaction of host defences, beetle arrival in response to pheromones, host nutritional quality, and competition. Relative reproductive success influences population size and has genetic consequences, which jointly feed back into host acceptance mechanisms (green).

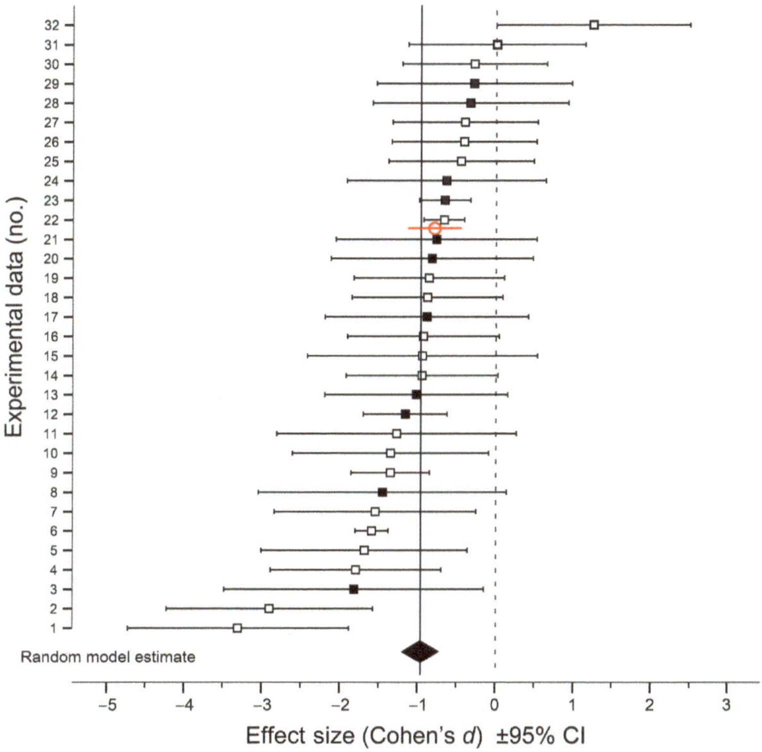

K.F. Raffa *et al.*, Fig. 6 Tree protection efficiency of antiattractant dispensers in *Ips typographus* and *Dendroctonus ponderosae*, tested by a meta-analysis comparing treated and untreated plots from a range of experimental results published until 2011. The efficiency is gauged by the effect size, where the difference between treatment and control is scaled by division of their pooled standard deviations, with a strong effect often considered to be at $|0.8|$ or larger (Nakagawa and Cuthill, 2007). Thus, the overall effect is here rather large with a mean of ca. −1, as shown by the 'diamond' on last row giving the mid-point and width of the 95% CI of grand mean effect size. The graph is an ordered plot of individual study effect sizes and their ±95% confidence intervals; known as a 'Forest plot'. *Vertical line*: ——, estimated overall effect size; - - - -, effect size = 0 or no effect. Numbers on the dependent axis correspond to the 'Expt nr' in Appendix of Schlyter (2012). ■, *I. typographus*; □, *D. ponderosae*; ○, data point summarizing a recent 7-year study of high-mountain whitebark pine protection against *D. ponderosae* (Perkins et al., 2015). *Redrawn from Schlyter, F., 2012. Semiochemical diversity in practice: anti-attractant semiochemicals reduces bark beetle attacks on standing trees—a first meta-analysis. Psyche J. Entomol. 2012, Article ID 268621.*

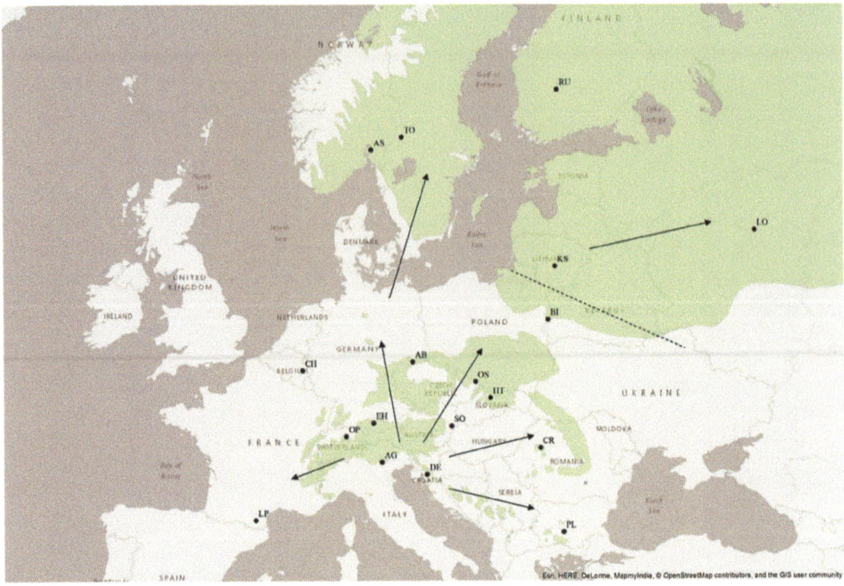

J.K. Janes and P.D. Batista, Fig. 1 Map of the phylogeographic dispersal of the eight-spined spruce bark beetle throughout Europe. *After Stauffer, C., Lakatos, F., Hewitt, G.M., 1999. Phylogeography and postglacial colonization routes of* Ips typographus *L. (Coleoptera, Scolytidae). Mol. Ecol. 8, 763–773.*

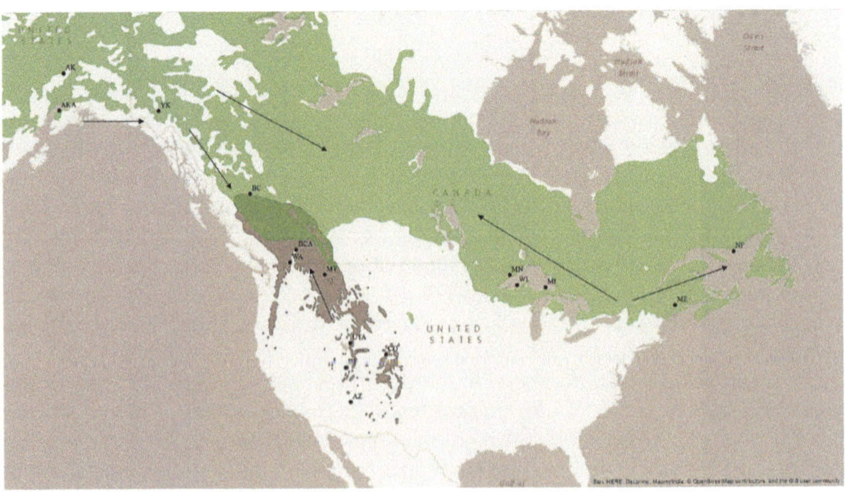

J.K. Janes and P.D. Batista, Fig. 2 Historical movement of the spruce beetle throughout North America. *After Maroja, L.S., Bogdanowicz, S.M., Wallin, K.F., Raffa, K.F., Harrison, R.G., 2007. Phylogeography of spruce beetles (*Dendroctonus rufipennis *Kirby) (Curculionidae: Scolytinae) in North America. Mol. Ecol. 16, 2560.*

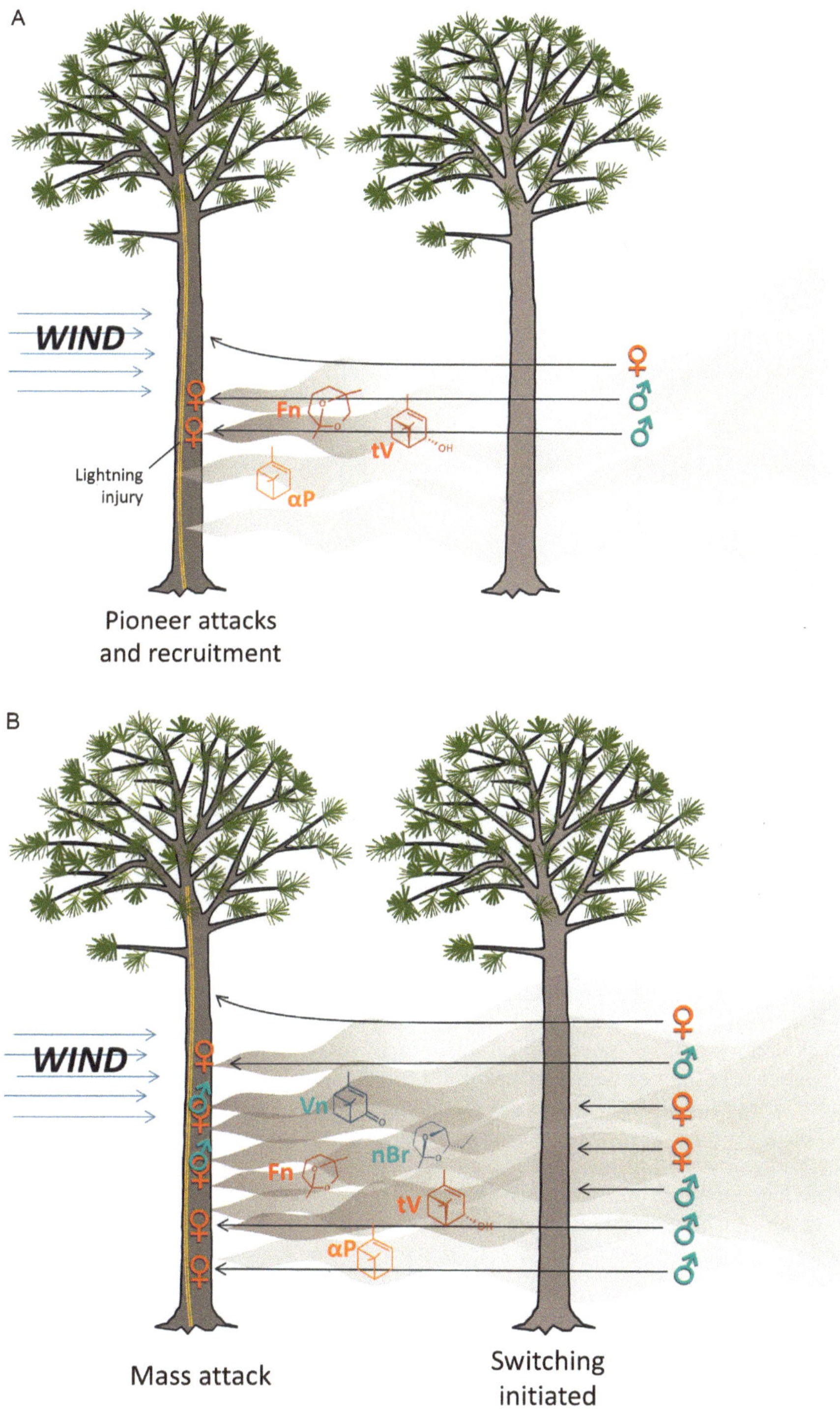

B.T. Sullivan, Fig. 8—Cont'd

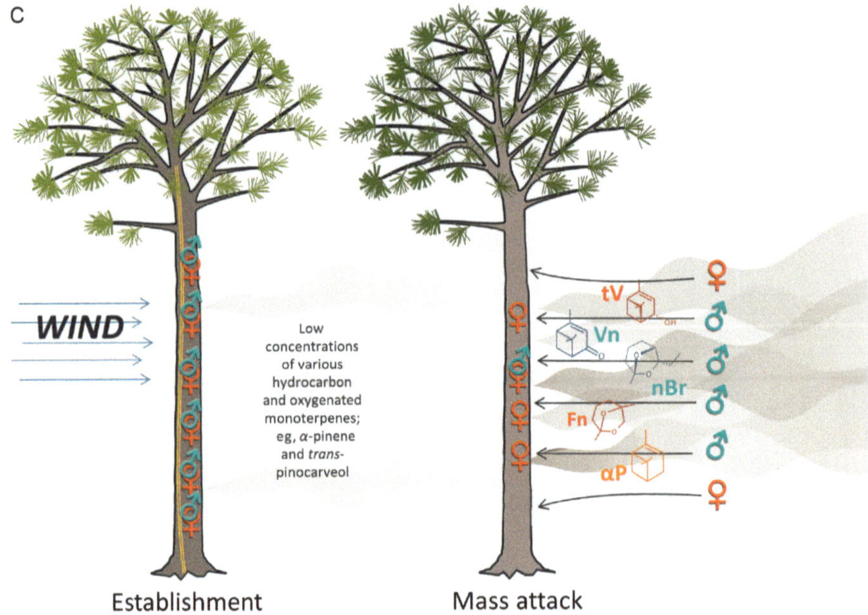

B.T. Sullivan, Fig. 8 (A) Stage I: Initiation of a *D. frontalis* infestation. Dispersing pioneer female beetles locate a lightning-weakened (or otherwise susceptible) pine apparently by random searching. They then release frontalin (Fn) and *trans*-verbenol (tV) which function synergistically to attract conspecifics of both sexes. Host-produced α-pinene (αP) from the lightning wound (and later from the females' entrances once they penetrate the phloem) may act as a synergist that duplicates the effects of the *trans*-verbenol. Attacking females establish nuptial chambers if resin defences are sufficiently low; otherwise they may expel resin from the gallery entrances to form pitch-tubes while continuing to release aggregation pheromone. *endo*-Brevicomin (not shown) from the first arriving males synergizes attraction to the pheromone-releasing females. Further arriving females, deterred by high concentrations of frontalin odours, tend to land above and below the area of initial female attack at mid-bole. (B) Stage II: Mass attack and initiation of switching of attack focus to an adjacent tree. Large numbers of beetles of both sexes arrive on and attack the original pine in response to the aggregation attractant of the first and subsequent arrivers. Males locate attacks of females and pair with them; they assist the female with clearing the gallery entrance while they simultaneously release both verbenone (Vn) and *endo*-brevicomin (nBr). These two compounds indicate to mate-seeking males on the bark surface that a female in an entrance has already paired, in which case a searching male must either remove the occupying male forcefully or continue looking for an unpaired female. At high concentrations (ie, with a high density of males), *endo*-brevicomin may also deter beetle landing, as may verbenone. However, male-produced *endo*-brevicomin strongly synergizes responses of flying beetles to solitary female attacks within a distance of at least several metres of the releasing males, and this aids in synergizing responses of beetles to "pioneer" female attacks on surrounding trees. Paired females on the original tree continue to release frontalin (Fn) and *trans*-verbenol (tV) at reduced levels, and

B.T. Sullivan, Fig. 8—Cont'd synergistic α-pinene continues to be released from damaged host tissue. A plume of attractant and synergists encompasses downwind, unattacked trees; upwind-orienting (anemotactic) females land on them and may begin to bore if the semiochemical concentration of the engulfing plume is sufficiently elevated. These attacks initiate the shift of the focus of attack to the downwind tree. (C) Stage III: Termination of attack on the original tree and re-focus of attack onto the "switch tree". After the original tree ceases to produce a constitutive resin response and has reached its carrying capacity of parent beetle pairs, each pair plugs its gallery entrance with frass and begins to extend the gallery in the phloem while laying eggs. This event coincides with a substantial reduction in semiochemical release (particularly of beetle pheromone components) from the original tree. Nonetheless, the tree continues to release odours for many weeks, including a variety of possibly inhibitory/repellant oxygenated monoterpenes generated from spontaneous and microbially mediated oxidation of the dead host's residual resin monoterpenes. The foliage of the tree crown begins to fade at this time. Meanwhile, a mass attack proceeds on the new focus tree as it had on the original tree, with its resulting semiochemical plume likewise being capable of starting attacks on nearby, particularly downwind trees. Parent adults reemerging from the original tree (and later, the emerging, developed brood) may participate in mass attack of new focal trees, and thus these beetles may initiate and then maintain a continuous cycle of new attacks by both re-emergent parents and newly emerged brood beetles that can propel a steadily growing infestation and continue to consume trees over a period of months.

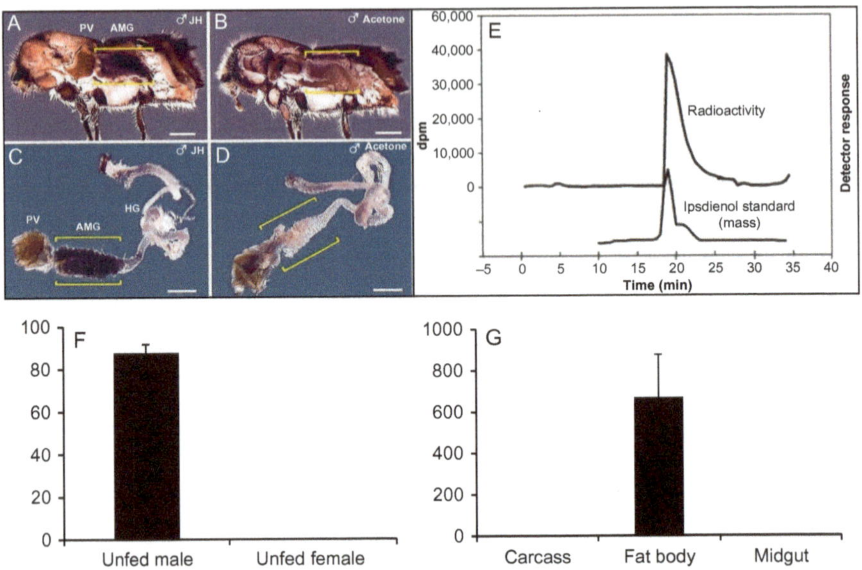

C. Tittiger and G.J. Blomquist, Fig. 3 Localization of pheromone production in *I. pini* and *D. ponderosae*. (A, B) In situ hybridization of JH III-induced HMGR expression in *I. pini* with the sagittal view exposed showing the proventriculus (PV) and anterior midgut (AMG) of a JH III-treated (A) and acetone treated (B) male. (C, D) Isolated alimentary canal isolated from a male *I. pini* that had been treated with JH III (C) or with acetone (D). Radio-HPLC (E) of pentane–diethyl ether extract of midgut tissue from five fed male *I. pini* incubated with [1-^{14}C] acetate confirms that ipsdienol is produced de novo in midgut tissues. (F, G) *exo*-Brevicomin is found in unfed male, but not female *D. ponderosae* (F), and localizes to the fat body (G). *Figures reproduced from Hall, G.M., Tittiger, C., Andrews, G.L., et al., 2002. Midgut tissue of male pine engraver,* Ips pini, *synthesizes monoterpenoid pheromone component ipsdienol de novo. Naturwissenschaften 89 (2), 79–83; Song, M., Delaplain, P., Nguyen, T.T., et al., 2014.* exo-*Brevicomin biosynthetic pathway enzymes from the Mountain Pine Beetle, Dendroctonus ponderosae. Insect Biochem. Mol. Biol. 53 (0), 73–80; and Song, M., Gorzalski, A., Nguyen, T., et al., 2014.* exo-*Brevicomin biosynthesis in the fat body of the mountain pine beetle,* Dendroctonus ponderosae. *J. Chem. Ecol. 40 (2), 181–189.*

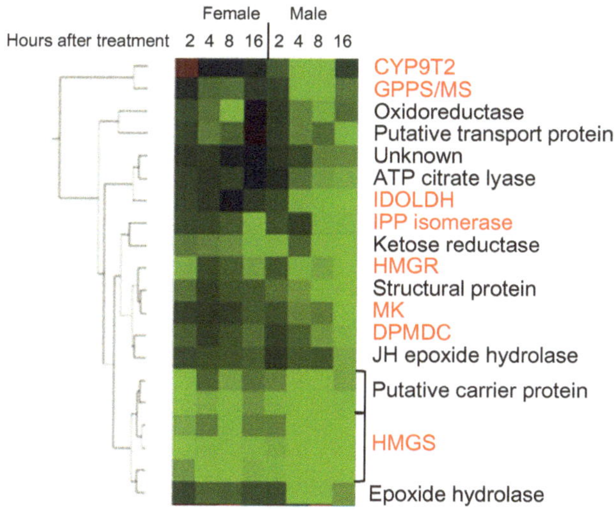

C. Tittiger and G.J. Blomquist, Fig. 5 Coordinated gene expression in *Ips pini* midguts. Heat map of genes induced upon topical treatment with JH III. Each column represents average relative expression ratios for each time point. Genes encoding mevalonate pathway and pheromone biosynthetic pathway enzymes are noted in *red. Adapted from Keeling, C.I., Bearfield, J.C., Young, S., Blomquist, G.J., Tittiger, C., 2006. Effects of juvenile hormone on gene expression in the pheromone-producing midgut of the pine engraver beetle,* Ips pini. *Insect Mol. Biol. 15 (2), 207–216.*

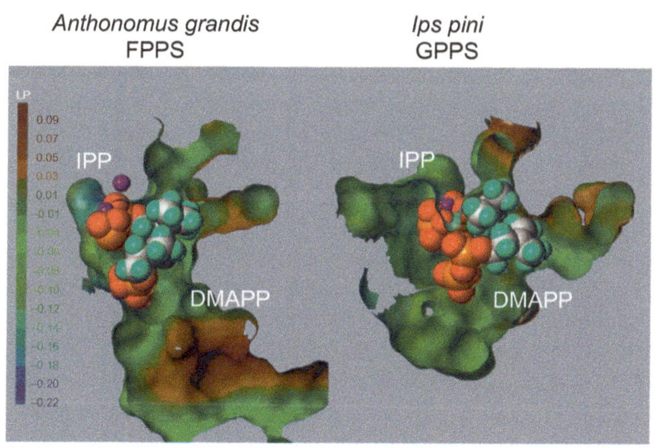

C. Tittiger and G.J. Blomquist, Fig. 6 Molecular models of prenyl transferase substrate binding sites. Cut-away views of the substrate binding domains for FPPS from the cotton boll weevil (*Anthonomus grandis*) and GPPS/MS from *Ips pini*, shaded to show relative hydrophobicity (*brown* to *blue* corresponding from most to least hydrophobic). IPP and DMAPP substrates are shown in space fill, hydrocarbon portions in *blue* and *white*, phosphate portions in *orange* and *red*. Mg^{2+} cofactors for each enzyme are in *purple*. The substrate binding domain for GPPS/MS is clearly smaller than that for FPPS. *Adapted from Gilg, A., 2004. The isolation and characterization of geranyl diphosphate synthase from the pine engraver, Ips pini (Coleoptera: Scolytidae). Dissertation. Biochemistry and Molecular Biology, University of Nevada, Reno Ph.D. 157 pp.*

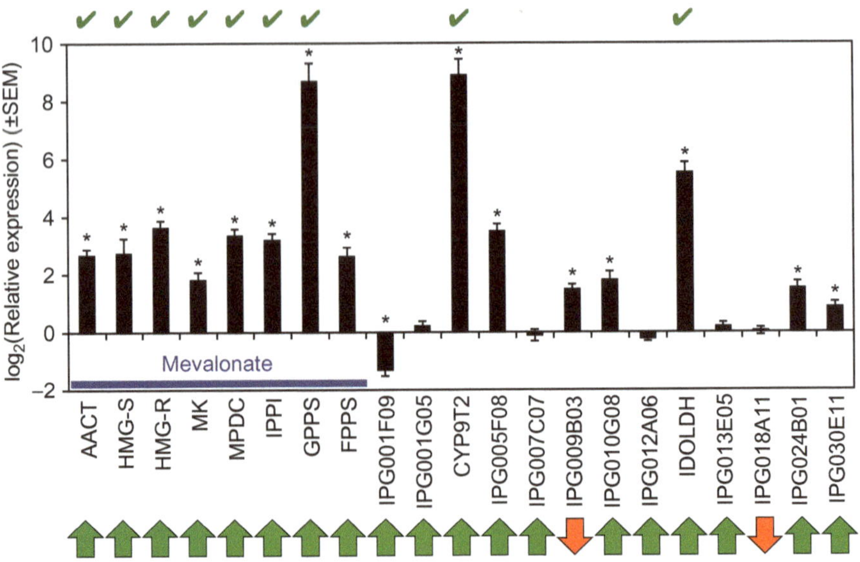

C.I. Keeling, Fig. 2 Candidate genes in *Ips pini* pheromone biosynthesis. qRT–PCR analysis of relative basal mRNA levels in midguts between sexes (males/females) of acetone and juvenile hormone III-treated beetles. Expression values are $\log_2$ for the ratio of the transcript levels of acetone-treated males over acetone-treated female beetles. *Asterisks* indicate significant difference between sexes. Checkmarks indicate genes with confirmed role in ipsdienol biosynthesis. Blue line indicates genes of the mevalonate pathway involved in the early steps of ipsdienol biosynthesis. *Arrows* indicate the direction of the significant change in transcript levels when males are treated with juvenile hormone III, compared to acetone-treated males. *Adapted from Keeling, C.I., Bearfield, J.C., Young, S., Blomquist, G.J., Tittiger, C., 2006. Effects of juvenile hormone on gene expression in the pheromone-producing midgut of the pine engraver beetle,* Ips pini. *Insect Mol. Biol., 15, 207 216.*

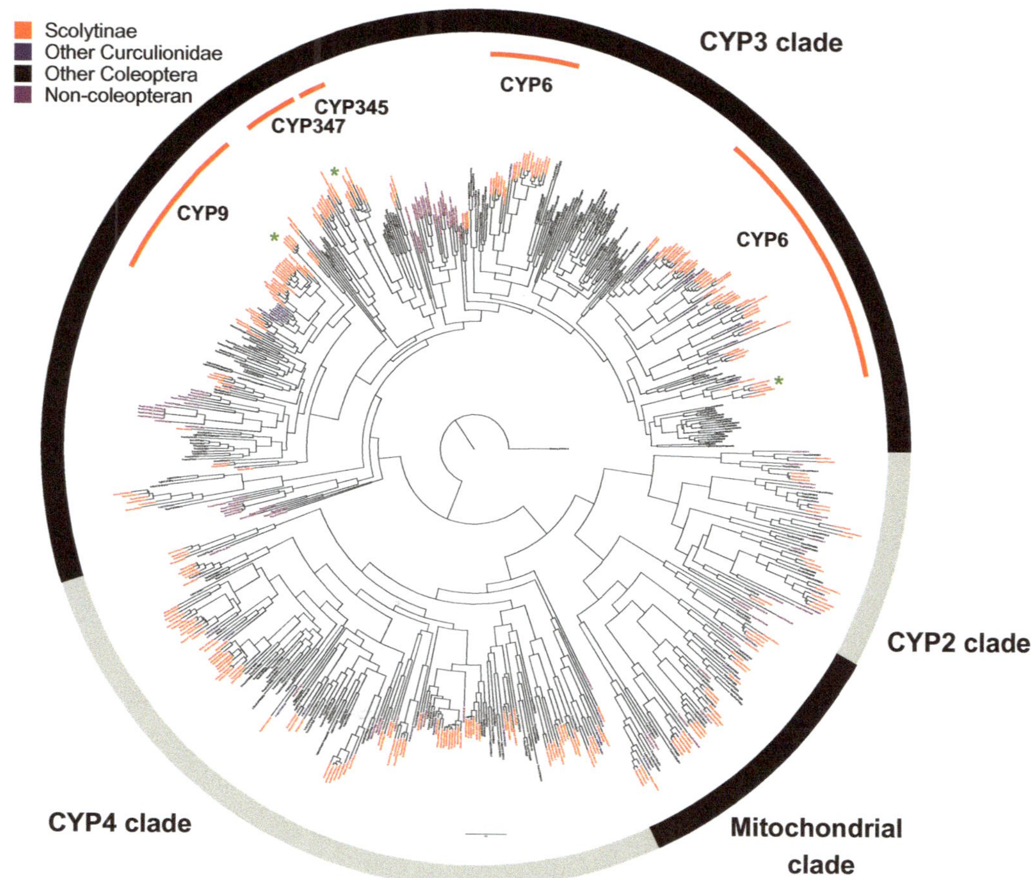

C.I. Keeling, Fig. 3 See legend on next page.

C.I. Keeling, Fig. 3 Cytochrome P450 proteins in bark beetles and other insects. Phylogeny of P450s from bark beetles, other beetles, the honey bee, and the silk moth. An alignment was prepared with MAFFT version 7.123 (Katoh and Standley, 2013) and the phylogeny subsequently prepared with FastTree 2 version 2.1.6 (Price et al., 2010) and displayed with FigTree version 1.4.2 (http://tree.bio.ed.ac.uk/software/figtree/). Proteins less than 200 amino acids were excluded from the alignment and phylogeny. The phylogeny was rooted with human CYP3A4. Abbreviations: Agla, *Anoplophora glabripennis*; Amel, *Apis mellifera*; Baen, *Brassicogethes aeneus*; Blon, *Brontispa longissima*; Bmor, *Bombyx mori*; Cmon, *Cryptolaemus montrouzieri*; Darm, *Dendroctonus armandi*; Dfro, *Dendroctonus frontalis*; Dhel, *Dastarcus helophoroides*; Dpon, *Dendroctonus ponderosae*; Drhi, *Dendroctonus rhizophagus*; Dval, *Dendroctonus valens*; Dvir, *Diabrotica virgifera virgifera*; Hham, *Hypothenemus hampei*; Icon, *Ips confusus*; Ipar, *Ips paraconfusus*; Ipin, *Ips pini*; Ldec, *Leptinotarsa decemlineata*; Pdiv, *Phyllopertha diversa*; Pjap, *Propylea japonica*; Pstr, *Pissodes strobi*; Tcas, *Tribolium castaneum*; Tmol, *Tenebrio molitor*; Tyun, *Tomicus yunnanensis*. The functionally characterized bark beetle P450s discussed in the text are highlighted with green *asterisks*.

Kolibri 360 GmbH
Ferdinand-Jühlke-Straße 7
99095 Erfurt, Deutschland
produktsicherheit@kolibri360.de